Sampling Opinions

An Analysis of Survey Procedure

Sampling Opinions

An Analysis of Survey Procedure

FREDERICK F. STEPHAN

and

PHILIP J. McCARTHY

GREENWOOD PRESS, PUBLISHERS
WESTPORT, CONNECTICUT

Library of Congress Cataloging in Publication Data

Stephan, Frederick Franklin, 1903-1971.
  Sampling opinions.

    Reprint of the ed. published by Wiley, New York,
  in series: Wiley publications in statistics.
    Bibliography: p.
    1. Public opinion polls. 2. Sampling (Statistics)
  I. McCarthy, Philip John, 1918-   joint author.
  II. Title.
  [HM263.S84 1974]   301.15'42   73-6395
  ISBN 0-8371-6901-1

Originally published in 1958 by John Wiley & Sons, Inc.,
New York, Chapman & Hall Limited, London

Reprinted with the permission of Mrs. Eda A. Stephan and
Philip McCarthy

Reprinted in 1974 by Greenwood Press, Inc., 51 Riverside
Avenue, Westport, Conn. 06880

Library of Congress catalog card number 73-6395
ISBN 0-8371-6901-1

Printed in the United States of America

10 9 8 7 6 5 4 3 2

Committee on Measurement of Opinion, Attitudes, and Consumer Wants

of the

National Research Council and the Social Science Research Council

To

EDA and MARY ANN

Foreword

This book shculd interest a wide variety of intelligent readers who wish to know more about one of the most important tools of modern investigation, namely, the sample.

Whether one is a social scientist adding to our knowledge of what makes society and its citizens function, or a practitioner applying this knowledge in market research or merchandising, or just a citizen alert to the world around him, the problems dealt with with such competence in this volume can be of abiding concern.

We have long needed a practical, wise exploration such as this one, which uses a rich number of fresh concrete examples from actual research experience to illuminate important problems. Some of these problems have hitherto been discussed, if at all, in treatises inaccessible to the non-mathematician.

The studies that led to this book originated in a concurrence of interest, shortly after World War II, expressed by the National Research Council, for the natural sciences, and the Social Science Research Council, for the social sciences. A joint committee was appointed, called the Committee on the Measurement of Opinion, Attitudes, and Consumer Wants. It included mathematicians, statisticians, social scientists, engineers, and leading practitioners of opinion research. The Committee sponsored studies of interviewing and respondent panels, as well as this study of sampling, and it reviewed other problems related to their projects. (Its membership and that of its advisory subcommittee on the sampling study are given on page v.) The Committee can assume neither credit nor responsibility for the conclusions of the authors, since they were left free to prepare and publish their results as they saw fit. The Committee, having finished its work, disbanded. It did so with a hope that it had set in motion lines of research that will be developed by a succession of research workers and organizations, inspired by the pioneering work that has been done in this field and by the growing importance of dependable reports and analyses of the thoughts and wishes of people concerning the complex issues of these momentous times.

Our democratic society and free competitive economy are expressions of the beliefs and wants of the American people. Indeed, our whole way of life is supported and maintained by certain fundamental attitudes in which there is general agreement. These attitudes are freely fulfilled in the daily activity of the people acting individually and in association with one another in a great variety of groups. Equally vital to the maintenance of our way of life is tolerance of divergent opinions, open discussion of issues, open-mindedness to innovation, withholding of judgment in disputes pending the outcome of fair trial and establishment of adequate proof, and responsiveness of public officials and other leaders to prevailing views of the majority, moderated by due consideration of minority interests. This way of life, so maintained, we deem vastly superior to any other we know or can imagine; it commands our love, our loyalty, and our labors.

There are many ways in which the attitudes of people are expressed. One is by public statements voiced by leading citizens in the press, in the halls of legislatures and councils, and at public meetings. These representatives succeed to the extent that they in turn have had expressed to them the thoughts and feelings of the people for whom they speak. Another way is by voting, especially when contending candidates espouse opposite sides of some clearly defined issues, or when questions are put to the voters in a referendum. A third way is by interview, the forte of the local political district leader canvassing his district, the newspaper reporter or commentator preparing his story, the elected official returning to his constituents to learn what they think and want, and other men in their special roles in business, labor unions, and even the arts who seek renewed understanding of the people they serve.

In the space of a generation we have developed new means of learning about opinions, attitudes, and consumer wants. The methods of interviewing and of measuring and analyzing attitudes have been greatly improved, and research centers have been established to provide efficient organizations capable of surveying attitudes on a wide variety of issues across the nation and abroad. The reports of their work as it accumulates give very impressive evidence of the important contribution these new developments are making to our national life.

Not all reports are equally dependable. If the studies have been produced by careless or biased interviewing, by inaccurate analysis, or by slanted writing, they may be worse than useless. Also if what they present as the attitudes of all the people or of a clearly defined group are actually the views of one faction or even a collection of interviews in which one faction has had more than its share of representation, the "finding" may be misleading and harmful. This possibility has called forth an examination of the *sampling*.

Various methods of sampling have been invented to insure the selection of a truly representative cross section of the population to be surveyed. These methods in turn have been questioned and examined to determine their reliability. The seriousness of the sampling problem has increased with the growing importance of surveys and the improvement of the other methods which are incorporated in them.

The present book is a report of an examination of sampling methods, undertaken as a result of uncertainty about the sampling that has been and is being done in surveys of opinion. It goes far in explaining sampling practices and marshaling the evidence available on the accuracy of procedures that have been commonly used. It also probes the difficulties of appraising sampling methods and shows how they are intertwined with other parts of survey taking. And in accomplishing these ends it happily manages to combine clarity of presentation with sound and scholarly judgment.

SAMUEL A. STOUFFER

Harvard University
April 1958

Preface

This book is written for those who are concerned with methods of studying the opinions, attitudes, needs, and wants of various parts of the population. Over the past thirty years many surveys of opinion and studies of consumer demand have been made. Such surveys are being taken at the present time in great numbers. Important decisions in business and in government frequently hinge on their results. Important theoretical advances in psychology and the social sciences have been based upon data provided by sample surveys. Frequently the same survey serves both operational and scientific purposes.

In using data provided by sample surveys, or in appraising conclusions that other people have drawn from such data, we must know the accuracy or trustworthiness of the data. The accuracy of the final results of a sample survey is conditioned by all aspects of the survey process: the definition of the population to be surveyed, the design and execution of a sampling procedure, the plan for obtaining information from individuals who have been drawn into the sample, and the techniques and line of reasoning to be used in processing and analyzing the data. Consequently in the past thirty years there has been increasing concern about the theory and operation of the various components of a survey.

During this period there has been a magnificent development of mathematical theory to explain the operations of certain types of sampling procedures and to aid in designing more efficient sampling methods and estimation techniques. These developments in mathematical statistics have been set forth very ably in a series of books which we have not attempted to parallel or even summarize. Instead we shall refer the reader to them at several appropriate places in this volume. There are also available a large number of books which treat the problems involved in obtaining information from individuals who have been drawn into a sample and the problems involved in analyzing such data. These problems are discussed in terms of such topics as: research design; wording of questions; construction of scales; selection, training, and supervision of interviewers; techniques of interviewing; analysis of data; and the like. Again we have not attempted to parallel or summarize this material, but have drawn attention to it in appropriate places.

Unfortunately these studies of sampling methods and of the other aspects of survey design are incomplete in a number of ways. In particular, there is relatively little information about the interrelationships between the sampling procedure and the other portions of the entire survey undertaking. In this book we have attempted to supply part of what is lacking by emphasizing these interrelationships between sampling and the other components of survey design. We have also attempted to take a more general approach to sampling than is customary in the standard treatises by including discussions of the problems of sampling that are raised when the methods do not conform to the mathematical concepts underlying the standard theory or when an investigator starts with a sample of data provided by natural processes not under his direct control.

This volume is the result of studies that were started some years ago under the auspices of the former joint Committee on the Measurement of Opinion, Attitudes, and Consumer Wants appointed by the National Research Council and the Social Science Research Council. The authors had excellent cooperation from, and are deeply indebted to, the members of the Committee on Measurement of Opinion, Attitudes, and Consumer Wants and the members of the Advisory Subcommittee.

The book consists of three parts. The first part is introductory. Its purpose is to explain the great variety of sampling methods that may be encountered by anyone who attempts to read or use the results of opinion studies, and to describe the more or less essential aspects of any sampling operation and the survey within which it is embedded. The aim here is to analyze survey operations into what appear to be their principal parts and to consider some of the interrelations among these parts. All this is done without recourse to mathematics, in the hope that it will permit a larger group of readers to understand the analysis and to gain a definite understanding of the way in which various kinds of sampling operations actually take place in surveys of opinion and related subjects.

The second part is devoted to an examination of empirical studies that were undertaken by the authors and the results that they assembled from studies undertaken by other people. In these studies it became evident that few, if any, simple generalizations could be made about sampling procedures and their results. It also became evident that there are no generally adequate procedures for determining the accuracy of data that result from sampling surveys. The types of research necessary for developing adequate evaluation procedures and the importance of applying these procedures in the future are discussed in both the first and second part.

The third part will represent for most readers the beginning of a synthesis of what has gone before. It is an attempt to describe and discuss in general terms the problems that arise in actually designing a sample survey and putting it into operation. Even a person who does not have direct responsibility for designing or operating a survey will find it important to understand the problems faced by those who do. This will help him reach a sounder judgment of the dependability of the results obtained from such surveys.

The authors are quite aware of the fact that this book is not a *final* answer to any of the major questions. It represents a progress report and a description of the present state of many of the problems which have been considered, or which should be considered, by people with serious interests in the results of sampling surveys of opinions, attitudes, needs, and wants.

Statisticians will find much in this book that is not discussed in statistical treatises. They may be impatient with the verbal discussion of problems that can be expressed more succinctly and precisely by use of mathematics. They should not, however, overlook the fact that there are in this field important elements of the sampling problem which have not yet been treated adequately by mathematical theory and which seem to offer very serious difficulties to anyone seeking at this time to treat them mathematically. The authors wish to make it clear that they regard the development of mathematical theory for all aspects of survey operations as an extremely important goal for mathematical statisticians and survey statisticians alike. At the same time they feel that preoccupation with mathematical models now feasible will detract attention from many important problems not yet incorporated effectively in these models.

The authors wish to acknowledge their indebtedness for the advice and materials they have received from a large number of individuals and organizations. In addition to the contributions definitely recognized in the text, there are many others which contributed to the development of the general conclusions and analysis. It is our hope that this book will encourage many organizations and individuals to conduct research projects on the problems of sampling discussed here and also on related problems. It is our further hope that, when they have done so, they will publish their results fully and frankly.

This work was conducted under a generous grant from the Rockefeller Foundation to the Social Science Research Council. None of the conclusions or statements made in this book is to be attributed to any organization or any person other than the authors. The authors are responsible for any errors that may occur in the text.

FREDERICK F. STEPHAN
PHILIP J. McCARTHY

March 17, 1958

Contents

PART I. THE NATURE AND ROLE OF SAMPLING

PART II. EMPIRICAL STUDIES

PART I

The Nature
and Role of Sampling

CHAPTER 1

The Value of Information
Obtained from Samples

1.1 THE FUNDAMENTAL QUESTION

From time to time some event precipitates a widespread public discussion of the value and fallability of information obtained from samples. Some of the recent debates have centered on pre-election surveys of voting intentions (1, 2), Dr. Kinsey's studies of sexual behavior (3, 4), the failure of midseason crop forecasts to predict the fall harvest (5, 6), and discrepancies among estimates of unemployment and other economic conditions (7). In the future there will be similar discussions about other deviations of sample information from what is expected or known through more thoroughly trusted sources.

More frequently, and in a less spectacular manner, questions arise about the scientific conclusions that may be drawn and the practical actions that may be taken on the basis of information obtained from samples. Before valid conclusions can be drawn and sound decisions made, such information must be appraised for its adequacy, quality, and accuracy. Frequently, a group turns aside in its deliberations to examine the dependability of a set of data bearing on its problem. This happens in court proceedings, in commission hearings (8), in scientific meetings (9), and in business conferences. It also occurs repeatedly in an individual's study of a problem.

Anticipating this scrutiny by the users, the people who collect information must wrestle with the difficulties of obtaining dependable data, including the choice of various techniques to improve their chances of getting good samples. In this choice, they are concerned with problems closely related to those faced by scientists who try to select typical specimens and material for observation, comparison, dissection, and experiment. Their efforts are also quite similar to those of business executives, government officials, engineers, and others who seek to apply the results of past experience to new problems. Considered planning of sampling operations and careful scrutiny of the information they produce are only a particular form of prudent action.

This prevalent concern about information obtained from sampling is, then, the fundamental question to which our book is devoted:

How can we improve our knowledge of sampling so that we understand it better, use it more effectively, appraise its results more accurately, and apply these results to greater advantage in scientific work and in practical affairs?

Our study of this question will be centered in the use of sampling to obtain information about a broad class of human variables for which there seems to be no suitable common name. These variables include what is referred to by psychologists and social scientists as: beliefs, opinions, expectations, interests, intentions, tendencies, attitudes, predispositions, desires, wants, needs, satisfactions, values, utilities, prejudices, stereotypes, responses, reactions, choices, decisions, and other similar terms.

Many of these words are used outside the field of social science with some variation of meaning, but their general import may be expressed simply. *We are concerned with observations and measurements of what people think, how they feel, what they want, and what they are ready and likely to do with respect to some particular problem, object, or situation.* The object may be a political candidate, a public issue, a commercial product, or some combination of people and things considered together (for instance, automobile traffic). The problem may center in the objects or in the individuals whose thoughts, feelings, and wants are directed toward the objects.

Attitudes and wants cannot be observed directly. They must be inferred from what people do and say. For example, people may say they expect prices to decline during the next six months, or that they want the government to reduce taxes, or that they would like to buy a color television set. They may change their buying behavior, cast their votes preponderately for one candidate, and place orders for television sets. Underlying such statements and actions are the psychological variables that guide and motivate behavior. These variables introduce a degree of consistency in behavior that makes it possible to understand the responses of people, to depend on them, and to satisfy them.

The scientific study of human variables has made notable progress but is still in the pioneering stage. Experimentation with human subjects is necessarily restricted by the danger of harming them.

Their cooperation is essential. Even with full cooperation, it is not yet feasible to observe and measure most human variables. Few such variables are well defined. Even these few are more difficult to measure than stature, weight, temperature, pulse, breathing rate, hearing acuity, eyesight, and the like, but some of them may be no more difficult than certain laboratory tests now being used successfully for medical purposes.

When measurements must be obtained from every member of a large group or population, the size and difficulty of the job become overwhelming. It is natural, therefore, to consider taking surveys by sampling methods. This reduces the burden of work and still permits the results to have a satisfactory degree of accuracy.

1.2 LINKAGE OF PRACTICAL AND SCIENTIFIC INTERESTS

Information obtained from samples of people is valuable for practical purposes if it contributes to the effectiveness of the decisions and actions used in attaining practical objectives. Every human being takes some actions in his personal affairs, in his work, or in other activities for which he needs to know a few people quite well, many others less well, and large classes or groups of people in general terms. Such knowledge can be obtained only to a limited extent by personal contact and direct observation. Hence we have to use many indirect means, among which is the sampling of individuals who are expected to be typical of the larger groups to which they belong.

Sampling is not only valuable but inescapable for scientific research on human behavior. Scientific generalizations can seldom, if ever, be based on a study of all the behavior or all the persons to which they apply. Moreover, any practical problem can become the object of scientific study. Hence the practical and scientific applications of sampling are thoroughly interrelated, and it is advantageous to develop them together.

There are many examples in which sampling has been employed to aid the solution of linked practical and scientific problems. We may expect even more in the future. For example, public attitudes toward monopoly and "big business" have been powerful political forces in the past. They are also of scientific interest to economists. A recent sample survey provides measurements and descriptions of the attitudes of Americans toward big business (10). Other sample surveys have attempted to measure the spread of ownership of corporations among millions of stockholders (11, 12).

Annual surveys of the expenditures and savings of American families are conducted for the Federal Reserve Board to provide information that is useful in connection with banking operations and policies (13). Similar studies conducted by the Bureau of Labor Statistics (14, 15) form one basis for the determination of changes in the prices of those consumer goods usually purchased by workers in business and industrial enterprises. In some instances wages are adjusted periodically to the changes in the index.

Likewise, government loans and price-support policies for agricultural products depend on data collected by sampling, as do the forecasts of crops and livestock production (16, 17, 18). Economists are interested in data of the foregoing types for research purposes as well as for immediate practical use.

Similar ties between practical and scientific interests in information appear in studies of voting (1, 19, 20, 21, 22, 23), marriage and family life (24), reading and the use of libraries (25, 26), the use of telephones, radio, and television (27, 28), morale and performance of soldiers (29), behavior of people in disasters (30, 31), rumor and propaganda (32), personal influence (33), and many other subjects.

1.3 PRACTICAL IMPORTANCE OF OPINIONS, ATTITUDES, AND WANTS

Democratic government is based on the principle that the opinion and desires of all citizens should determine important decisions. This principle is followed whenever major questions are submitted to discussion and vote. Not only are candidates for office elected by the people they are to represent, but many questions are submitted directly to the electorate, such as changes in the form of local government, issuance of bonds for major public purposes, and extraordinary proposals for legislation. The practice of settling questions by vote is also carried into the relations of employees and their employers. Thus balloting may determine the union that will represent employees in collective bargaining or may decide that no union will represent them. A strike is seldom called without a vote of union members. Farmers vote to determine whether there will be government control of prices and production for wheat, cotton, and other crops. The decisions of many groups of people are made by voting: the directors of corporations, governing bodies of private associations, juries in court proceedings, and property owners affected by proposed developments. In each instance voting is a means of taking account of the interests, opinions, and wishes of each member of the group.

It is not feasible to submit every question to a vote or to seek public approval for every act of responsible officials. As an alternative to voting, the attitudes and desires of interested people may still be taken into account through petitions, public hearings, discussion in the press, individual conferences, and other similar means. In some court and commission proceedings the opinions, understandings, beliefs, desires, and wants of people in various groups may be sought out by means of surveys and tests. Examples of this recognition of the importance of personal factors in what are frequently complicated and difficult problems of policy or justice appear in litigation about trademarks (34); actions brought by the Federal Trade Commission (35), the Food and Drug Administration (36, 37), and the Attorney General of the State of New York (38); the allocation of radio channels for broadcasting by the Federal Communications Commission (39); and a motion to move a criminal trial to another county on the grounds that a fair trial was not possible where public feeling was strong against the accused (40). Though the status of surveys of opinion as legal evidence is still far from settled, these instances reveal the extent to which courts and regulatory bodies need dependable information on the opinions and attitudes of representative consumers, typical citizens, and specific populations (41).

The Second World War produced several notable examples of the importance of information about the thoughts and feelings of various groups of people. The Norwegian "government in exile" in England is reported to have taken no important actions without first consulting the "underground" in Norway. Even more difficult obstacles stood in the way of determining the attitudes and morale of soldiers and the populations of enemy countries, particularly in the case of Japan. However, considerable success was achieved in doing just this. A remarkable series of studies of American soldiers was made by the Research Branch, Information and Education Division of the Army. The results have been analyzed thoroughly and reported in detail in another example of the linking of the interests of social science and practical administration in substantially the same problems and data (29). Though apparently no question was ever put to a formal vote, the soldiers' views about such crucial problems as the order in which they were to be discharged were sought out and used in the selection of the plan adopted by the Army.

A great many examples of lesser importance can be found in the work of market research agencies whose studies of consumers' wants and preferences among products guide the business decisions of firms. By their choices of what they will buy, consumers do effectively cast their votes for what can be produced and sold. This in turn influences the growth and even the survival of various sections of the economy.

The evidence provided by these and other examples is clear. With the possible exception of totalitarian and monolithic states, the personal factors of opinion, morale, attitudes, and wants have a profound influence on the life of a nation. Without a large measure of acceptance and cooperation, no actions that touch large numbers of people can be completely successful. Further, the wants and interests of people press for satisfaction. When these wants go unmet, they create serious problems and tend to bring on changes. In view of their great practical importance and scientific interest, it is remarkable that so little thoroughly scientific effort has been directed toward improving our meager understanding of these personal opinions and interests (42, 43).

1.4 IS IT NECESSARY TO MEASURE
HUMAN VARIABLES?

Granting then that it is important to know about opinions, attitudes, and consumer wants and to take them into account in making decisions and acting upon them, is it not sufficient to have a verbal description of them in general terms? Is it necessary that they be measured? Must they be measured accurately? Moreover, is it really possible to measure these personal factors?

A full answer to these fundamental questions is a proper task for some other book. Certainly many needs for this information can be satisfied adequately by descriptions of a verbal nature. Even when numerical measurements are made, the meaning of the measured variables must be described in words. But this is not always enough.

The widespread use of voting is an indication of the necessity of measuring the direction and strength of opinion. In small groups the chairman or leader often sums up a discussion and states the "sense of the meeting," subject, of course, to correction when others do not agree. In large groups, however, and especially those in which it is not possible for many to speak, some other method of arriving at the aggregate reaction to a question is necessary. Voting provides an explicit measurement based on counting the number of persons who support each side in a contest. It also induces those who have not decided previously to take sides or abstain from the voting. To do this many individuals compare the strength of their motivations in each direction and take the action supported by the stronger motivation. This implicit process of measurement is necessary if the individual is to make a choice among the alternatives open to him. If he cannot compare and measure, he cannot choose.

Measurement of attitudes by voting does not make any distinction between different degrees of favorable and unfavorable opinion. There are, of course, some exceptions such as the voting of juries on the degree of guilt of a person found guilty of murder. Practical expediency leads to a simple counting of votes, quite as much as does the political philosophy of equality.

The importance of considering the opinions back of the votes as variables, that is, as differing in degree over a range of possible positions, has been recognized for a long time. One of the early formal statements of this point was made by Rice (44) in 1928 for research in political science. At about the same time Thurstone applied the methods of psychophysics to the measurement of attitudes. He presented individuals with a set of statements, each appropriate to a possible position on a scale of variation in favorableness, so that a person could choose the statement or statements best approximating his own position. By analyzing the internal variation of responses to the statements, he was able to find a measure of the relative distance between any two statements on the scale (45).

Further advances have been made by many individuals in testing the extent to which a set of statements or responses to questions behave as if they belonged to a single scale and for measuring more than one dimension of variation in attitude. Methods of analyzing a set of attitudes into "common factors" and "latent variables," which in some sense can represent opinions more simply than the original data, have been developed by Thurstone (46), Guttman and Lazarsfeld (47), Coombs (48), and others (49, 50, 51). These methods of measurement and analysis arose in response to theoretical and practical needs, and they are finding new application as they become more familiar to research workers.

In measuring attitudes, we must determine not only finer gradations of the strength of opinions and differences in position on an issue but also changes over time and from group to group. In experiments and tests of the effect of various factors on attitudes, it is necessary to measure them both before and after the factors in question have been introduced. Thus in tests of the effect of information on attitudes in the Army (52), in an educational program about the United Nations (53), and in a semi-public test of the atomic bomb (54), the effect was measured as the difference between determinations of attitude made before and after for the same sample of people or for comparable samples. Similarly the trend of opinion over a period of time has been measured by determining the attitudes of a population at successive intervals during the period. Some examples center in a political campaign (20, 22), public reactions to telephone service (55), consumer attitudes (56), and opinion trends in wartime (57).

The measurement of opinion and other related personal variables is still a crude and primitive affair. So too is the measurement of wants and what economists call "utility." There is no reason why this kind of measurement should not be improved rapidly by research and experimentation. Some notable attempts have been made in this direction both by direct experiment (58) and by the clarification of the theoretical problems involved (59).

Since this book is not devoted to the problem of measurement of individuals, but nonetheless must incorporate such measurement in the larger survey systems in which sampling is employed, we shall assume that *acceptable though imperfect methods of measurement are available for use and that their accuracy and other properties are known to a fair degree of approximation. We further assume that the errors of measurement cannot be ignored in the analysis of the data.*

1.5 WHAT IS ACCURACY WORTH?

Whenever information about attitudes is obtained and put to use, it is expected to produce some valuable result. The expected result may be the improvement of human relations in a factory or office; it may be the solution of problems arising in the work of a school, a church, or community organization; it may be the clarification of poorly understood phenomena in a branch of science; or it may be the acceptance or rejection of hypotheses about human behavior. Often it is merely a contribution to the knowledge and understanding of the thinking of a large population on important issues of the day. Whatever its purpose, we can see that the value of obtaining the information depends on how well it serves its purpose. If it corrects and makes precise the information that people had beforehand, its value may be great. If it is false and misleading, actually reducing the accuracy of what is known, it may be worse than useless. Usually we may expect that the value of the results will depend on the relative accuracy of the new and prior information, as shown in Table 1.1.

This crude outline can be elaborated but, even as it stands, we see that the value of information depends not only on its accuracy but also on the accuracy of the information which was already available and with which it is likely to be combined. Moreover, it depends on

TABLE 1.1

Value of New Information in Relation to Its Relative Accuracy

General Situation	Value of New Information
Previous information accurate New information accurate	Only as a check or confirmation
Previous information accurate New information inaccurate	No gain and possible great loss
Previous information inaccurate New information inaccurate	No gain and possible loss
Previous information inaccurate New information accurate	Possibility of gain, depending on knowledge of the extent to which the new information can be trusted
Previous information inaccurate New information extremely inaccurate	Possible loss which may be small if the new information does not seem plausible and hence is not used

the degree of confidence placed in it. These relationships can, to some extent, be formulated in mathematical terms and used in a quantitative analysis of the performance of sampling systems (cf. 60, 61).

Perhaps no more dramatic instance of the importance of accuracy in the measurement of opinion can be cited than the pre-election polls in 1948 (1). The division of the vote in that election was such that a relatively small change could mean either the election of Mr. Dewey, who was defeated, or a great landslide of electoral votes for Mr. Truman, who was elected. Many people who were confident in their anticipation that Mr. Dewey would be elected took actions that had to be retracted hastily after the election. An analysis of the polls was made after the election by a committee of the Social Science Research Council. It was found that the polling methods used as a basis for pre-election reports of the apparent division of voters' preferences for presidential candidates were not accurate enough to provide a clear and dependable indication of how the election was likely to go. Moreover, the reports were presented to and understood by people who read them to be more accurate than they actually were.

In other instances, the requirements for accuracy of attitude surveys may not be so stringent. The results may be subjected less frequently to a definite test such as an election. In each survey, however, the actual gains and losses that occur, whether they are known or not, are dependent in part on the accuracy of the information and in part on the extent to which the users of the information know its accuracy.

1.6 THE USEFULNESS OF SAMPLING

Any attempt to measure attitudes over a large population presents a dilemma. Since accuracy is important, there is pressure to measure the attitudes of each member of the population and to do so very carefully. Since the money and other available resources are usually limited, there is pressure to restrict the measurement to a few members of the population and to do the measurement of each individual as cheaply as possible. These two tendencies are incompatible. Hence some compromise is necessary, and sampling provides one part of the solution of the problem.

The usefulness of sampling lies not in the reduction of the amount of work that must be done—if it did, then each further reduction, even down to zero, would make it still more useful. Instead, it lies in the working out of the best feasible compromise between the need for accuracy and the need for economy.

During the 1920's, the *Literary Digest* conducted a number of polls by mail on various public issues. One of the issues, for example, was the prohibition of the production and sale of alcoholic beverages. No national popular vote was ever taken on this issue, and consequently the polls of the *Literary Digest* on this controversial question provided information of great public interest. The *Literary Digest* conducted its polls by mailing ballots to millions of people. In the 1936 election ballots were mailed to some ten million, and about two million, or one-fifth, were returned. We do not know the accuracy of the poll on prohibition. Even the *Digest's* election polls do not help answer this question, for the 1932 poll was in error by less than two percentage points whereas the 1936 poll was in error by nineteen percentage points. Whatever the accuracy of the *Digest's* polls, it could have been improved greatly. A change in method of sampling would have cut the number of ballots sharply and would have improved both their effectiveness in determining the attitude of each individual who filled out one of the ballots and the representativeness of the sample of ballots returned. Instead of ten million ballots, some 100,000 might well have been sufficient even to provide separate tabulations by states.

The usefulness of sampling in improving the accuracy and in reducing the cost of large data-collecting operations was demonstrated in a number of large-scale surveys during the 1930's (62). Its application has been extended to many subject matter fields during the past decade or more (63, 64). Although these past and current uses of sampling are quite impressive, it is clear that they reveal only a few of the possible uses of sampling that will be found and cultivated in the next decade or more.

1.7 NEEDED RESEARCH ON SAMPLING

Sampling is frequently used in studies of human behavior and in the measurement of attitudes as only a part of a much larger operation of obtaining and applying information. These applications of sampling may be experiments with a few subjects, national surveys, intensive studies of groups of people over a long period of time, and other research enterprises. Whatever their nature, such complex undertakings pose questions about the appropriateness of the sampling, its effects on the other parts of the whole process, and the effect of the other parts on it, as well as specific problems about the internal working of the sampling operations themselves. These questions can seldom be answered by direct tests or observations of the sampling but often they can be answered to a degree by indirect tests, by analysis of the sampling as it occurs, by application of general statistical theory, by use of the results of other kinds of statistical research, and by special arrangements within the entire system of data production.

Research on the problems of sampling is needed not only for the improvement of particular surveys but also for building up an increasingly effective body of theory and know-how for use in the future. Although the two purposes are not distinct, the second and more general purpose is often neglected. The influences of the financing of studies and the incentives of research personnel are much stronger in the first direction than in the second. This is not the best balancing of objectives from a broader and longer-range point of view. Many studies now being made are less effective than they might well have been if past studies had placed more emphasis on the more fundamental research problems of sampling. By leading to improved methods, such research increases the accuracy of measurements and of the analytical results that are derived from them. In this way it adds to the value of attitude measurement for practical and scientific purposes.

1.8 SOME PERSISTENT PROBLEMS OF SAMPLING

A number of similar problems arise repeatedly in the use of sampling. They may be summarized, perhaps too simply, in the following terms.

1. There is a pervasive problem of discovering how we can best gauge the dependability and the accuracy of information collected by various methods of sampling. Sometimes we can learn by our own experience with similar operations, providing they are tested by later experience which reveals their accuracy. Most of the time we must draw on more remote experience and on general knowledge to infer their degree of accuracy. There does not seem to be any particular device or principle for assessing a method of sampling or the results produced by it, but the general scientific approach that combines painstaking observation and theoretical analysis, each checking and stimulating the other, would appear to be the most effective way to make the appraisal.

2. There is a generic problem of selecting the best number of people to form a sample and the best distribution of these people throughout the population being studied. The best number depends so closely on the distribution throughout the population that there is only one problem here, not two.

3. Since many bodies of data come to the user from sources that obtain information as a by-product of natural processes of selection, there is a perplexing problem of understanding how the selection occurs and how it affects the data. This problem arises in the use of letters, diaries, complaints, petitions, and conversation (mass observation) as a source of data on attitudes and other personal variables.

4. There is a sharp difference of opinion on the degree to which the operation of a survey should be left to the judgment of interviewers and other rank and file observers in contrast to strict control of all steps by specific instructions and by central direction. This reliance on the judgment of interviewers involves more than expediency and compromise with practical difficulties. It is also an attempt to utilize the interviewers' resourcefulness in adapting to local conditions and in obtaining the cooperation of potential respondents.

5. Another problem for research is the evaluation of the relative usefulness of sampling procedures that differ so much from each other that they are not immediately comparable. This is a particular

form of the problem, appearing in economics and in the theory of decision making, of comparing the worth of alternatives.

6. There is a practical problem for which research is needed, namely, how can the difficulty and expense of obtaining information from samples be reduced through cooperation between survey organizations. This cooperation could occur in the preparation of basic materials and the utilization of the materials of one survey or study in other studies. There are other ways in which the costs, both overhead and field, can be spread and reduced. As a subsidiary problem there is the need for better cost analysis.

7. There is a problem common to the producers and the users of information: What should producers report and users demand, in addition to the basic survey information, to insure that it will be used wisely and to enhance the future value of similar data? Some description of the methods that were used in obtaining the information and of the losses or limitations that resulted from difficulties encountered in applying the methods is needed for intelligent understanding and use, but what and how much is needed is seldom clear. Research on the effectiveness of applications of data to practical and scientific uses could provide some answers to this question.

8. A great deal remains to be learned on the interrelations of the sampling operations and the other operations in a large and complex survey or experiment. This is particularly true for the measurement of attitudes and similar variables, for they involve the cooperation of the people on whom the observations are made and their results will be distorted by some sampling methods in certain situations. We need to know how much the observations are distorted in various circumstances and also how the methods influence the use of the data.

9. Some attention should be given to problems of informing potential users of the results of sampling about the principles of sampling and the possibilities of applying it in new ways. The usefulness of sampling depends to some extent on the prior acquaintance of many social scientists and leaders with the possibilities of sampling in problems similar to those they are facing. This problem involves a study of the effectiveness of taking courses in sampling, reading available books, seeking the advice of consultants, and communicating with people who are concerned with similar work.

10. Although a great number of systems of theory aimed at understanding and controlling sampling operations have been published in recent books and monographs, an extensive range of theoretical problems still remains to be studied. Hence, a final persistent prob-

lem is to guide the development of theory into important questions that remain unanswered and to facilitate the progress of the theory. Applications of sampling have stimulated most of the principal advances in sampling theory, and new applications are likely to do this in the future. A lively interchange of information among theorists and practical men should be very fruitful to both, even though there are difficulties in finding a common language and in understanding one another.

These problems deserve the attention of statisticians and social scientists. What is sketched out quite meagerly in the rest of this book should be replaced as soon as possible by more solid results.

In this chapter, we have recognized the universal importance of information about people and particularly what they think, how they feel, and what they want. We see that this calls for the measurement and surveying of many human variables—of the kind that are called opinions, attitudes, expectations, wants, and the like—and that the value of such measurements depends on their accuracy and the extent to which their quality is known to those who use them. Sampling contributes to the value of the results by permitting a reasonable balance between the need for accuracy and the cost of getting it. Research on problems of sampling, in turn, leads to better sampling and hence increases the value of the information-producing operation. Problems about sampling can only be answered in relation to the whole operation of which they are a part and in relation to the uses to which the information will be put. A broad and common interest in these problems of sampling is shared by practical men and scientists since they use the same data and methods to a great extent.

REFERENCES

1. Frederick Mosteller, Herbert Hyman, Philip J. McCarthy, Eli S. Marks, and David B. Truman, *The Pre-election Polls of 1948*, Bulletin 60, Social Science Research Council, New York, 1949.
2. Norman C. Meier and Harold W. Saunders, editors, *The Polls and Public Opinion*, Henry Holt & Company, New York, 1949.
3. Alfred C. Kinsey, Wardell B. Pomeroy, and Clyde E. Martin, *Sexual Behavior in the Human Male*, W. B. Saunders Company, Philadelphia, 1948.
4. William G. Cochran, Frederick Mosteller, and John W. Tukey, *Statistical Problems of the Kinsey Report on Sexual Behavior in the Human Male*, The American Statistical Association, Washington, D. C., 1954.
5. Charles F. Sarle and Others, "Discussion of Congressional House Committee Report of the Investigation of the Federal Crop Reporting Service," *Agric. Econ. Res.*, 5 (1953), 25–40.

6. U.S. House of Representatives, 82nd Congress, Committee on Agriculture, *Crop Estimating and Reporting Services of the Department of Agriculture,* U.S. Government Printing Office, Washington, D. C., 1952.

7. U.S. Bureau of the Census, Special Advisory Committee on Employment Statistics, *The Measurement of Employment and Unemployment by the Bureau of the Census in Its Current Population Survey,* Washington, D. C., 1954.

8. W. Edwards Deming, "On the Presentation of the Results of Sample Surveys as Legal Evidence," *J. Am. Statist. Assoc.,* 49 (1954), 814–825.

9. Sidney J. Cutler, "A Review of the Statistical Evidence on the Association Between Smoking and Lung Cancer," *J. Am. Statist. Assoc.,* 50 (1955), 267–282.

10. Burton R. Fisher and Stephen B. Withey, *Big Business as the People See It,* The Survey Research Center, University of Michigan, Ann Arbor, 1951.

11. Lewis H. Kimmel, *Share Ownership in the United States,* The Brookings Institution, Washington, D. C., 1952.

12. Burton Crane, "U.S. Stockholders Top 10 Million," *The New York Times,* July 24, 1956.

13. U.S. Federal Reserve Board, *Fed. Reserve Bull.* (monthly), Washington, D. C.

14. U.S. Bureau of Labor Statistics, *Family Spending and Saving in Wartime,* Bulletin 822, U.S. Government Printing Office, Washington, D. C., 1945.

15. U.S. Bureau of Labor Statistics, "Consumer Expenditure Study, 1950: Field Methods and Purposes," *Mon. Labor Rev.,* 72 (1951), 56–59.

16. Charles F. Sarle, *The Theory of Sampling as Applied to Crop Estimating,* Division of Crop and Livestock Estimates, U.S. Dept. of Agriculture, Washington, D. C., 1929.

17. F. Yates, "Crop Estimation and Forecasting," *J. Minist. Agric.,* 43 (1936), 156–162.

18. U.S. Bureau of Agricultural Economics, *The Agricultural Estimating and Reporting Services of the United States Department of Agriculture,* Miscellaneous Publication No. 703, U.S. Government Printing Office, Washington, D. C., 1949.

19. Daniel Katz and Hadley Cantril, "Public Opinion Polls," *Sociometry,* 1 (1937), 155–179.

20. Paul F. Lazarsfeld, Bernard R. Berelson, and Hazel Gaudet, *The Peoples Choice,* second edition, Columbia University Press, New York, 1948.

21. Angus Campbell and Robert L. Kahn, *The People Elect a President,* The Survey Research Center, University of Michigan, Ann Arbor, 1952.

22. Bernard R. Berelson, Paul F. Lazarsfeld, and William N. McPhee, *Voting,* University of Chicago Press, Chicago, 1954.

23. Angus Campbell, Gerald Gurin, and Warren E. Miller, *The Voter Decides,* Row, Peterson & Company, Evanston, 1954.

24. Samuel A. Stouffer and Paul F. Lazarsfeld, *Research Memorandum on the Family and the Depression,* Bulletin 29, Social Science Research Council, New York, 1937.

25. National Opinion Research Center, *What . . . Where . . . Why . . . Do People Read?,* Report No. 28, Denver, 1946.

26. Angus Campbell and Charles A. Metzner, *Public Use of the Library and Other Sources of Information,* The Survey Research Center, University of Michigan, Ann Arbor, 1950.

27. Alfred Politz, *BMB Evaluation Study,* a report submitted to the Broadcast Measurement Bureau, New York, 1947.

28. *A Study of Four Media,* conducted for *Life* by Alfred Politz Research, Inc., New York, 1953.

29. Samuel A. Stouffer and Others, *The American Soldier, Vols. I, II, III,* Princeton University Press, Princeton, 1950.

30. Irving L. Janis, *Air War and Emotional Stress,* McGraw-Hill Book Company, New York, 1951.

31. Charles E. Fritz and Eli S. Marks, "The NORC Studies of Human Behavior in Disaster," *J. Soc. Issues,* 10 (1954), 26–42.

32. Leon Festinger, Dorwin Cartwright, and Others, "A Study of a Rumor: Its Origin and Spread," *Hum. Relat.,* 1 (1947), 464–486.

33. Elihu Katz and Paul F. Lazarsfeld, *Personal Influence,* The Free Press, Glencoe, Ill., 1955.

34. Frank R. Kennedy, "Some Legal Aspects of Sampling," *Industr. Qual. Contr.,* 7 (1951), 24–27.

35. Eugene P. Sylvester, "Consumer Polls as Evidence in Unfair Trade Cases," *Geo. Wash. Law Rev.,* 20 (1951), 211–241.

36. Frank R. Kennedy, "Sampling by the Food and Drug Administration," *Food Drug Cosmet. Law J.,* 6 (1951), 759–774.

37. U.S. Food and Drug Administration, *Notices of Judgment under the Federal Food, Drug, and Cosmetic Act, March 1955, Cases 4321–4340,* U.S. Government Printing Office, Washington, D. C., 1955, Case 4327, pp. 301–305.

38. Lester E. Waterbury, "Opinion Surveys in Civil Litigation," *Publ. Op. Quart.,* 17 (1953), 71–91.

39. U.S. Bureau of the Census, "FCC Radio Survey," FCC *Docket* 6741, Washington, D. C., 1945.

40. Julian L. Woodward, "A Scientific Attempt to Provide Evidence for a Decision on Change of Venue," *Am. Sociol. Rev.,* 17 (1952), 447–453.

41. Joel Dean, W. Edwards Deming, Irston R. Barnes, Albert E. Sawyer, Martin R. Gainsbrugh, and Edward F. Howrey, "The Role of Sampling Data as Evidence in Judicial and Administrative Proceedings," *Current Business Studies,* No. 19, Society of Business Advisory Professions, Inc., New York, 1954.

42. Frederick F. Stephan, "Sampling in Studies of Opinions, Attitudes, and Consumer Wants," *Proc. Am. Phil. Soc.,* 92 (1948), 387–398.

43. Frederick F. Stephan, "Measuring Opinions, Attitudes, and Consumer Wants," *Mech. Engng., N. Y.,* 70 (1948), 432–433.

44. Stuart A. Rice, *Quantitative Methods in Politics,* Alfred A. Knopf, New York, 1928.

45. J. P. Guilford, *Psychometric Methods,* second edition, McGraw-Hill Book Company, New York, 1954.

46. Leon L. Thurstone and E. J. Chave, *The Measurement of Attitude,* University of Chicago Press, Chicago, 1929.

47. Samuel A. Stouffer, Louis Guttman, Edward A. Suchman, Paul F. Lazarsfeld, Shirley A. Star, and John A. Clausen, *Measurement and Prediction,* Vol. IV of Studies in Social Psychology in World War II, Princeton University Press, Princeton, 1950.

48. R. M. Thrall, C. H. Coombs, and R. L. Davis, editors, *Decision Processes,* John Wiley & Sons, New York, 1954.

49. Bert F. Green, "Attitude Measurement" in Gardner Lindzey, editor, *Handbook of Social Psychology*, Vol. I, Addison-Wesley Publishing Company, Cambridge, Mass., 1954.

50. Paul F. Lazarsfeld, editor, *Mathematical Thinking in the Social Sciences*, The Free Press, Glencoe, Ill., 1954.

51. Matilda White Riley, John W. Riley, Jr., and Jackson Toby, *Sociological Studies in Scale Anolysis*, Rutgers University Press, New Brunswick, N. J., 1954.

52. Carl I. Hovland, Arthur A. Lumsdaine, and Fred D. Sheffield, *Experiments on Mass Communication*, Vol. III of Studies in Social Psychology in World War II, Princeton University Press, Princeton, 1949.

53. Shirley A. Star and Helen MacGill Hughes, "Report on an Educational Campaign, The Cincinnati Plan for the United Nations," *Am. J. Sociol.*, 55 (1950), 1–13.

54. Leonard S. Cottrell, Jr., and Sylvia Eberhardt, *American Opinion on World Affairs in the Atomic Age*, Princeton University Press, Princeton, 1948.

55. C. Theodore Smith, "Seeing Ourselves as Others See Us," *Bell Teleph. Mag.*, 30 (1951), 26–37.

56. George Katona and Eva Mueller, *Consumer Attitudes and Demand*, 1950–1952, The Survey Research Center, University of Michigan, Ann Arbor, 1953.

57. Hadley Cantril, "Opinion Trends in World War Two," *Publ. Op. Quart.*, 12 (1948), 30–44.

58. Frederick Mosteller and Philip Nogee, "An Experimental Measurement of Utility," *J. Pol. Econ.*, 59 (1951), 371–405.

59. Milton Friedman and L. J. Savage, "The Utility Analysis of Choices Involving Risk," *J. Pol. Econ.*, 56 (1948), 279–305.

60. Russell L. Ackoff and Leon Pritzker, "The Methodology of Survey Research," *Int. J. Op. Att. Res.*, 5 (1951), 313–335.

61. Irwin D. J. Bross, *Design for Decision*, The Macmillan Company, New York, 1953.

62. Frederick F. Stephan, "History of the Uses of Modern Sampling Procedures," *J. Am. Statist. Assoc.*, 43 (1948), 12–39.

63. Statistical Office of the United Nations, *Sample Surveys of Current Interest*, Statistical Papers, Series C, Nos. 2–7, 1949–1955.

64. C. A. Moser, "Recent Developments in the Sampling of Human Populations in Great Britain," *J. Am. Statist. Assoc.*, 50 (1955), 1195–1214.

CHAPTER 2

Some General Questions
about Sampling

2.1 THE PURPOSE OF THIS CHAPTER

The purpose of this book is to help readers to a better understanding of sampling so that they may gauge its dependability and use it intelligently. We shall therefore start with an attempt to answer some of the broad questions which anyone might want to ask before he looks at details. The answers will be drawn from a wide range of facts and from some of the thinking that is developed in a more specific form throughout the book, particularly in Part III. These answers set forth assumptions that are often left unstated in discussions of sampling. We hope that they will tend to prevent some of the confusion and misunderstanding which might arise here as it has elsewhere. Our subsequent discussions of details will be seen in better perspective and be set in proper relation to each other if this chapter succeeds in this purpose.

2.2 WHAT ARE WE TRYING TO SAMPLE?

By our title and introductory chapter, we have directed this book at the measurement of opinions, attitudes, and consumer wants. The reason for linking sampling problems to this field of measurement is that special problems of sampling are developed, or at least raised to major importance, by the conditions under which such measurements are made today. When greater progress has been made on the problems of measurement, these distinctive features of sampling human variables may tend to disappear. When they do, the problems of sampling can be handled on a broader and more satisfactory basis throughout. Until that time, however, we believe it will be necessary to give special attention to sampling that involves human responses of the kind commonly referred to as opinions, attitudes, wants, beliefs, prejudices, preferences, intentions, expectations, and even those more elusive and sustained responses designated as satisfaction and morale.

"But," we may ask, "How can you possibly sample such vague and elusive things as these? How can you get a hold on them and measure them? Really, now, aren't you just sampling people?" This is an important point and if we ignore it we shall surely find ourselves confused before long. The answer will be useful in the chapters that follow.

Actually, in most surveys not one but several samples are taken. For example, a typical survey in a city may take in succession: (a) a *selection of small areas* or districts within the city, such as city blocks or parts of blocks, drawn from maps and Census reports; (b) a *selection of names* of persons to be interviewed, obtained by visiting each dwelling in the areas in sample a, listing all the people who live in it, and determining by some standard rule those who should be interviewed; (c) *the sample that consists of people who are actually interviewed*, excluding those whose names were in the sample drawn in b but who could not be seen, refused to be interviewed, or were omitted for other reasons; and (d) *the sample of information* that is recorded by the interviewers on the basis of the answers they received when they asked questions of the people in sample c.

Thus there are at least four distinct samples, each derived from the one preceding. They are: a sample of areas, a sample of names, a sample of persons, and a sample of their responses. They represent stages in the progress of the survey, beginning when we take the first step toward the interviews that are desired and ending when the recorded information is assembled in the survey offices. Other samples may be identified in the survey but the samples that are listed here show that we do not merely draw a sample of people. We make successive selections until we finally obtain *a sample of results or measurements*. These results form the final sample; they are a sample of *data*. All the survey findings and conclusions are derived from them.

Something more must be said in answer to this question about what is sampled. It might seem that what we obtain is *all the data* from the persons interviewed. We see readily that some areas were not selected and some people were not interviewed, but we may wonder how any answers to questions could have been left out of sample d, assuming of course that the interviewers faithfully recorded the answers to every question that they asked. To understand the sampling that produced d, we must consider the different answers that might have been given if the interviewer had come a few minutes or hours earlier or later, if the weather had been different, if the interview had not proceeded exactly as it did, or if any of a number of other factors had interfered in a different way.

The notion of possible answers is difficult to grasp and somewhat complicated, but it is not unlike our conception of what would happen if one of us wrote a long letter to a friend, then set it aside, and tried to write the same letter over again. The second letter would differ somewhat from the first. So would a third and a fourth letter. If a person could continue this process without becoming tired or careless and without remembering more than he did when he sat down to write the first letter, then he would produce a large stack of letters, each of which he could have written the first time and sent to his friend. Only by chance did he write the first letter the way he did.

The notion of a sampling of possible answers is also similar to making a number of measurements on an object, any one of which might just as well have been the first or the only measurement to be made. We can look at any of a number of clocks to tell the time or any of a number of thermometers to tell the temperature, and each will give us a slightly different result. The conception of a much larger body of possible data that might be obtained under the same conditions is just as important, though not as familiar, in interviewing as it is for repeated measurements in those fields of science in which the use of instruments naturally leads to this broader conception of data.

The sampling of possible responses is important because the variation it introduces into survey results is too great to be ignored. Hence we aim this book at the sampling of data that are subject to a substantial degree of inaccuracy in the measurement processes by which they are produced. We anticipate that great improvements will be made in the accuracy of such measurements in the future. As they come, the problems of sampling will be changed. In the meantime it is necessary to make a clear distinction between a sample of *people* and a sample of the *data* or the information that is obtained from them.

2.3 JUST WHAT IS SAMPLING?

The basic idea in sampling is simple: (*a*) we seek information about a whole class of similar objects, (*b*) we examine some of them, (*c*) we extend our findings to the entire class. This fundamental aspect of sampling is quite familiar. All our experience with the world is of this *partial* nature. We learn about people and about physical objects by accumulating experience. We assume that they will continue to be in the future approximately what they have been in the past. It is in this way, for example, that the common conceptions of national and regional character are formed from a few observations. New Englanders are believed to be cold and reserved. Southerners are easygoing and very hospitable. Scotsmen are thrifty. We can extend the list as we please.

A conception of sampling that applies to all such generalizations from limited experience is too broad to be useful for our present purposes. Something more is needed to make it definite and fully serviceable. It is necessary to restrict further the scope of the notion of sampling. We shall define it in the following way:

Sampling is the use of a *definite procedure* in the selection of a part for the express purpose of obtaining from it descriptions or estimates of certain properties and characteristics of the whole.

Although a sampling procedure may depend on expert knowledge and skill, ordinarily it can be expressed in formal instructions that are communicated to another person who then carries out the procedure. The procedure is usually one phase of a larger and more complex activity directed toward obtaining information. The sampling procedure itself may be simple or complex. It may be quite distinct and separate from the other phases of data collecting, or it may be combined with them in such a way that no one can separate it out completely. This emphasis on a definite communicable procedure does not, of course, draw a sharp line between sampling and similar activities. It does, however, make the first step necessary for sound analysis of the reliability of sampling, and it opens the way for further restrictions that can produce still sharper definitions.

This definition of sampling is necessarily somewhat arbitrary. Moreover, it is not, and cannot be, perfectly adequate to distinguish sampling from other procedures. Modern sampling technology includes some procedures in which the sample that is drawn does not necessarily represent or resemble closely the whole from which it is derived. Estimates are made from it, however, by methods that undo the distortion intentionally introduced in the sample selection process to achieve greater economy or accuracy. By the combination of procedures, adequate accuracy and representativeness are obtained in the final estimates. This inevitably entangles sampling with *estimation* and *analysis,* so that it is impossible to draw a clear and sharp distinction between them. However, estimation methods are not completely intertwined with sampling procedures, for they often make use of material provided by sources other than the sampling.

The definition of what is sampling and what is not had best be left a bit flexible and further distinctions made explicitly when they are needed. Future developments of sampling are bound to raise new questions of definition and make all neat logical definitions impractical to use.

2.4 WHAT CAN WE ACCOMPLISH BY SAMPLING?

It is obvious that by sampling we can obtain important information we could not get otherwise, unless we took pains to examine all the persons or objects to which our findings and conclusions will apply. However, there is more to the story than just this. If sampling is a more or less formal, deliberate, and communicable procedure, then it may offer several opportunities not offered by a procedure of complete examination.

1. We can plan or adjust the procedure so as to obtain the information more efficiently, more conveniently, at lesser cost, and to greater advantage in other ways, until a practical limit is reached for such improvements.

2. We can learn something about the dependability of the information from the procedure itself, in addition to what we can learn from other sources.

3. We can gain certain additional advantages in cost, accuracy, and extra information by employing a series of sampling operations of a sufficiently similar character, standardizing some parts of the procedure and using the same personnel and resources for more than one operation.

In general, sampling makes feasible many studies that could not be conducted on the basis of a complete census or comprehensive survey. It also makes the results available more quickly and often yields results of higher quality. On the other hand, in many surveys there are serious limitations that result from the restricted size of the sample. Fewer comparisons may be made between parts of the whole because the sample does not provide enough items in the smaller parts for accuracy. Nonetheless, sampling may also permit an extension of the scope of a survey and thus permit more comparisons than a complete canvass. Fundamentally, sampling widens the range of choice among ways of conducting a quest for information. Since it need not exclude the possible choice of a 100 per cent sample or complete coverage, it does not curtail the possibilities from among which we can choose the most advantageous.

2.5 WHERE CAN WE USE SAMPLING PROCEDURES?

There would seem to be no general limit to the possibility of using sampling in any situation in which we can make measurements, but there may be obstacles in particular cases. In some instances it is necessary to obtain complete coverage because each measurement is, or may be, of the utmost importance. Such instances do not occur very often in the field of human response. Severe practical difficulties beset some sampling operations. Suspicions may be aroused if questions are only asked of a few persons and not of everyone. Persons selected for the sample may feel that they are imposed upon because they are requested to give something others are not required to give. In emergency situations it may be difficult to follow any definite procedure and information may then be obtainable only by expedient measures and by improvising. People who are to use the information may be prejudiced against conclusions that come from anything short of complete coverage. These are all special situations. They are not often insurmountable.

Sampling studies have been made of a great variety of people: illiterate populations, college graduates, soldiers, civilians in occupied countries, customers, employees, voters, subscribers to publications, young people, and the aged. They range from the people who work in a single building to samples of an entire nation. The use of sampling is being extended to groups not previously sampled.

The practice of sampling is likely to become the most common means of obtaining information by more than casual observation. It will become more convenient and more valuable as it is better understood.

2.6 HOW IS SAMPLING DONE?

Some samples are drawn by an expert who looks over a body of material and selects from it a portion that seems to be representative of the entire lot; for example, a scientist selects specimens to be exhibited in a museum or a research report writer selects "typical" examples of the verbatim remarks made by people when they answered the interviewer's questions.

Sometimes interviewers are given general instructions about the districts in which they are to interview and about the number of men, women, older and younger people, people in different economic levels, and other broad classes that they are to seek for interview. They are also instructed in making what are regarded as representative selections within such groups but are left free to approach and interview the particular persons they think are appropriate. This procedure is called "quota sampling." It will be described in greater detail in subsequent chapters, especially Chapters 3 and 12.

A considerable variety of procedures is based on selection of names from directories, registration records, and other lists of residents, employees, stockholders, members, etc. The ration lists in European countries have proved to be an excellent source of such samples. Some samples are selected from maps or lists of areas. These methods and others are often combined in more complex methods. After names or areas have been selected, the next step consists of visiting the persons who are selected, calling them to a meeting, sending them questionnaires through the mail, or using some other means of obtaining from them the data needed to complete the sampling operation. Difficulties that arise in these stages lead to various attempts to overcome the problems of finding people who are hard to reach and enlisting the cooperation of those who are reluctant. Like any complicated operation, sampling often involves a thoroughly planned and well-controlled program of work. Hence it is usually done by a group or organization rather than an individual. A description of the general and specific methods used is given in Chapter 3 and in Part III.

2.7 WHAT IS THE BEST METHOD OF SAMPLING?

There is no one particular method of sampling that is uniformly the best for all purposes and all situations. Nevertheless it is possible to make a rational choice of a method that is close to being the best, considering the information that is available in advance, the principles of rational choice that are generally accepted, and the restrictions that are imposed on the sampler. The problem is not unlike that of choosing the best house to live in. Many houses can be eliminated as not suitable or as inferior to others on almost every score. The final choice depends, however, on the needs and tastes of the people who are to live in it, the rent or price, and other considerations of like character.

It is no longer necessary to choose sampling methods imitating those used by someone else. It is possible to design excellent procedures according to general principles and to fit these procedures readily to the particular needs of the prospective user.

2.8 HOW CAN WE JUDGE THE DEPENDABILITY
OF A SAMPLE?

There are several ways to judge dependability: (a) by the past performance of the survey organization in similar studies or by other comparable experience, (b) by independent tests and audits, (c) by analysis of internal evidence and of the operating records of the survey, and (d) by the application of well-established principles and doctrines which establish presumptive judgments of dependability. All these methods are incomplete; a combination of them may be more nearly adequate than any one alone. They can be applied, of course, with different degrees of stringency ranging from irrelevant criteria to the most exacting tests.

The record and reputation of the organization is a helpful indication but it may be misleading. We should ask, "How often and how severely has its work been put to a conclusive test?" and "Do the results of these tests really apply to the present situation and the sample with which we are concerned?" The *Literary Digest* engaged in large-scale polling for twenty years before its reputation for polling was ruined in 1936. Its methods had worked fairly well in the Presidential elections of 1932, and though they had been criticized previously for having a Republican bias, the extent of their error in 1936 (19 percentage points) came as a great surprise. Elmo Roper's forecasts of the Presidential vote had been highly accurate from 1936 to 1944 (1.5, 0.5, and −0.2 percentage points for the Democratic share of the vote for the President), but in 1948 his error was −12.3 percentage points. Gallup's and Crossley's errors in 1948 were more than double their errors in 1944 and were consistently in the Republican direction. Should we expect the experience prior and subsequent to 1948 to be a fair indication of probable future accuracy or should the experience of 1948 be taken as a better indicator? Should these errors in 1948 be charged to the sampling procedure, to other parts of the survey, or to changes that occurred between the polling and election and to discrepancies between the intentions expressed by potential voters on the one hand and their actual behavior on election day? This question will be explored further in Chapter 7 and elsewhere in this book, but the general difficulty of depending on reputation and the record of past performance is evident.

No one would doubt that the competence and past achievements of an organization are valuable guides in assessing its current work. The difficulty is that a change may occur in some of the essential conditions that contributed to the success of its work in the past. Without any slackening of its efforts it may run into a period of poor results until it manages to discover and correct what is wrong. Experience is a good guide to point out possible sources of trouble but a poor guarantee of safety.

Independent tests are perhaps the most effective way to determine the dependability of samples, but they are not always feasible. They are often very expensive since, to be effective, they must do the job they are testing much better than it was done in the sampling operations. This may be feasible when highly accurate information becomes available some time after the sampling has been completed. For example, when the annual income tax tabulations are completed, they can be used to check previous estimates based on sample surveys of income (1, 2, 3). Unfortunately, few if any such tests can be made on opinion and attitude surveys except those provided by election returns (4, pp. 41–51). Response errors as such can be tested against independent records and by reinterviewing a subsample of individual respondents (5, 6, 7). The quality check conducted immediately following the 1950 Census of Population (8) included an elaborate program of such tests. Testing and calibrating should be practiced more frequently, as they are in manufacturing and in scientific measurement.

The internal evidence of dependability may take many forms. There are well-developed techniques of discovering from an examination of schedules certain kinds of repeated "cheating" or gross deviation from instructions on the part of interviewers. The estimation of the sampling variability or stability of the data by methods like those that are utilized in Chapter 10 is another kind of internal evidence. The biases of measurement and sampling are not likely to be found by this approach, and many kinds of factors that weaken dependability usually escape it. Hence this approach can only supplement the evidence we should seek from the other sources.

Often the most important consideration in determining how far to trust a sample is the procedure by which it was obtained. We may judge what can be expected to result in any instance by examining (a) the nature of the procedure, (b) our knowledge of the situation in which it was used, and (c) the record of experience with the procedure in other studies under similar circumstances. This kind of knowledge about samples and the procedures that produce them is built up both by the analysis of experience and by progress in theory. In time it becomes incorporated in a body of doctrine or "know how" which is shared by people actively engaged in such work.

These formulations of doctrine may be too restricted and exclude practices that are quite dependable when they are applied competently under appropriate circumstances. Some of the rules and doctrines may actually be false. Critical examination of them in the light of fresh experience, experimentation to test them, and utilization of the related scientific generalizations will ultimately correct whatever errors are embodied in these principles. Their value lies in the fact that until better knowledge is at hand they provide guidance to the taking and using of samples. They make it possible for us to act in an intelligent manner and not just blindly on hunches and custom.

Samples are like medicine. They can be harmful when they are taken carelessly or without adequate knowledge of their effects. We may use their results with confidence if the applications are made with due restraint. It is foolish to avoid or discard them because someone else has misused them and suffered the predictable consequences of his folly. Every good sample should have a proper label with instructions about its use. This is why many surveys now include rather full descriptions of the methods that were used and the limitations of the results.

REFERENCES

1. Selma F. Goldsmith, "Appraisal of Basic Data Available for Constructing Income Size Distributions," in *Studies in Income and Wealth,* Vol. 13, National Bureau of Economic Research, New York, 1951.
2. George Katona, *Psychological Analysis of Economic Behavior,* McGraw-Hill Book Company, New York, 1951.
3. Robert Wasson, Abner Hurwitz, and Irving Schweiger, "An Appraisal of Field Surveys of Consumer Income," in *Studies in Income and Wealth,* Vol. 13, National Bureau of Economic Research, New York, 1951.
4. A. Angus Campbell and George Katona, "The Sample Survey: A Technique for Social Science Research," in Leon Festinger and Daniel Katz, *Research Methods in the Behavioral Sciences,* The Dryden Press, New York, 1953.
5. John A. Clausen, "Studies in the Postwar Plans of Soldiers: A Problem in Prediction," in Samuel A. Stouffer and Others, *Measurement and Prediction,* Vol. IV of Studies in Social Psychology in World War II, Princeton University Press, Princeton, 1950.
6. Leslie Kish and John B. Lansing, "Response Errors in Estimating the Value of Homes," *J. Am. Statist. Assoc.,* 49 (1954), 520–539.
7. John B. Lansing and Stephen B. Withey, "Consumer Anticipations: Their Use in Forecasting Consumer Behavior," in *Studies in Income and Wealth,* Vol. 17, National Bureau of Economic Research, New York, 1955.
8. A. Ross Eckler and Leon Pritzker, "Measuring the Accuracy of Enumerative Surveys," *Bull. Int. Statist. Inst.,* 33 (1951), pt. 4, 7–25.

CHAPTER 3

The Variety and Characteristics
of Sampling Procedures

3.1 INTRODUCTION

Our primary concern throughout this book is the use of sampling to obtain information relating to the distribution of various measures of attitudes, opinions, consumer wants, and related variables among the members of a specified human population. Many different procedures have been suggested and used for choosing samples of persons from such populations. Some of them are widely known and well understood. Many of them can only be described in vague and superficial terms. Some require a rather technical description to make clear the manner in which they operate to produce samples and their relative strengths and weaknesses for this purpose. This chapter will describe some of the types of sampling that are frequently reported and discuss certain of their characteristics. It will also outline a broad classification scheme and then indicate the general features that appear to be common to all sampling operations.

Sampling methods have frequently been labeled in papers and reports with very general titles such as "representative," "random," "quota control," "purposive," "judgment," "stratified," "area," "precision," "unbiased," "probability," and the like. Actually, such general labels have little meaning, either because there is no commonly accepted and precise definition of the term applied or because there exist many variations of the general method, both in concept and application. For these reasons it is important that survey operators provide adequate descriptions of the procedures they use. Moreover, each description should include an account of the way the sampling was actually done and how it differed from the way it was planned. Even here, however, we must acknowledge that there is an arbitrary element in stating what constitutes an adequate description, since it is never possible to set forth every relevant detail of a sample survey.

3.2 REPRESENTATIVE SAMPLES

The first aim of most sampling procedures is to obtain a sample of people that will *represent* the population from which it is selected. In other words, it is expected that results obtained from the sample will agree "closely" with the results that would have been obtained had the entire population been subjected to study. This same idea has frequently been stated by saying that a sample should be a miniature population or universe.

Many different procedures, arising from intuition or common sense, have been suggested and used for obtaining representative samples. Thus it seems fairly obvious that the sample should be scattered evenly throughout the population being sampled. A simple illustration of such scattering is the sampling of cars which cross a bridge during a fixed interval of time, every tenth car being drawn into the sample. Bowley obtained such a scattering of his sample in 1912 when he selected every twentieth household in Reading by counting down the street lists in the directory. Another intuitively appealing method of obtaining representative samples is to select the sample in such a manner that it *must* agree with the population in certain respects. Illustrations of specific applications of this type will be given later in this chapter.

The foregoing remarks have been phrased as though the quality of representativeness were absolute, that is, as though any given sample of people could be classified as either entirely representative or entirely non-representative of the population from which it is drawn. Actually, this is not the case. In the first place, a sample may represent the population with respect to one characteristic or variable and may not represent the population with respect to other variables. For example, suppose that in selecting a sample of pupils from a school, only those pupils in the first grade were taken. Such a sample would certainly be very far from representing the school population with respect to age. However, the fraction of pupils in the first grade having blue eyes might be very close to the corresponding fraction in the entire school. Consequently we see that representativeness must be specified not only with respect to a population but also with respect to one or more variables. In the second place, it is clear that representativeness is also a matter of degree. A sample selected from the children in each of the first three grades would be more representative of the school with respect to age than a sample selected only from the first grade, though neither sample would be as representative as the sample made up of all the blue-eyed boys in all the grades in the school.

The problem of the relative nature of representativeness is particularly marked when human populations are being sampled for the measurement of opinions, attitudes, and consumer wants. The difficulties arise because in practice we must first obtain a sample of individuals in order to obtain a sample from the population of all opinions or attitudes on a specified topic, even though it is only the population of opinions in which we are interested. This means that a sample of persons may be quite representative according to some individual traits but still not yield a very representative sample of opinions when the measurements have been made. Moreover, it is a rare survey in which only one variable is studied. Consequently, it is necessary to balance the requirements for representativeness, one variable against another.

In view of these considerations, it is quite clear that the term "representative sample," by itself, can never be given a precise meaning. Accordingly, we offer the following definition:

A *representative sample* is a sample which, for a specified set of variables, resembles the population from which it is drawn to the extent that certain specified analyses that are to be carried out on the sample (computation of means, standard deviations, etc., for particular variables) will yield results which will fall within acceptable limits set about the corresponding population values, except that in a small proportion of such analyses of samples (as specified in the procedure used to obtain this one) the results will fall outside the limits.

Any individual or agency is, of course, free to adopt its own standards of representativeness, and these may well vary from investigation to investigation. However, it is only when these standards are set forth in detail, together with the reasons for assuming that the sample conforms to them, that the term takes on a precise meaning. Thus the mere statement or claim that a sample is representative of a population tells us nothing. This is not necessarily a condemnation of the sample itself but only of the incompleteness of the description of the sample.

3.3 SAMPLING FROM LISTS OR OTHER NATURAL ORDERINGS OF POPULATION ELEMENTS

A sample can be drawn from a list by taking items from the list at regular intervals—say, every fifth or every twenty-third or in general every kth, where k is some integer that will yield a sample of the right size. When every kth element is taken from a list or from some other form of natural ordering, we say that a *systematic sampling procedure* is being used. Systematic sampling has a strong intuitive appeal, for it spreads the sample elements uniformly over the population. This appeal is particularly strong when the ordering is a geographic one, but it also exists in many other cases. Thus the sales-girl who hands out a questionnaire to every tenth person making a purchase spreads her sample out evenly over all her customers with respect to the order in which they appear at her counter.

Even with a simple procedure such as systematic sampling, intuition is not always an adequate guide to the degree of representativeness that may be achieved by a given sample. For example, there may be a relationship between the order of occurrence and an element's characteristics that will produce unexpected or "queer" results. As a simple example, consider the following. During World War II, there was a research agency connected with the Army which was engaged in finding out how the soldiers felt on particular subjects. When a member of this organization went into a camp, he obtained a list of the soldiers in the camp from which to draw a sample. An obviously convenient procedure was to draw a systematic sample from this list. In one particular instance in which such a sample was being drawn, some of the characteristics of those men drawn into the sample were examined. It was found that much too great a proportion of master sergeants was selected for the sample. Upon learning this, the surveyor had no trouble in finding the reason. The men in the camp were listed barracks by barracks. Within each barracks the men were listed in order of their military rank. Also there were the same number of men in each barracks. Unfortunately, the chosen sampling interval was such that it stepped off the same line numbers on list after list, and in this way it happened to fall again and again on lines that usually contained the names of master sergeants.

In addition to problems such as the above, many practical difficulties may arise in applying systematic sampling. For example, in sampling from lists or schedules, some of the records may be out of the file, in use, or waiting to be refiled and consequently may be overlooked entirely. Further accounts of such difficulties will be found in a paper by Stephan (1). Moreover, in any large-scale undertaking, problems of time and cost are of the greatest importance.

If the data are to be obtained not only from the list but from further inquiry about each case in the sample, new problems arise. It is one thing to select a systematic sample of 1000 names from a list, and it is quite another thing to obtain a sample of data by personal interviews with each of these persons when they are located in 700 or 800 towns scattered across the United States. These practical problems of time and cost require modifications in the selection procedure for their solution, and these modifications are usually designed to obtain some geographic clustering of the sample elements. In other words, one interviewer can obtain six or seven interviews in one town much more quickly than he can obtain the same number of interviews if they are scattered over two or three counties.

A relatively simple example of this type of sample design was used to select a sample of names from the list of subscribers to the magazine *Country Gentlemen,* each selected subscriber to be visited for personal interview. Stripped of details, the procedure was as follows. First a systematic sample of names was selected from the list. Since the names were arranged in a geographic fashion, this imposed a geographic spread on the sample. The size of this sample was determined by the number of *different geographic areas in which it was feasible to conduct interviews.* After these names had been selected, an automatic procedure was devised to associate with each of them an additional group of names such that (*a*) the total sample was of the required size and (*b*) the additional names were "geographically close" to those originally chosen. Most samples used in personal interview surveys are necessarily subject to similar considerations, except when the population is confined to a very compact geographic area. The use of mail questionnaires avoids this problem of "scattering," but as is well known, mail questionnaires present other problems.

3.4 RANDOM SELECTION

Some of the difficulties that may arise in the use of systematic selection can be traced directly to the fact that the inclusion or exclusion of any given element may not be independent of certain characteristics of the element which are associated with its position or order in the list. The same statement can of course be made even more strongly in instances where *self-selection* is operative (e.g., where an individual decides whether or not he will respond to a mail questionnaire), where *convenience selection* is used (e.g., where those elements are taken which are conveniently located with respect to the sampler), or where *judgment selection* is used (e.g., where the sampler decides whether or not an element will be included in the sample).

It is therefore natural that we should look for a method of selection which bears no relation to the properties or characteristics of individual elements, and in which no subjective influence is allowed to affect the determination of what elements will make up the sample. Such procedures can be obtained by considering the operation of games of chance, lotteries, and other related activities. A simple procedure for performing this type of selection comes to mind immediately. Suppose we assign a number to each population element and write the numbers on identical cards. Having done this, we shuffle the cards thoroughly and draw "blindly" a number of cards equal to the required sample size. The population elements with the corresponding numbers will then complete the sample. It is quite clear that this process removes all subjective influences from the selection procedure, and also that the characteristics of individual elements have no bearing on whether or not they are included or excluded from the sample.

When such a selection procedure as this is used, we say that every element in the population has an equal chance of being drawn into the sample and, furthermore, that each of the possible sets of elements of the required size has the same chance of being selected. We call it *random selection.*

Even when systematic selection may be assumed to give a representative sample, it is sometimes preferable to use random selection. For example, suppose a garage wished to survey customers' attitudes toward repair work done in the shop. This could be done by questioning every fifth customer. However, this might allow shop employees to figure out in advance which jobs would be followed up, and consequently to give these jobs special attention. The use of random selection would remove this difficulty.

Random selection is one example of a method which conforms to theoretical principles from which deductions can be made about the performance of the method. In this instance, the principles assume that there is assigned to each population element, in advance of the draw, the same probability of being included in the sample. Provided that a mechanical procedure can be devised for selecting elements in accordance with this principle, probability theory can then be helpful in predicting how the method will work in practice and thus assessing the representativeness of the chosen sample.

If random samples are drawn repeatedly from the population, estimates made from the samples will vary from time to time around an average value. Probability theory permits us to deduce how closely these estimates will agree with the results that would have been obtained if the entire population were subjected to the same measurement and estimation procedures. A similar approach can be made to systematic sampling, but the methods are necessarily more complex (2).

3.5 WHY COMPLEX SAMPLING PROCEDURES MAY BE REQUIRED

The simple procedures of random and systematic selection described in the preceding sections can frequently be applied if lists of population elements exist. Even when such lists are not available, there may still be a possibility of using these procedures, provided the population is of relatively small size and is confined to a relatively compact geographic area. For example, we might walk along every street in the area that contains the population and draw every kth dwelling unit into the sample. However, even here we must proceed with care in the application of the scheme. Some of the dwelling units may be missed because they are located in alleys, in the rear of stores, and so on. Moreover, it is necessary to specify exactly what population is being sampled. If we are interested in dwelling units, then information can be obtained from any responsible person in each of the selected units. On the other hand, if we are interested in the population of individuals living in the dwelling units, there is the further problem of obtaining respondents from among the people living in the selected dwelling units. If we always take the person who comes to the door, the sample will contain too many housewives and other persons who are usually at home. In this instance it is possible to use random selection to obtain one person from each dwelling unit, or we may wish to interview every person in the selected units. Probability theory and experience must serve as the guides in deciding what is to be done.

In the event that the population is large and scattered over a wide geographic area, it is frequently too costly in terms of time, money, and personnel to carry out a procedure such as that outlined above. Consequently, some other form of sampling must be devised. The general outlines of the procedures that are currently being used under such circumstances are relatively easy to describe and understand. We shall now consider a number of illustrative examples. For the sake of simplicity, we shall confine our attention to city sample designs. The procedures that will be described can be carried over, with little change, to the sampling of populations which are located in rural areas or in areas of both urban and rural population.

3.6 QUOTA SAMPLING

For a long time the dominant form of sample design in opinion surveys has been "quota sampling." This procedure was developed by Cherington, Roper, Gallup, and Crossley, whose sample surveys of opinion became widely known after the presidential election of 1936, and by other market research and opinion survey organizations whose studies received less publicity. Its basic features have been widely adopted, partly in imitation of the more widely known surveys and partly because of their attractiveness with respect to costs and convenience. In recent years it has been superseded in many surveys by various forms of "probability sampling."

As with many other sample designs, a quota sampling procedure starts with the premise that a sample should be scattered over the population and should contain the same fraction of individuals having certain characteristics as does the population. Census data and other information are used to divide the total population into a number of mutually exclusive and exhaustive sub-populations or *strata* and to determine or estimate (depending on the adequacy of these external data) the fraction of the population in each of these strata. The total sample is then allocated among these strata—geographic regions, age and sex groups, racial groups, and the like—in proportion to their real or estimated size.

The imposition of controls on the sample, which are external to the survey, is carried out through stratification in many other types of sample designs. Therefore, the mere statement that a quota sample is a "stratified" sample is not sufficient to distinguish it from other sampling procedures. Many other considerations have to be raised. What strata were used? How were the strata defined? How was the sample allocated among them? How was sampling carried out *within the strata?*

The final step in quota sampling is to apportion the sample among the interviewers and tell each the number of persons he is to obtain in each one of the strata. Thus he is told to obtain so many men and so many women, so many persons of high income, and so many of low income, and similarly for the other characteristics used to control selection. Interviewers are free to choose the particular respondents they will interview, subject only to the restriction that the quota requirements be fulfilled. A detailed analysis of such a sampling operation is given in Chapter 12.

If it were true that the interviewers selected their respondents in a purely random fashion within each stratum or "quota control" group, then quota sampling could be labeled as stratified random sampling. Actually, it is extremely difficult to see how we can regard interviewers as completely random selecting devices. There are many ways in which the conscious or unconscious likes, dislikes, and habits of the interviewers may influence their selection of respondents. For example, an interviewer may be attracted or repelled by the appearance of a house; he may concentrate his work in certain sections of a city and miss others; and he may fail to reach certain kinds of people because he selects those who are most convenient to interview.

This lack of full control has long been recognized as a fundamental weakness of quota sampling by both proponents and opponents of the method. It has led to various modifications in the method intended to make the interviewer operate like a mechanical selecting device. This may be done by specifying more definitely the procedures he is to follow in selecting people to interview, or by narrowing the group from which he makes his selection to such an extent that the subjective element can have little effect. The instructions issued to interviewers caution them against neglecting the lower education and income groups, request them to make their interviews in the home or on the farm, stress the necessity for making evening calls, and so on. Many of these points are covered in the manual prepared by the National Opinion Research Center (3) and in similar manuals used by other survey organizations. The group from which the interviewer selects his respondents is usually narrowed by telling him to obtain interviews only on specified blocks, in specified sections of a city, etc.

There are many different ways in which the foregoing considerations, and others that have not been mentioned, may be translated into a definite quota sampling procedure. Therefore it is not sufficient simply to state that quota sampling was used in a survey and expect anyone to have more than a very general idea of how the sample was drawn. A substantial portion of the design must be described before we can even state that two such procedures are roughly similar. A number of specific quota procedures will be described and referred to in Chapters 8, 10, and 12.

3.7 PURPOSIVE SELECTION OF AREAS

There are many situations in which it may be necessary to conduct interviews in a relatively small number of compact groups or areas. That is, it may be impossible to identify individual population elements, whereas groups of elements can be specified (workers in a plant, members of social or professional organizations, people living in defined geographic areas, and the like); also, the geographic "clustering" of sample elements will tend to reduce the amount of time and the number of personnel required for obtaining a given number of interviews. Under these circumstances, we may wish to conduct the sampling in two stages. A sample of groups or areas is chosen first, and then a sample of elements is taken from each of the selected groups.

There are many ways in which this sampling of areas, or groups, can be accomplished; at this point we shall present a brief description of one approach which is closely related to the fundamental premise of quota sampling, namely, that forced agreement between sample and population on a number of characteristics or *controls* is desirable. The particular example to be considered ultimately ends with the assignment of quotas to interviewers. However, prior to the determination of these quotas, a sampling of areas was carried out by what we may call *purposive selection*.

In making predictions for the 1948 presidential elections, the *Philadelphia Evening Bulletin* used the following sample design (4). Philadelphia has a total of 52 election wards, each of which is divided into a number of voting divisions. After each election the Registration Commission publishes detailed reports on voting and registration by wards and divisions. The *Bulletin* selected 25 wards from the city which, when combined, voted in the same proportions as did the city as a whole, over the three election years 1942, 1944, and 1946. From each of the selected wards, two divisions were chosen which, when combined, voted over the three election years in exactly the same pattern as the ward itself. The total number of interviews to be obtained was then allocated among the divisions in proportion to the total voting population. The interviewers were given maps showing the boundaries and the number of interviews to be obtained from each division, and the sampling proceeded from this point in a fairly typical quota fashion. The interviewers were not given specific addresses within the divisions but were instructed to scatter the calls, one or two to each block. White-Negro quotas were assigned to each division and checks were maintained on the sex and economic composition of the sample.

A procedure very similar to this (at least as far as the selection of areas is concerned) was used by the American Institute of Public Opinion (AIPO) to design state samples in connection with the 1944 and 1948 presidential elections. It has been described by Benson,

Young, and Syze (5) who refer to it as a "pinpoint" sample. The current method of "precinct" sampling that is used in Gallup election polls appears to be a development of "pinpoint" sampling.

It is quite clear that many variations of this fundamental approach are possible—depending on how the areas are defined, how the *control* variables are chosen, and how the sample of individuals or households is taken from each of the selected areas. Therefore it is not sufficient merely to state that a purposive sample of areas has been taken. The details must be specified so that others will know what was done.

3.8 SELECTION OF AREAS IN ACCORDANCE WITH A PROBABILITY MODEL

When a list of the population elements is available, the methods of random and systematic selection may be used, and they may be made to conform to theoretical models from which deductions can be made about the long-run performance of the method. We may well ask whether such models exist in the more complex situation where population lists are not available. As has been shown by the work of the Indian Statistical Institute, the Bureau of the Census, the Bureau of Agricultural Economics, and Iowa State College, and many other agencies too numerous to mention, the answer must be in the affirmative. In the material which follows, we shall speak of a sampling method as conforming to a probability model if it is possible, in advance of the actual selection of the sample, to assign to each possible set of population elements a known probability. These probabilities give the fraction of times that each possible set would be drawn if the sampling procedure were repeated an indefinitely large number of times.

These concepts can be illustrated by considering a probability model design which might be used in a city. We could first obtain maps and divide the city into a large number of smaller areas, or *primary sampling units*, as they are usually called. These primary sampling units might be blocks, portions of blocks, or any other areas which seemed desirable. The areas must be mutually exclusive, must be easily identifiable, and must include between them all the population. From the totality of these units a sample of units is selected. In the event this is done by simple random selection, the probability that any given unit will be in the sample is simply n/N, where N is the number of units in the city and n is the number of units to be drawn into the sample.

Once the primary sampling units have been chosen for the sample, field workers are sent out to prepare a list of every dwelling unit located within their boundaries. Then a random sample of dwelling units is selected from this list. If, for example, a 10 per cent sample of dwelling units is to be selected from within each primary unit, then the probability that any given dwelling unit is included in this sample is equal to $(n/N) \times (1/10)$—i.e., the product of the area selection probability multiplied by the dwelling unit selection probability. If all the eligible persons within each chosen dwelling unit are to be included in the sample, this probability is also the probability that a specified individual will be chosen; but if further sampling is performed within the dwelling unit, the number obtained must be multiplied by still another factor to determine the probability for an individual. A step-by-step account of such a procedure as this for obtaining a city sample has been provided by Kish (6).

As far as practical application is concerned, this procedure accomplishes the following things: (a) it makes it possible to state in advance the probability that any given individual will be included in the sample and the relation of this probability to the probabilities for other individuals, without requiring preparation of a complete list of all population elements, thus conforming to our concept of a probability model sample; and (b) it clusters the selected respondents from a geographic point of view and thus effects economies in the amount of travel required to contact the selected respondents.

Once a probability model of this kind is set up, a mechanical procedure is devised to select elements from the population in accordance with the model. This is frequently done by means of random-number tables (7). It designates the elements that are to be included in the sample. As the final step, interviewers are sent to obtain information from the predesignated respondents.

Sample designs of this general type have frequently been referred to as "area samples" because we first select a sample of geographic areas. However, the term is not particularly apt because it is possible to make use of areas without having a probability model sample. An example of such a sample design was given in the preceding section. Also, we can apply similar techniques to groups of population elements which are not defined on an area basis. Thus a sample of workers in a particular occupation could be obtained by first selecting a sample of firms that employ such workers and then choosing a sample of workers from within each firm.

3.9 MULTI-STAGE SAMPLING PROCEDURES CANNOT BE CLASSIFIED SIMPLY

As illustrated in the preceding sections, many sampling operations are carried out in successive stages, a member of the defined population being obtained only at the final stage. This means, among other things, that an extremely large number of sampling procedures are possible in even a simple situation. For example, Table 3.1 shows

TABLE 3.1

POSSIBLE COMBINATIONS OF SAMPLING PROCEDURES IN A THREE-STAGE SAMPLING OPERATION

Sampling Stage	Unit of Sampling	Procedures That Might Be Used in Selection
1	Blocks	*a.* Random
		b. Systematic
		c. Purposive
2	Dwelling units within blocks	*a.* Random
		b. Systematic
		c. Selection left up to interviewer
3	Adults from within dwelling units	*a.* All adults interviewed
		b. Random selection of a single adult
		c. Selection left up to interviewer (subject to quota restrictions)

what might have to be considered in drawing a sample of adults from a city.

Any of the procedures given for one of the three stages may be used in combination with any of the procedures given for another stage, and thus it is possible to conceive of 27 different sampling methods in this situation. Moreover, some of the procedures that we might wish to consider have not been included in this table. Further variations could be obtained by imposing external controls on the sample. Thus we might wish to stratify blocks with respect to geographic location, or with respect to average rental, or with respect to estimated size.

This example emphasizes the point made in the opening portions of this chapter, namely, that general titles can tell us little concerning the actual procedures used in selecting a sample. It is essential that the producers of survey information provide a complete description of sample design, and that the consumers of such information expect this description as a matter of course. Otherwise there is no basis on which to evaluate the accuracy and reliability of the data.

3.10 THE ACTUAL APPLICATION OF THE PROCEDURE

We have already noted that a sample of individuals is important only insofar as it leads to a sample of data. Two surveys may use the same type of sampling procedure but apply it in different ways and so get markedly different samples of data. They may use identical sampling procedures to select persons, and yet the ultimate results may not be comparable for a variety of reasons. There may be essential differences in the methods that are used to obtain information from the sample elements such as mail questionnaires, personal interviews, group interviewing (where sample elements are called into a central location and fill in their own questionnaires), or telephone interviews. There may be important differences in the extent to which information is actually obtained from the intended sample. Thus some of the individuals may refuse to be interviewed; it may be impossible to contact others without a prohibitive expenditure of time and money; or persons sent a mail questionnaire may not respond. These latter eventualities often mean that we do not obtain information from an important fraction of the people in the sample originally selected. Consequently our advance predictions about the behavior of the sampling procedure may no longer be correct. All these points must be included in the evaluation and comparison of sample survey results. They will be treated in some detail in Parts II and III of this book.

3.11 A BROAD CLASSIFICATION OF SAMPLING OPERATIONS

The kinds of sampling that have been described in this chapter are those that have been distinguished and discussed most frequently. A review of the reports of sampling surveys and studies of attitudes, opinions, and wants reveals many instances in which the processes by which the data were obtained do not seem to correspond to any of these methods. It would be fruitless to set up a large number of specific classifications to include the many varieties of sampling, but several broad classifications are useful. They will be illustrated by a discussion of three examples: (A) classification by the method of selection, (B) classification by the type of contact or connection between sampler and sample, and (C) basis for cooperation or compliance of persons in the sample.

A. The Method of Selection

The methods of selecting samples of people and of data may be grouped into four broad classes:

1. Taking what is readily available.
2. Expedient choice.
3. Selection constrained by group quotas, matching, and other purposive controls.
4. Probabilized selection.

1. Taking What Is Readily Available. In this class the sample is scarcely selected at all. It is merely discovered or picked out of the immediate environment. Although this may seem a careless or slipshod method, it is not necessarily so. Its application often consists of a prolonged and careful search for available cases. This is essentially the method used in a study of letters received by members of Congress (8). It is the method by which Thomas and Znaniecki obtained a large collection of letters written by and to Polish immigrants in America to form a principal part of their pioneering study of attitudes at the time of the First World War (9). Another similar application occurs in the interrogation of prisoners of war and refugees (10, 11), in analyses of captured correspondence (12), and in studies of excommunists such as Gabriel A. Almond's study, *The Appeals of Communism* (13). It is also the method that must be used in the studies of the reactions of civilian populations to major disasters. Thus the sampler may have no other choice than to take the sample that is offered to him ready-made by the force of circumstances.

In other instances, however, the sample seems not so much a rare opportunity or scarce set of data as it does a windfall or a convenient by-product that frees the user from more costly or arduous methods of obtaining his sample. Thus the selection of readily available cases is the method used by innumerable psychologists and sociologists who have made studies of attitudes by giving questionnaires to the students in their classes or members of other readily available groups (14). It is the method used in some of the voluntary panels of consumers, correspondents, and crop reporters, in which people are accepted who have been recommended, are willing to cooperate, and are qualified according to some general criteria.

2. Expedient Choice. In this category would be placed methods which recognize that the elements of a sample should be "chosen" from among all elements in a defined population but which permit this choice to be made in the most expedient manner. This may not be very different from taking readily available persons, but it does include active steps to make more members of the population available or it rejects some that are readily available. For example, a questionnaire may be published in a newspaper with a request that everyone cut it out, answer the questions, and then mail it to a central office. A card may be attached to a product so that the users of the product can report their reactions to it.

People have been approached for their opinions when they are assembled at meetings, county fairs, conventions, women's club meetings, bus stations, hospital clinics, and other places. There are two selective processes that are usually involved: first, the sampler's selection of the place and the occasion, and second, the processes of attraction that bring people to it. These processes introduce obvious possibilities of bias and distortion that may make it impossible to obtain a fair sample. On the other hand, certain types of places and occasions will produce mixtures of persons who are widely dispersed with respect to place of residence and ordinary daily activities. Dr. Kinsey's sampling methods appear to have been of this kind. He traveled about seeking all who would agree to be interviewed; he attempted to interview all the members of various groups; he even interviewed many hitchhikers. The results of his sampling procedure may be fairly representative of the population he was interested in studying in some characteristics but quite different in some others (15). How good they were is still debatable.

An element of expediency appears even in some of the more rigorous methods of sampling. For example, some surveys exclude people who live in institutions, are migrating, or are homeless. The selection is thus confined for the most part to private households when it is believed that the results are substantially the same as would be obtained by more refined and difficult methods attempting to cover the whole population. Relative to the other problems involved in the survey, the distorting influence of such exclusions is regarded as negligible. This assumption may be unwarranted. Some test of it is always needed.

3. Selection Constrained by Group Quotas, Matching, and Other "Purposive" Controls. There are methods of selection in which the taking of readily available individuals or the active search for individuals is regulated by definite instructions that aim to make the whole sample similar in a number of ways to the population that it is to represent. The instructions may be negative so as to avoid certain kinds of unrepresentative persons; or they may be positive and seek to provide specified numbers of persons of particular kinds. They may specify that no children are to be included and that specified numbers of men and of women are to be obtained. They may warn interviewers against taking people who are too readily available, such as their friends, members of their own households, or persons who all live in the same district of a city. The instructions may be given in such general terms as "Interview people who are regarded in their communities as typical citizens" and "Try to pick a representative sample with the right proportion of rich and poor, old and young, well educated and poorly educated, and members of the important religious and social organizations."

The indefiniteness of these general instructions, plus the latitude that is left by even the more specific instructions of a positive kind, leads many statisticians to call these methods "judgment sampling." Judgment is used in many phases of sampling operations, but here it is used rather freely in the actual determination of the individuals to be included in the sample. Sometimes the element of judgment is subordinated in some degree to definite rules. For example, a set of cities or counties or other areas may be selected provisionally and then tested by comparing its average values on each of several population characteristics with the averages for the whole population. Next some of the areas are removed and others added until the set of areas compares quite closely with the whole population in these characteristics. Some judgment is involved in the choice of areas to be removed and added but the entire procedure could be reduced to a set of formal rules that involve no judgment in their application. Several decades ago this method was tried under the name of "purposive sampling" with somewhat disappointing results.

Some more recent developments have shown the usefulness of exercising judgment in the stages of sampling prior to the final choice of the sample rather than in the actual choice itself (16). The selection by a group of judges or experts of "typical" cities, or regions, or districts within a city, or individuals is still and will continue to be a common form of sample selection. It aims to use what is known by the best-qualified persons in putting together a better sample than could be obtained without that knowledge. Undoubtedly there are many instances in which the neglect of expert knowledge, or indeed of any pertinent information, will seriously weaken the sampling operation. Expert judgment has limitations in the selection of samples, however, and even some biasing tendencies and defects. Therefore this class of methods is often based on a somewhat optimistic appraisal of the effectiveness of the expert. It may also overestimate

;he effectiveness of the control variables on which the sample is matched to the population.

4. Probabilized Selection. In methods of this type chance is deliberately given a principal role in the determination of the individuals who are to go into the sample. Since, in the simpler varieties of this method, every individual in the population has the same chance of getting into the sample, this may seem to involve no selection at all. In fact it almost seems like an instance of the class discussed first, in which the sample is not chosen by the sampler but just happens. With these methods, however, the sampler does actually accomplish certain definite purposes by delegating the selection to a definite system of chance. Under some specific conditions, his action could be fully justified by the recently developed theory of games and statistical decision theory. It is "blind selection" only in a restricted sense, for ordinarily the sampler knows quite a bit about what he is doing. By utilizing known facts and principles, he can *channel* the operation of chance and gain the advantages it offers while avoiding the losses and disruption that it might otherwise introduce. Probabilized selection is favored by people who are relatively pessimistic about the effectiveness of choice by experts and relatively optimistic about rigorous methods. It opens up, as none of the other methods does, the possibility of analyzing the operation of the sampling procedure in terms of probability theory and of applying the theory of mathematical statistics to the design, operation, and analysis of the results of a specific sampling procedure.

When we attempt to classify a particular selection procedure, the line of division between these categories will be found to be blurred. Many surveys use a mixture of two or more types. The classification scheme should be useful, however, to bring some order into the variety of methods that are used.

B. Type of Contact or Connection

The two other examples of principles for classifying various sampling operations stem from and qualify the foregoing system for categorizing selection methods. The first of these concerns the nature of the contact or connection between the sampler and the individuals in the sample. This contact is the means by which the final sample of data is extracted from the sample of persons. We might make a distinction between obtaining the data: (*a*) through other persons, (*b*) through communication in writing, (*c*) through direct contact by talking, and (*d*) through observation of behavior. This classification would not be wholly satisfactory, however, and hence a somewhat more extended one will be outlined. It follows a scale of increasing involvement of the sample of persons in the process of obtaining information about them. We distinguish the following classes of contacts:

1. Indirect connection.
2. Semi-direct contact.
3. Direct impersonal contact.
4. Direct personal contact.
5. Participating contacts.

This classification is based on the relation between the sampler and the sample of persons. It generalizes the answers to the question, "How did the sampler get the data from the persons in the sample?"

1. Indirect Connection. Ordinarily this type of contact occurs without the knowledge of the persons who are in the sample. It taps their attitudes and wants through knowledge and information which already exist in the minds of people who know them, or which exist in records in the possession of other persons. For this reason this type of contact usually does not offer any opportunity to improve the records and knowledge, or to discover and correct misunderstandings. In some recent studies, for example, the people in the sample were asked how their friends and relatives voted in an election. They had no opportunity to ask their friends and relatives but could only reply out of knowledge they had gained previously in conversation.

The use of letters and documents is also an example of indirect connection. Sometimes, of course, it is believed that such records provide more accurate and franker replies than could be obtained by direct questioning. However, the record may be one that tends to conceal the "true" intentions and attitudes of the people who contributed to it. For example, the record is not assumed to be a frank one in "content analysis" when we attempt to see behind the mask of propaganda. In this type of contact the possibility of using standard questions and measuring devices is severely limited.

2. Semi-direct Contact. In this second type of contact, there is some degree of "individualizing" of the contact. The most meager step in this direction may be a request to each individual for his permission to use existing records or to furnish records in his possession pertaining to his own opinions, experiences, and actions. Records requested might be correspondence, diaries, professional and institutional records, recordings of interviews and private meetings, etc. The essential point is that the individual usually has some control over the material and may refuse to assent to its inclusion in the sample.

Of course his control may not be fully effective. The material may be used in an anonymous manner or in such summary form that the objections are overcome. There are important questions of legal and moral rights in such instances that may be more prominent in the future than they have been in the past.

A second feature of semi-direct contact is the opportunity afforded for supplementing the recorded material with explanations, descriptions of changes that took place after the time covered by the record, and personal interpretations of a stated position. The individual concerned may also provide clues to the location of other material related to the original records. Such opportunities for further communication will often lead to more direct contacts, though they may remain on a relatively impersonal and indirect basis. Even though such subsequent contacts are developed, the contact can be regarded as being primarily semi-direct so long as the principal data are those that were recorded in some form for another purpose. If they are shaped to the purposes or circumstances of the current inquiry, then the type of contact changes.

3. Direct Impersonal Contact. The third type of contact involves a request to specific individuals, or to all members of some population collectively, to furnish information about their own individual attitudes or wants. Ordinarily the individual or organization making the request is known to, or is at least identifiable by, the members of the population.

The responses are influenced by the respondents' attitudes toward the source of the request as well as the indicated use to be made of the results. The request may be ignored or it may be definitely refused. Questionnaire surveys that are broadcast to widely dispersed groups may find the percentage of refusals and other failures to reply going well above 75 per cent. This is very likely to produce a strong biasing of the selection process so that the sample of replies is quite different from the replies that would have come if all the questionnaires had been returned. Thus, even though the sample of persons selected to receive the requests may have been closely representative of the population in the characteristics to which the questionnaire applies, the actual sample of replies is not representative of the characteristics or of the desired responses concerning them. It might be observed that a low rate of response is not inevitable. Many questionnaire surveys are made in which 90 per cent or more of the sample of persons is covered successfully.

4. Direct Personal Contact. The fourth type of contact is one of person-to-person communication. Ordinarily it is an interview, but it could be conversation and correspondence, or a combination of the two, over a period of time. An example of this combination is the contact of a biographer with a person whose biography he is writing. The biographer has an opportunity to go beyond the examination of records, letters, and writings, and beyond collateral interviewing of persons who have known or been associated with the biographee. He can arrange for a period of personal contact for the mutually recognized purpose of improving the biography. An interesting example in early studies of attitudes is the analysis of an immigrant autobiography by W. I. Thomas and F. Znaniecki (9). There are many studies of a similar nature.

Interviewing takes many forms. It may be done by telephone, by visit to the home of the interviewee, by appointment at the office of the interviewer, or casually in some public place. It provides a full opportunity for supplementing each major question with further questions designed to clear up ambiguities, fill gaps, explore the background of important material, check interpretations, and otherwise enrich the data by using the responses to develop further questioning. If carried far enough, the respondent may even be able to contribute to the determination of the questions that should be asked in order to develop the essential information. This of course requires that the interviewer give the interviewee a reasonably complete account of the purposes and problems of the survey.

It should be noted that direct and personal contacts are not necessarily "face-to-face." Moreover, not all face-to-face contacts are direct and personal. The mere fact that a person hands a questionnaire to the people in the sample and receives it when it is filled out is not sufficient. Even though something is said about the purpose of the study and the method of completing the questionnaire, this kind of contact has only a remote and modest relation to the answering of the questions. It is not essentially different from direct impersonal contact, even though the spoken words may be important as a means of increasing cooperation.

5. Participating Contacts. In some studies the persons in the sample take an active part in the prosecution of the study. The method of participant observation is a well-known example. In contacts of this type the participant may only be a passive subject during part of the data-collecting process. However, his connection with it is such that his performance as a subject is influenced substantially by his participation in the planning of the study, in contacts with other subjects, or in similar phases of the work.

The manner in which information is obtained about his attitudes, vants, and reactions is usually altered considerably by this connection. He is not merely better informed about the purposes and methods of the study, but his responses are likely to be formalized to some extent by his knowledge of the theoretical implications of the study and of the analytical scheme according to which it will be carried from data to conclusions. Some of the recent studies of community life in new housing developments and of preferences with respect to various features of residential construction are examples of this kind of contact. So too are some recent studies of rumor and organization.

C. Basis for Cooperation or Compliance

The preceding principle of classification of sampling operations is largely concerned with the degree of personal involvement of the subjects in the production of the final sample of data. It measures the opportunity they have to determine what data are to be obtained, to formulate it for the record, and to participate actively in the research undertaking. It necessarily involves the particular means of communication and measurement, and it goes beyond these means to the character and limitations of the communication involved in eliciting responses on the one side and responding on the other. Communication, when fully developed, requires quite as much initiative on one side as the other. Some cooperation is usually necessary, even when the process involves little or no relation beyond the implicit linkage established by a researcher in discovering and gaining access to records from parties other than the persons to whom they refer. We may find it useful to include a third principle to classify sampling operations by the influences that enable the sampler to obtain from other people the sample of data, including his ability to motivate individuals to provide information about themselves or about other persons in the sample. This somewhat elusive facet of sampling is essential to the operation of obtaining a sample of data.

The influence may be simply that of calling on the good nature of people to help a stranger, or it may be strong pressures that compel replies. It may be exerted directly through the contact or indirectly by calling on the prestige and authority of some generally influential agency. Some of the interviewing of civilians in occupied countries by survey organizations that work for the occupying forces are examples of the latter type of influence.

The nature of the pressure, and the personal reactions of individuals to various kinds of pressure, not only affects the degree to which the sample of persons is covered but determines the kind and amount of distortion of the data that will result from this aspect of the sampling. If strong influence is brought to bear on the entire sample of persons solely to induce a small number of reluctant respondents, the biases produced in willing respondents may prove a heavy price to pay for the cooperation of the unwilling. Even a very friendly approach or the awarding of money and prizes may sometimes have undesirable effects on the accuracy and completeness of the data obtained.

The sample of data will also be affected by the motivations of the people in the sample, whatever the kinds of influence used to obtain the data. Thus in indirect contacts the motivations that lead the originating persons to write the letters they wrote, to keep a diary and write in it what they did, to submit for recording the statements and other information they gave all have an effect on the subsequent processes of sampling and measuring attitudes, opinions, wants, and related variables.

D. Other Principles of Classification

There are other principles of classification that are useful in grouping together the many varieties of sampling operations that have been employed in the study of human populations. For example, the sampling may be done at one time, at successive times, or continually. It may also be classified according to the technical means used to record the responses of persons in the sample, such as by standard questionnaires, speech-recording devices, interviewers' notes, and self-written replies. It may be done publicly or with various degrees of privacy and anonymity. All such means are part of the production of data and have a characteristic effect on it. Selection of what to put into the record, losses from faulty memory, misunderstandings, self-consciousness about the contact, outright deception, and many other processes are involved in the recording of data.

No attempt will be made to develop the classification scheme to include these other phases, but their effects and the problems of controlling them deserve attention in a thorough study of the operations of sampling and in valid assessment of the reliability of the results they produce.

3.12 INTERRELATIONS OF SELECTION OF PEOPLE AND OTHER PARTS OF SAMPLING OPERATIONS

Within a sampling operation, the selection of a sample of people is a step toward obtaining the sample of data. Sometimes this step can be planned and taken with little reference to how the data will be obtained from the sample of people after it has been chosen. Usually, however, the two steps are closely interrelated. A few simple examples will serve to illustrate these points.

A. An Audience

A theatrical producer, who does not take the critics to be infallible, wishes to find out from the first-night audience whether his play will be a success. He can judge the answer from the frequency and vigor of the applause (readily available, indirect), but this may not be a satisfactory sign. He can ask his friends who are in the audience (readily available, personal). He can listen to the conversation in the lobby during the intermission (readily available, indirect). This conversation, however, may reveal only the reactions of those people who are most prone to express their impressions of the play. They may be people who make snap judgments and change them quickly.

He can ask some of the people who are not talking (expedient choice, personal). If he tries this, how should he pick out the people to talk to? He can also talk with some of the people who remain in their seats when the others go out to the lobby. Their views may be different. Will the front rows suffice as a sample or should he include people sitting farther back? Will it be necessary to select people who represent each price of ticket? Are the reactions of the audience at intermission time a dependable indication of what their reactions will be after they have seen the whole play?

He can have assistants do some of the interviewing. The assistants may obtain the opinions of a greater part of the audience by giving people a brief questionnaire to be filled out and collected (impersonal contact). Will this be as good as talking with them? Or would it be better to ask people to fill out cards, giving their telephone numbers, so that he can telephone to a random sample of them on the following day? Is the first-night audience sufficiently representative of the people who will purchase tickets for subsequent performances?

These are a few of the questions that might be considered even in so simple a survey as obtaining the reaction of a theatre audience. There are other questions of the same kind to be asked regarding a survey of reactions to a program on television or a political debate on the air. Some of them would still have to be answered if it were possible to get the whole audience to write down its reactions to the play at the end of the show. They are a mixture of questions about selecting people, making contact with them, and influencing them to give their candid opinion about the play. Most sampling problems are just such a mixture.

B. A Library

The trustees of a public library wish to improve the service it performs in its community. Two of their number spend an afternoon at the desk talking with the people who come to charge out the books they wish to take home. Is this a good sample? One of the other trustees challenges the report. He claims that the wants of people who come in the afternoon differ from those who come in the evening. Many of the afternoon users are school children. The night-time users tend to be people who work during the day. He also points out that a special exhibition at the library on that particular afternoon may have attracted some types of people more than others. Moreover, he recalls that it was raining.

Another trustee thinks that people should be interviewed in their homes where they may feel freer to express their criticisms. Still another remarks that there may be many people who are not likely to be interviewed at the desk because they seldom use the library. They might be interested if new services were offered to attract them. It becomes clear that people are likely to be interviewed in proportion to the frequency with which they use the library, or even more specifically, the frequency with which they bring books to the desk to be charged out. Should the sample be one of people charging out books, or of people entering the building, or of card holders, or of people who reside in the neighborhood of the library, or of all people who read books, including those who buy them or get them from rental libraries? How should their opinions and wants be ascertained? Is there a best way to sample for this purpose? Should a sample of the community be taken and then the data for each person be given a weight corresponding to the likelihood that he will use the new services if they are offered? What advice can we give the trustees on these questions?

C. A Store

The owners of a large department store wish to find out what their customers would like to buy but do not find in the store at the present time. Where and how should the owners ask them? What sample of customers is needed? Is this problem any different from that of the library? Does it differ from that of the theatre?

D. A Nationwide Survey

To help make orderly plans for the future and take prudent action in the regulation of their affairs, both the agencies of government and the private enterprises of our economy need accurate information about the wants of consumers and their expectations about spending their incomes. The consumers in turn need some assurance about the income they may expect to receive before they can make wise decisions about spending and saving.

A survey organization conducts an annual survey covering a sample of the population to obtain some of the information that is needed. Should it ask a few questions of a great number of people or more questions of fewer people? How can it make sure that the answers it gets truly represent what people think and what they plan to do? How closely can we expect their future behavior to follow the plans they now have in mind? Should the interviewers speak only with the people who are employed or should they also talk with other members of the families who may have some influence on how the income is spent? Should they interview in the home or at work or somewhere else? Could they get satisfactory information by mail or by telephone?

These and many other questions arise in a national survey. They are complicated by the wide scattering of the population and the long distances that we must travel to reach a national sample. Must it be spread over the entire area of the United States in the same way that the population is spread over that area or can it be concentrated in ten or fifty or a hundred places? If it can be concentrated, how should we choose the cities, towns, and rural areas in which the sample of people will be located? What is the most economical way of choosing the sample and surveying it? When the survey is completed, how much confidence can we place in its results? How much can an organization rely on the survey in determining its policies? How much can a family rely on it in deciding whether to make a major purchase now or wait until later to buy?

E. An Experiment

In the summer of 1946 the effect of the atom bomb on naval vessels was tested at Bikini (17). Prior to the test there was a wide range of opinions about the ability of ships to withstand the effect of the tremendous explosions that had shown their force in the first test in the American Southwest and then at Hiroshima and Nagasaki. Although the Bikini test was conducted under conditions of great secrecy, enough information was released to the public to make a major addition to its knowledge of what the atom bomb could do and how it might affect world peace. At the least the test had great potentialities in provoking serious discussion and changing attitudes.

In effect, therefore, not only was the Bikini·test an experiment in the vulnerability of naval vessels, it was an experiment in the reaction of the public to terrific demonstration of power and danger. What effect would it have on public attitudes, both toward the atom bomb and toward the prospects of future warfare? What would be its effect on the outlook of people toward Américan policy in international relations?

The natural way to study this experimentally induced change is to measure attitudes before and after the experiment. How long before and how soon after the event should this be done? Would the effects be anticipated by some people as a result of advance publicity concerning the test and the question of what the bomb might do in future conflicts? Would the effects occur immediately after the event or would they develop more gradually? Would they be the same in all parts of the country or would they be greater in areas that might some day be the targets of enemy bombing? Would the public reaction be an increased demand for air raid shelters and other defensive measures? Would it be for the abandonment of naval shipbuilding programs and the use of the funds for defense programs? Would there be a call for international control of atomic weapons or even for world government? Would the further development of atomic weapons be stimulated or restricted? What features of the test and the publicity about it would prove the most influential in producing the changes that occur?

In order to answer such questions as these, what kind of interviewing and how extensive a series of contacts with a sample of the population are necessary? How should a sample of persons be drawn for a study of this kind? Should the same people be interviewed before and after the test or should a fresh sample be interviewed after it occurred? Would the previous interview tend to alert the people who were in the first sample to the coming event and thereby modify their reactions? Would they remember what they said at the first interview and try to be consistent with it in what they said in the interview after the event? What is the best way to conduct a study of this kind and how do we decide the adequacy of its results as scientific data to be used in drawing inferences about the processes of public reaction to major events?

There are countless other examples that might contribute to the picture of the variety of methods of sampling, their close association with the purposes they are to serve, and their involvement with methods of obtaining information from individuals in the sample. The questions that have been listed are difficult to answer. There is no completely satisfactory answer to most of them. None the less it is very worthwhile to consider them carefully and come to a reasonable conclusion on each in order to increase the value of the data that are to be obtained. Fortunately, both experience and some general principles help.

3.13 COMMON CHARACTERISTICS OF SAMPLING OPERATIONS

It is possible to learn about sampling and to understand how it works by accumulating experience with actual samples. It is also possible in this way to make relatively sound judgments about the dependability of some sets of sample data.

Unfortunately learning by experience does not proceed very fast in the case of surveys. They are not made frequently enough to give very many examples of how they work out. We know all too little about their accuracy. We know much less about the conditions that affect their dependability. Taking a survey of a human population is usually quite complicated. It involves a great many factors that affect the results, each of which is difficult to observe. The approach to sampling wisdom by the long road of experience is all the more disappointing because there are so few opportunities to determine just how well the samples we encounter on the way actually do turn out.

There are some exceptions in which a check on accuracy is available. The surprises that have occurred during the past twenty years in the election polls are an example. We would progress more rapidly in learning about sampling from actual experience if other surveys were subject to similar tests.

Faced by the paucity of such direct tests of survey results, we must turn to other means of appraising the dependability of both the output and the methods of survey operations. One way is to look into a survey and see "how it works." We must examine each operation more carefully. We must identify the particular parts that it includes and discover how each affects the dependability of the whole operation. We must then form a comprehensive judgment for each survey from the specific information we obtain in this way. It is an overwhelming job to do this afresh for each survey we encounter.

The task becomes feasible only when the multiplicity of methods and situations can be reduced to a much smaller number of typical systems or to what we shall call "models" in the discussion that follows. To start in this direction it is necessary to determine what features are common to all sampling operations, or at least to a considerable proportion of them. We need to have a fairly clear understanding of what these common features are and of what is meant by the words that are used to refer to them. In the course of clearing up our vocabulary, we shall acquire sharper tools of thought with which to describe, analyze, appraise, compare, and generalize the sampling operations we examine. One result we obtain in the end is a set of rather general principles of sampling and some familiarity with the ways in which they can be used. The following list presents and discusses briefly some of these common features of sampling operations. This list is not meant to be exhaustive. Furthermore, many of the separate items are related to one another. These relations will be treated extensively in the chapters that follow.

A. Sampler, Situation, and Purposes

1. The Sampler. A natural place to start is with the person or organization doing the sampling, or at least with the persons who are contemplating a possible sampling operation. For simplicity we shall refer to the *sampler,* leaving it understood that several persons may be sharing this role. We shall use the term even if a number of persons have a part in the sampling operation but no one of them directs the operation or is involved to any greater extent than the others.

2. The Situation. Next there is what we may call the *circumstances* or the *situation,* namely, all those aspects of the environment within which the sampler will perform the sampling operation and which may affect the operation to an important degree. It includes other people. It includes the resources of funds, equipment, information, transportation, services, and other means the sampler needs to perform the survey or study of which the sampling is a part. It also includes aspects of the survey that are wholly distinct from the sampling. It must include the customary behavior of people and various existing relationships between them that both make possible and limit the activity of the sampler. It is the context of his own activity.

3. Purposes. There are in the situation, and perhaps in the sampler's mind, one or more *purposes* to be served by the survey. The sampling may not be viewed broadly enough to link it directly to these purposes, but it is at least connected to them through intervening sets of purposes, proposals, specifications, aims, and objectives. The most immediate of these may be merely a request for a sample of a certain kind or for data on a certain problem. There may on the other hand be a complete and detailed program of what is wanted and how it should be obtained.

Here at the beginning, then, is a sampler operating in a situation under the ultimate direction of certain purposes. The producer of the play was such a sampler. So too was the board of trustees of the library. The group that attempted to measure the public reaction to the Bikini experiment was also a sampler operating in a much larger and more complex situation and with far more serious purposes. We can now be more specific about the various details of a sampling operation and describe them in whatever form we find them.

B. The Population, Selection Process, and Operations

1. The Population. All sampling operations aim to produce information about some *population* of persons larger than the sample that is actually selected and measured. The population may be only vaguely defined by the purposes of the survey, and this vagueness may not be cleared up at any point in the survey process. Much scientific information is about populations that are only partially defined. This is true, for example, of psychological experiments on the factors that change attitudes. These experiments are almost always performed with a group of subjects that is not representative of the population of the country in which the experiments are done or of any similar population, however defined.

It is more important in most surveys than in such experiments to determine what population is to be represented and to regulate the sampling to that end. Often there are several populations, each to be represented by some of the data that the survey produces. A sample of the opinions of voters on public issues may also be directed toward all persons who were eligible to vote, including those whose views on the issues were not strong enough to induce them to take the trouble to cast their ballots. It may even extend to include persons who would make themselves eligible if they were more deeply involved in the issues and political contests that are to be decided. Whether there is one population or a set of several populations, the sampling is directed toward the production of measurements and estimates that will represent the populations to which they refer.

2. Selection Process. All sampling involves some process of selection by which one or more samples are designated in the population. It may be merely the act of taking what is readily available, or it may be an arduous series of technical procedures. It may consist of taking a sample from one population for use in obtaining a sample of another population; for example, a sample of persons who are 21 years old or older is selected and interviewed in order to obtain a sample of persons eligible to vote. The persons who are found to be not eligible, on the basis of the information they supply in the interview, can then be separated when the data are analyzed from those who are found to be eligible, leaving the sample that was desired.

Sometimes it works the other way. Studies of family expenditures were made in 1936 on samples of the native white population. Later an attempt was made to use the data as a sample of the entire population. Although several small samples of Negro and foreign-born persons were available, they were not sufficient to represent this part of the larger population; thus the studies were not fully representative.

The methods and techniques of obtaining samples from populations are many. In any sampling operation it is important to have a clear and definite understanding of the methods that are being used and to take account of their characteristics in appraising the data produced. This will be called the "sampling procedure" or the "selection procedure." When all the related steps, such as the actual activities of the sampler to apply the procedure to a population are also included, it will be called the *selection process*. For example, the sampler may make a public appeal for cooperation, write letters to people who will be visited by an interviewer, send interviewers back when their previous efforts have been unsuccessful, substitute new names for people who are missed, etc.

3. Operations. All sampling operations have their own *operating problems.* Some of the circumstances may interfere with operations. Storms, heavy snow, muddy roads, floods, and other results of unfavorable weather are one kind of hindrance to interviewing in the field. Reluctance and outright refusal to be interviewed are hindrances on the human side. The difficulty of recruiting and training personnel for dispersed and usually temporary jobs is another problem. Many methods of sampling involve technical operating difficulties of their own.

C. Measurement and Estimation

1. Measurement Processes. The measurement of groups and individuals is not simple. In the broader sense of the term it involves identifying those people who are of the kinds that are being studied, such as persons eligible to vote when the survey is one of the formation of voting decisions. It involves counting, observing, interviewing, or otherwise measuring the people who are selected for the sample. These operations we designate by the general term "measurement," sometimes with emphasis on the methods or procedure, sometimes on the actual performance of the operation, and sometimes on both. In certain instances, the terms "process of measurement" or "measurement process" will be used. It is just what is necessary, starting with a sample of people, to obtain from them a corresponding sample of information in the form of numerical quantities or the equivalent.

2. Group Measurements. The purposes always include, or at least imply, a need for information about a class or population of people. Ordinarily what is needed is not information about any individual by himself, but information about the number of individuals of different kinds, the average values of certain measurements, the proportions of persons in various categories, and measures of the relationships of two or more variables in the population. These are properties or characteristics of groups, or of the whole population, not of individuals.

Group measurements are derived from individual measurements but they cannot be obtained simply by measuring only one individual. There may be no possibility of finding an individual whose measurements are exactly the same as those of the population. Thus there is no individual who is 55 per cent Republican in the same sense that the voting population may be.

The measurement of opinions, attitudes, and consumer wants is still quite primitive but, within this limitation, the purposes of a survey may well become translated into a definite request or specification, "Obtain a reasonably accurate estimate of the prevalence of X attitude in the population P." It may include many such simple requests, and the specifications may also call for various combinations and subdivisions of the variables. The survey will often require the collection of other information needed for subsequent analysis and interpretation of the principal variables, and it may also make use of information obtained from other sources.

3. Measurements of Individuals. The information about a group or population of people must be obtained by measuring individuals and combining the measurements statistically. The individual characteristics that are counted or measured are often called "variables." The term "variable" is used to designate not only information that may take any of several numerical values but also information that takes the form of non-numerical categories such as "yes" and "no" or "favors" and "opposes." When there are two categories, they can be treated as if they were numbers by assigning the value 1 to one of them, it does not matter which, and the value 0 to the other. If there are more than two categories, then a somewhat more complicated but definite numerical coding can be made.

4. Estimation. Sample data seldom are suitable indicators of the population variables when taken raw, just as they stand. Usually they must at least be converted into averages, proportions, and percentages, as well as certain measures of relationship. Often more complex statistical methods are used to derive from the data the estimates that are appropriate to the population. Even the estimation of frequencies—counts of particular kinds of individuals in the population—may be complex. In the simplest instances they are made by multiplying each count or frequency recorded from the sample by a suitable constant factor. Thus if the sample is 2 per cent of the population, each sample frequency may be multiplied by 50 to bring it up to an estimate of the frequency in the whole population. In other instances they require extensive computations, and they may even be performed by electronic computers. These procedures and others like them will be referred to as "methods of estimation." Estimation may involve the use of data on several variables other than those that are estimated and even some use of non-numerical information and judgment. When the method is considered to include the closely related phases of the sampling and measurement processes, it may be termed the *estimation process.*

D. Appraisal and Gain

1. Accuracy, Representativeness, Use, and Appraisal. Every sampling operation is likely to include some judgment or appraisal of the *accuracy* and *representativeness* of the data it produces. It is also likely to involve, as the final stages, certain steps necessary to apply the data to the purposes of the survey. This may be little more than a brief period of thinking about them on the part of the sampler himself, or it may involve extensive discussions, comparisons, analyses, and interpretations. It may involve the preparation of reports for publication or presentation orally. There may be subsequent periods of re-examination of the whole survey and the data it produces in the light of information that comes to hand at a later date.

2. Costs, Values, and Gain. In any situation the sampling operation requires some effort and involves some *costs.* The situation also offers some means of placing a *value* on the information about one or more variables. Were it not for this value, sampling would not ordinarily be undertaken unless the costs were strictly negligible. The value arises out of the uses to which the information may be put and tends to depend both on the accuracy of the information produced by the survey and on the accuracy of information that was available prior to the survey. If the effect of the sampling on the whole survey, including all its purposes, is evaluated accurately in terms of the value of the data produced and the costs properly attributable to the production of the data, then the best method of sampling is that which produces the greatest *gain* (value less costs), given the circumstances and the other phases of the survey.

3. Risks and Chance of Gain. There is an element of risk in all sampling operations. Although the use of sampling to obtain information of a certain type in a particular situation may be quite advantageous in the long run, it may turn out to be actually disadvantageous in occasional instances. This is to say that if we knew enough to evaluate an actual sample accurately, we might be correct in concluding that we were better off without it than with it. The only way that we can gain the long-run advantages of sampling is to accept the risk of the occasional losses. It may not even be possible to recognize the losses as such when they do occur. When it is clear that a loss has occurred in sampling, there is no sufficient reason to discard the sampling method or even to terminate a series of sampling operations that may be in progress, unless it is judged that this will avoid additional losses that are greater than the long-run gains that may be expected from a continuation of the sampling. In this respect sampling is not essentially different than many other kinds of enterprise.

E. Summary of Common Characteristics

There are other common features of sampling, but at this point it may suffice to summarize those that have been brought to the fore. We start with a sampler operating in a situation under the general guidance of certain purposes which may be translated to some degree into operating specifications. The sampler is engaged in obtaining a sample of information in the form of measurements of certain variables or their equivalent from which, by suitable estimating procedures, will be produced a set of estimates of the values of the variables that are characteristic of the population. He may seek to obtain the greatest possible gain by accepting certain costs necessary to get information of greater value, the value being dependent on the uses that will or can be made of the estimates and on their accuracy and representativeness relative to the population or populations involved in their use. In the course of producing the data, various operating problems will be encountered and use will be made of some method or methods of selection to obtain an actual sample from the particular population that is sampled. This sequence of the selection process, the measurement process, and the estimation process, in relation to the situation and its operating problems, may be thought to be the *survey process*. These terms emphasize what is common to sampling operations.

Much that is not common is of equal importance. Samples differ in many ways. They vary greatly in size, ranging from as few as ten or twenty persons to millions of people and from a single variable to hundreds of variables for each person. They may be confined to a single locality or occupation or age group. They may cover the full range of kinds and conditions of people found in a national population. They differ in the form of contact of the survey personnel with the people in the sample.

They vary in the simplicity or subtlety of the measurements that are made and in the accuracy of the data that result. They differ in the amount of detail and explanation that is needed with each measurement to aid in the interpretation of the data. Likewise they vary in the number of variables for which measurements are made and in the extent to which the survey is concerned with the interrelationships among these variables.

The operating conditions vary greatly too. There may be great haste to complete a survey in a few days or it may be spread over weeks or even over months. The budget may permit the employment of very competent personnel and ample expenditures for facilities and preparations. It may require severe economy and taking greater risks for want of adequate resources.

The survey may turn out fortunately, by chance, or it may make the sampler a victim of an adverse turn of fate. The external conditions, such as weather, the political and economic situation, shifts of opinion, and the moods of the public, are always different from one sample to another. The timing of various steps cannot always be adjusted prudently in relation to the timing of future events that may interfere.

In spite of these differences and uncertainties, however, it is possible to draw general conclusions about sampling procedures from an examination of past experience, especially on points of similarity among surveys. It is possible, by analysis of the way in which surveys work, to establish and use general principles. These principles constitute a working theory of sampling that can be employed to appraise the results of the sampling that has been done by others and to design better sampling operations for ourselves.

REFERENCES

1. Frederick F. Stephan, "Practical Problems of Sampling Procedure," *Am. Sociol. Rev.*, 1 (1936), 569–580.
2. Lillian H. Madow, "Systematic Sampling and Its Relation to Other Sampling Designs," *J. Am. Statist. Assoc.*, 41 (1946), 204–217.
3. National Opinion Research Center, *Interviewing for NORC*, University of Denver, Denver, 1945.
4. Frederick Mosteller, Herbert Hyman, Philip J. McCarthy, Eli S. Marks, and David B. Truman, *The Pre-election Polls of 1948*, Bulletin 60, Social Science Research Council, New York, 1949.
5. Edward Benson, Cyrus Young, and Clyde Syze, "Polling Lessons from the 1944 Election," *Publ. Op. Quart.*, 9 (1945), 467–484.
6. Leslie Kish, "A Two-Stage Sample of a City," *Am. Sociol. Rev.*, 17 (1952), 761–769.
7. The RAND Corporation, *A Million Random Digits*, The Free Press, Glencoe, Ill., 1955.
8. Hilda Hertzog and Rowena Wyant, "Voting via the Senate Mailbag," *Publ. Op. Quart.*, 5 (1941), 359–382 and 590–624.
9. W. I. Thomas and Florian Znaniecki, *The Polish Peasant in Europe and America*, 2 vols., Alfred A. Knopf, New York, 1927.
10. Alexander H. Leighton, *Human Relations in a Changing World*, E. P. Dutton & Company, New York, 1949.
11. Margaret Mead, *Soviet Attitudes Toward Authority*, McGraw-Hill Book Company, New York, 1951.
12. The United States Strategic Bombing Survey, *The Effects of Strategic Bombing on German Morale*, Vol. 2, U.S. Government Printing Office, Washington, D. C., 1947.
13. Gabriel A. Almond, *The Appeals of Communism*, Princeton University Press, Princeton, 1954.
14. Quinn McNemar, "Opinion-Attitude Methodology," *Psychol. Bull.*, 43 (1946), 289–374.
15. William G. Cochran, Frederick Mosteller, and John W. Tukey, *Statistical Problems of the Kinsey Report on Sexual Behavior in the Human Male*, The American Statistical Association, Washington, D. C., 1954.
16. Roe Goodman and Leslie Kish, "Controlled Selection—A Technique in Probability Sampling," *J. Am. Statist. Assoc.*, 45 (1950), 350–372.
17. Leonard S. Cottrell, Jr., and Sylvia Eberhardt, *American Opinion on World Affairs in the Atomic Age*, Princeton University Press, Princeton, 1948.

CHAPTER 4

Some Models for the Analysis
of Selection Procedures
and Measurement Processes

4.1 SAMPLING OPERATIONS AND MODELS

The preceding chapter was intended to improve understanding of sampling by abstracting, from the bewildering variety of instances of sampling, a number of features that are in some degree common to all sampling operations. When a new survey is encountered in the future, it is to be expected that the identification of these features and the description of the exact form that each assumes will help in analyzing the dependability of the results produced by the survey. There is, however, only limited value that can be attached to the deductions made on the basis of identification and description alone. Each characteristic feature of a sampling operation can appear in a wide variety of forms, and the various features can then be combined into a complete survey in a practically limitless number of ways. It would be impossible to find examples of more than a few combinations among the surveys that have been taken. Even for these few, the experience of using them would not always be adequate to provide an accurate guide for future use of similar sampling plans. Another approach is required.

Certainly some important conclusions can be found that apply to one or another large grouping of different sampling plans, grouped together because they are alike in certain essential characteristics, even though in other respects they appear to be quite dissimilar. Such essential similarities may not be readily apparent. Once they are discovered, however, they can be used to classify surveys or to associate them with a limited number of distinct "types" or species. Within each type or classification additional similarities in essential characteristics may be discovered, and the varieties as grouped together can be compared to see what essential differences remain to restrict the use-fulness of the classification.

The preceding step from description to classification is naturally followed by the selection for each class of one or more examples that represent it by portraying the common characteristics in the form and degree exhibited (with minor variations) by the members of the class. Study of the example may then take the place of separate study of the individual members of the class. If, for the purposes of a study, some common characteristics are important and others are not, an example can be chosen that portrays these important characteris-tics exceptionally well, even though it does not represent the class so well in other characteristics. It is necessary, however, that the ex-ample be representative of the class in essential interrelations among the characteristics as well as in the characteristics themselves. We are interested not only in what parts go to make up a survey opera-tion but in how they are put together and how they work.

We can go a step further. Taking advantage of the fact that the example need not be fully representative in all respects, we can con-trive somewhat artificial examples in order to enhance their effective-ness in representing the essential features we wish to emphasize. When we take this further step, we are making a *model* and using it to demonstrate certain features that are important. We can use a model to aid us in thinking more effectively about these features and in analyzing the entire operation. If we use the model properly, we may even be able to predict from the model how the operation will actually work out in practice.

The use of models is well developed in engineering and science. A small wood or metal object shaped like the hull of a ship, but with no engine or internal parts corresponding to those of an actual ship, is pulled through a long trough of water to determine how efficient a hull of this design would be in an ocean-going vessel. A small object shaped like an airplane or an airplane wing is fixed in the rapidly moving stream of air in a wind tunnel to determine how an airplane with a particular design feature would fly. These models are similar to what they represent in a few respects but not in many. None the less the data they produce, when used properly and with full allow-ance for the respects in which the model differs from the real object, prove in practice to be accurate and very valuable.

In addition to *physical* models, there are *conceptual* models con-structed out of ideas and relations among ideas. They are systems of thinking about the real systems they represent. Thus the physical model of an airplane can be represented by a blueprint or drawing. The model is now operated mentally. Relationships may be traced by measurements on the paper. We may even take one further step and free ourselves from any dependence on physical models and drawings. We can use instead the simple and powerful methods of mathematics to express the relations among various portions of the physical model. We then have a mathematical model. Mathematical models are being used in many practical ways, as well as in scientific theory, to an increasing extent. Many useful applications of mathematical models have been found in the social sciences (1, 2, 3).

Mathematical models for sampling procedures have proved especially helpful in exhibiting clearly their essential features, showing how they work, facilitating certain measurements and calculations, and providing a basis for comparing them. In view of the many excellent books on this subject (4, 5, 6, 7, 8, 9) we shall not attempt a review of this material here. Instead, the next section will set forth in very general terms conceptual models for some types of selection frequently used to choose samples of people for the measurement of opinion, attitudes, and consumer wants. A later section will be concerned with conceptual models of the measurement process. Then the models will be combined in an effort to describe the way in which sampling operations work and the ways in which they can be compared and appraised.

In summary, then, it is possible to use physical representations to demonstrate the nature and relations of systems of objects in the real world. We can even measure some of the characteristics of the objects and systems from their physical models. Maps and drawings serve a similar purpose, with more of the particular detail of reality eliminated in order to display the essential facts still more clearly. A further step takes us to sets of ideas that represent conceptually the system of objects and operations as it exists in the real world. When these ideas are developed in mathematical form, they permit logical deductions and calculations that would be much more difficult, if possible at all, in direct observation and manipulation of the real world.

The models do not represent reality in more than a few selected respects, and even for these features the representation may be quite imperfect. Hence conclusions drawn from models must always be checked in the actual world. Their approximate character as well as their oversimplified portrayal of what they represent must be taken into account.

The proper use of each model should be fully understood, and the following information should be clearly stated: (a) the features in which the model is representative and those in which it is not, (b) the part of the model that corresponds to each part of the reality that is so represented, (c) the scale relationships and the manner in which distortion can be adjusted, (d) the degree of accuracy maintained, (e) the basis for interpreting the results of working with the model, and (f) other facts to assure us that we can learn correctly and not be misled by the model. When used in this way, models are of great value.

4.2 MODELS OF SELECTION PROCEDURES

A primary requisite of a model for a sampling operation is that it set forth clearly the essential characteristics of the operation and the manner in which the different characteristics interact. Then we can attempt to isolate the key factors that are susceptible of numerical measurement and use these in deducing from the model predictions about the actual performance of the defined operation. In this section we shall present the very broad outlines of models for a number of situations. These models are intended only to illustrate a general method of approach. Each model will therefore contain only certain essential features of the procedures used to select samples of people, not the many other details of the entire selection process, the measurement process, or any other part of the survey. When they include such other parts, it is only for the purpose of making the example more concrete.

A. A Model of Selection by Taking What Is Readily Available

One of the convenient sources of information on opinion is the department of a newspaper devoted to letters to the editor. By accumulating clippings over a period of time, we can obtain a considerable amount of material on the opinions of persons living in the area served by the newspaper toward any specified timely topic. The opinions expressed in these letters would certainly be of interest in and of themselves, but suppose we wished to use them as a sample, on the basis of which to infer something about the distribution of opinion in the population of all people living in the area. What sort of model might be constructed to describe this situation, and what conclusions might we legitimately draw from the model?

To be more concrete, let us suppose that we wish to ascertain opinion toward a proposal to lower the voting age to 18 years. We might then think of the population as being divided into four groups.

1. Definitely in favor of the proposal.
2. Definitely against it.
3. Relatively indifferent to it.
4. Not aware that the proposal has been made.

The avowed purpose then of examining the published letters to the editor is that of determining the fraction of the population falling in each of the four groups.

Obviously each of these groups will have a different rate at which they send letters to the editor. Group 2 might be expected to have a relatively high rate, Group 1 a somewhat smaller rate, Group 3 a substantially smaller rate, and members of Group 4 will not write at all. They contribute at different rates to the daily flow of letters that reach the editor's desk. However, we must now recognize that the editor makes the final selection of letters that are actually published. Ordinarily he cannot publish all of them. Let us assume that the editor attempts to report both sides of the controversy equally well but, influenced by his own position in the matter, prints a few more letters in favor than he does of letters against the proposal. At times, other issues are being debated so vigorously that he receives an unusually large number of letters about them; consequently he may print fewer letters on each current issue, including the vote for 18 year olds. If we now determine the proportion of letters published in the newspaper that were in favor of the proposal, what can we conclude about the support and the opposition it has in the whole population?

We may formulate a model by conceiving the flow of information about reader opinions as a stream with three stages: (a) *The source.* Each person in the population can write a letter and thereby bid for selection of his letter for publication. The rate at which bids are made may be constant or variable over time and from person to person. For many persons it will be zero over their lifetime. (b) *Prior selection and diversion.* An intervening selection by the editor takes place and some of the bids are rejected. The letters that are printed are selected on the basis of certain properties. (c) *The final selection.* The surviving bids reach the sampler (the reader), who merely takes as his sample everything which comes to him in this convenient form within some convenient period.

Clearly, the model indicates that the information in the letters so selected provides no adequate basis for estimating the proportion of the population in favor or opposed to the proposal or the rates at which different groups originate bids. It does not even provide a basis for estimating the effect of the editor as a selection mechanism. Just as the water that flows in a river bed may be partly diverted, recombined, contaminated, purified, and modified in other ways before it reaches a given location downstream, so the flow of information may be modified in various ways. The sampler cannot be confident about his estimates of the sources.

Even though this model does not provide a sound basis for drawing many conclusions from the letters that appear in the newspaper, it does help us to understand how a process of selection can be severely distorting. It also points to the facts we need to know to draw any confident conclusions from the letters that are published. It suggests how little control of the sampling the sampler actually exerts. Each group has a different rate of writing to the editor. The editor prints a different proportion of the letters he receives from each group. That proportion depends on the number of letters he receives from each group, since it is aimed at equal space for both sides. We can see that if people in one group were stimulated by public leaders or an organized campaign to write more letters to the editor, it might have no effect at all on the proportion of letters from this group that would appear in the paper. The selection of the persons in the sample therefore is made primarily by the initiative of the persons who write and by the actions of the editor who determines which letters will be printed.

B. A Model of a Procedure of Selection by Searching and Matching

Instead of depending on people to "volunteer" by offering "bids" for selection, the sampler may take active steps to "enlist" people in his study. Thus he may go looking for persons of several kinds, keep track of the number found, and like a fisherman who catches his limit, stop taking people of a specified kind when he has reached the specified number. In this way he tries to match the sample to the population, at least in the categories he uses to determine the kinds of persons he seeks and also to some extent in categories that are closely associated with them.

As a simple case of sampling by searching and matching, we may consider the example of a quota interviewer who is told to obtain interviews with so many men and so many women, so many persons of high income and so many of low income, and similarly for the other characteristics used to control selection. Even though the quota interviewer may be constrained to do his searching in specified areas of a city and may be instructed to obtain interviews in the home, the ultimate choice of respondents is left up to his discretion. Some aspects of the actual search procedures used by quota interviewers are described in Chapter 12.

We may think of a model for such a searching and matching procedure in the following terms. An active sampler goes out to get a sample in much the same way as a housewife takes her list and basket to go shopping. He takes the initiative in seeking out persons beyond the group that is readily available to him. It is he who writes letters, or visits people in their homes, or finds them elsewhere, thereby originating the steps that lead to the selection of some of them for the sample. His efforts make more of them available, though many may remain still entirely out of reach of the selection process. He may reject some of the persons with whom he comes into contact as possible members of the sample, but in this model he has control of the rejection procedure and he has, whether he uses it or not, an opportunity to learn what effect it has on his sampling. Again we may think of several kinds of persons in the population and different proportions of each kind among those who are at least considered for possible selection. The sampler also has different rejection rates for each kind, possibly depending on how many persons of each kind he has already selected. If these rates are known or can be estimated with some degree of accuracy, it may be possible to use the model to predict how the sampler will function as a selective device under searching and matching constraints. When the sampling is done by many persons and in several steps, the model need only be elaborated to conform to these variations. This extension would be required to describe a large-scale quota operation where many interviewers are simultaneously carrying out the searching and matching procedure.

The American Youth Commission of the American Council on Education made a study of the conditions and attitudes of young people in Maryland in 1936. The report describes the search in vivid though not precise terms.

To reach these youth, the Commission's agents went anywhere and everywhere that young people were to be found. Most of them, 52 per cent, were interviewed in their homes. The next largest group, about 33 per cent, were found elsewhere in their neighborhoods—their clubs, community centers, street corners, swimming pools, public parks, dance halls, beer "joints," drug stores, and anywhere else that youth might congregate. . . . That they might not miss the other social extreme, the interviewers went to relief offices where youth along with adults awaited their turn at the dole, and to employment agencies where they awaited their chance at a job. And finally, to complete their invasion of the world in which young people live, the interviewers went to the places where they work. . . . Every kind of neighborood or area, every social and economic stratum, and every educational and intellectual level was proportionately represented in the final sample (10, p. 5–9).

Another kind of search procedure in which special efforts were necessary to gain the cooperation of the persons in the sample is described by Kinsey (11):

> The techniques of this research have been taxonomic, in the sense in which modern biologists employ the term. It was born out of the senior author's long-time experience with a problem in insect taxonomy. The transfer from insect to human material is not illogical, for it has been a transfer of a method that may be applied to the study of any variable population, in any field (p. 9).
>
> The difficulties . . . promised to be greater than those involved in studying insects. The gathering of the human data would involve the learning of new techniques in which human personalities would be the obstacles to overcome and human memories would be the instruments whose use we would have to master (p. 10).
>
> Lectures to college, professional, church, and other community groups have most frequently provided the entree to the better educated portions of the population. Hundreds of such lectures have been given. Perhaps 50,000 persons have heard about the research through lectures, and perhaps half of the histories now in hand have come in consequence of such contacts.
>
> Practically all of the contacts at lower levels, and many of those at other levels, have depended upon introductions made by persons who had previously contributed their own histories. One who has not already given a history is not usually effective as a "contact man." Contact men and women have often spent considerable time and gone to considerable pains to interest their friends and acquaintances (p. 38).
>
> In securing histories through personal introductions, it is initially most important to identify these key individuals, win their friendship, and develop their interest in the research. Days and weeks and even some years may be spent in acquiring the first acquaintances in a community. . . .
>
> The number of persons who can provide introductions has continually spread until now, in the present study, we have a network of connections that could put us into almost any group with which we wished to work, anywhere in the country (p. 39).

C. A Model of Selection Using Areas

To avoid neglecting some parts of the population that may be quite important for the purposes of the survey, a thorough search may be conducted aimed at making all persons equally and fully available for selection. When this is too difficult a task, we can confine the thorough search to a set of selected areas. Thus we reduce the amount of work necessary to find the people in each area who live there, work there, are present in the area at a specified time, or are attached to it in some other way. The selection procedure then starts with the choice of a sample of areas.

If all the persons associated with the chosen areas are then sought out and selected, the representativeness of the sample of people will depend on the selection of areas and not on any process of selection of persons within the sample areas. Sometimes the areas are selected by well-informed persons who pick those places which they think are representative of the whole population. They may choose one or more cities to represent a region or the urban population of the nation (12, 13, 14). They may select a number of blocks by chance from existing records and maps on which all the blocks are shown, or they may use some more highly controlled method of the kinds discussed later. Certain blocks may be rejected for special reasons and others not even considered at all.

An excellent example of choosing a representative city and then following an exhaustive search procedure is provided by a study of child-bearing in Indianapolis which included questions about the parents' desire for children and many related attitudes. The search was then conducted in the form of a complete survey of all white occupants of more than 100,000 dwelling units among whom only 2589 couples were found who were members of the particular population subgroup that was the object of the study, "couples with the following characteristics: husband and wife native white; both Protestant; married in 1927, 1928, or 1929; wife under 30 and the husband under 40 at marriage; neither previously married; residents of a large city most of the time since marriage; and both (at least) elementary school graduates" (15, p. 157).

The model for these types of sampling is obtained by dividing the population into parts in which each member of the population is associated with an area according to some rule. The selection of people is then made by choosing one or more of the areas that go to make up the entire territory occupied by the population. Since areas are easier to find in a search than people, and since the choice of a sample of areas reduces the subsequent work of searching for people, this method has great advantages. However, the selection of areas usually introduces offsetting disadvantages, as will be illustrated in Chapter 20. In a way similar to that of the preceding model, we think of each area as having a chance to get into the sample, depending on the method of selection the sampler uses. Each person's chance depends on the chance his area has of being chosen for the sample.

The search within the chosen areas should be a thorough and systematic canvass of the area aimed at finding every person in it, or at least every person in it who is a member of the population to be included in the survey. This may be done by going from one dwelling to the next along all the roads and streets and also looking for dwelling places that are not located on or near a road or are not readily visible from it. It may include other kinds of searching such as waiting in post offices or other public places to which people come and finding out from some of them what persons are yet to be seen. Thus all the persons who vote in an election could be seen at or near the polling places, except those who vote an absentee ballot.

D. A Model of Selection from a List

Sometimes the planners of a survey will find that some other agency has already conducted a survey of the defined population, or of a larger population that includes the desired population, and has prepared a list of the persons found. There are useful lists that have been produced in the course of legal or governmental functions such as censuses, tax lists, registrations of voters, property lists, registrations for rationing or for military service, and membership lists of various kinds. There are city directories, telephone directories, and other private compilations of the names of persons in various parts of the population.

When such a list is available, it may be used to produce a sample of names. The search then becomes one of finding the persons whose names have been selected from the list. The same problems arise for the selection of the names from a list that arose for the selection of areas in the preceding model. The lists may not be complete. Changes have occurred since the list was prepared. Certain addresses are inadequate to lead us to the location of the dwelling. The completion of the selection of persons by searching for the people whose names have been selected is therefore quite like the search of an area to find the people who are associated with it (16).

This model may be considered as a sampling of persons through a well-defined and readily available set of intermediaries, such as their names. The names on the list may be considered singly or in sets. For each name or set of names, there is a decision to include it in the sample or reject it. Then the persons whose names are included in the sample of names are sought out to obtain the sample of persons. Sometimes names may be selected conditionally and the final decisions about their inclusion made as the search progresses.

E. Combinations of Models in a Composite Model

The models that have already been described may be combined in various ways. For example, a survey of the interests of parents in plans for the development of the public school system may include the following selections:

1. Parents from a list of the children who are in school.

2. A typical area in which some one part of the future program of school expansion, such as a new junior high school, is most needed.

3. A number of families that have no children in school but do have younger children whose education might be affected by a change in the system.

4. Letters written to the school board by previous residents.

Another example of a combination of procedures is the selection, by one method, of a sample of smaller areas within each of the areas in a sample originally chosen by a different method. This is followed by the preparation of a list in each area for list sampling.

The actual technique of selection from a list or the selection of areas has not been discussed. Neither has the basis of selection that is involved in the first two models. This should now be done to make the models more definite.

4.3 ELEMENTARY TECHNIQUES FOR USE IN SELECTION PROCEDURES

The models of the previous section were presented in very general terms, and no attempt was made to specify just how the sampler decides which persons he will accept and which he will reject from among those he considers for possible inclusion in the sample of people. However, if a model is to be used for making predictions concerning the performance of the operation it represents, it is necessary to proceed to a stage in which the details are made more definite and certain aspects of them are made precise. We will not attempt to do this in full detail for a particular sampling operation. We shall do it partially and for some types of selection by describing several general techniques that may be used in a variety of operations. We shall also describe a physical model, a simple card game not connected with any survey, to illustrate the techniques. The reader can try working with this model, to see how the techniques turn out in practice. He will remember that only a few essential features need be similar in the card game model and an actual survey. Although it may seem quite artificial, it exhibits in the actual play how risks arise in sampling and how they can be reduced by certain methods of playing the game of "Sampling."

A. The Physical Model

Two players sit down with a deck of 52 playing cards. Since they are tired of all the games they know, they invent a new one which they play according to rather unusual rules. One player, *the dealer*, removes from the deck any 16 cards he chooses and sets them aside. In doing this he may look at the faces of the cards, but he does not permit the other player, *the sampler*, to see which cards he takes. Then he arranges the remaining 36 cards, *the population*, in any order he pleases and forms four piles of cards with 9 cards in each, all face down. The sampler may now take any 12 cards he wishes from the population but he must look at the faces of only the cards he takes. Then, counting kings, queens, and jacks as ten, and all the other cards according to the number of pips on them, he must estimate the total count of the cards in the population. His score depends on how close his estimate is to the actual count. The only question he faces is how he should pick the 12 cards from the 4 piles in order to get as good an estimate as possible.

At first glance, this game may seem to bear no relation to the problems that must be faced, for example, in drawing a sample of adults from the population of a city or nation, or in drawing a sample of workers from among all workers employed by a particular company. What, we may ask, corresponds to the dealer in these situations? Actually, any population we choose to study has characteristics determined by forces over which the sampler has no control. Thus such diverse forces as climate, geographical terrain, economic processes, and even public leaders have all played a role in setting up the population. We may therefore think of a composite force (nature, or *the situation*) as the dealer for the practical game of sampling. Sometimes the role of dealer is even more direct than this, as when a plant manager attempts to influence a group studying plant morale to talk only with those workers who will present the manager's point of view in a favorable light.

B. General Techniques of Selection

One possibility open to the sampler is just to take the nearest pile of twelve cards (ready availability). If the dealer wants to conform to the behavior that is expected of honest dealers, he may shuffle the cards and deal them out in the usual way. On the other hand, if the dealer wishes to be capricious in making up the piles he can do so. How much can the player tell about the population when he uses this technique of accepting as his sample whatever is offered by the dealer in the most convenient form?

The sampler may anticipate that the dealer will put high cards in two of the piles and low cards in the other two piles, in order to give a false picture of the population should he take most of his twelve cards from only one pile. The dealer is also free to choose which pile he puts closest to the sampler. To thwart such manipulations of the cards by the dealer, the sampler may take three cards from the top of each pile, getting close to the proper proportion of high and low cards. This is the technique of *proportional representation of parts or subdivisions*, but it still uses ready availability in selecting what is offered on the top of each pile.

After several trials of the game, the sampler may suspect that the dealer is putting high cards in the top of each pile, or low cards at the top, to mislead him again. To counteract this trick he may then decide to take the second, fifth, and eighth card in each pile. This is *systematic selection* or the technique of taking individuals at constant intervals. Of course, the dealer may anticipate this too and arrange the cards accordingly. Unless he behaves as an honest dealer should, he will make the game unfavorable to the sampler.

Finally the sampler will say to himself, "The only way I can prevent the dealer from taking advantage of me is to keep him ignorant of how I am going to pick my twelve cards. I'll just combine the four piles, shuffle the cards as any honest dealer would and then deal myself twelve." This is the technique of *random selection*.

Random selection follows no system except that of leaving the selection completely free of outside control by either the dealer or the sampler. Every possible set of 12 cards out of the 36 has the same chance of being chosen.

If the dealer refuses to permit the player to pick up the cards and shuffle them, the player can find another way to make a random selection. Thus he can prepare 36 cards, on each of which is written the pile number and order number of a single card of the pile. Next he can shuffle these cards, deal himself 12 cards, and then draw the corresponding 12 cards from the population. Likewise he could use other devices than cards to draw a random sample, such as dice, coins, and tables of "random numbers."

C. Method of Play Used by the Dealer

The game can work out differently if the dealer is not trying to oppose the sampler or if he does not arrange the cards to his own best advantage. Also after a while the sampler may learn something of the methods that the dealer is using to deal the cards and can take advantage of consistency in the dealer's behavior. The dealer can

rotect himself against the use of information he inescapably reveals
by changing his method from time to time or by using methods similar
to those of random selection. This is a subject that is considered in
detail in the "theory of games" (17, 18, 19).

In sampling, likewise, one of the major problems is that of how the
dealer, i.e. the situation, is arranging the cards, i.e. the population.
Any selective processes or rejection activities that stand between the
sampler and the population are analogous to the efforts of the dealer
to influence the success of the sampler in his estimating activities.

The dealer may, of course, be neutral or even very helpful to the
sampler. He may even make the estimation and selection almost fool-
proof. Raindrops contain very nearly the same proportions of hy-
drogen, oxygen, and other elements the world around, due to the selective effect of evaporation and the mixing operations of the atmos-
phere. A sampler who wishes to estimate this proportion is not likely
to go wrong in taking a sample of raindrops by any method, so long
as he does not contaminate it with other material. Nature has set up
the population of raindrops very favorably for him. However, the
water in lakes, rivers, and oceans varies greatly in chemical composition. A sampler can go wrong quite easily in taking a sample of
water from a river to represent all the water on the Earth.

Similarly the common language and customs of a people support
somewhat common ways of thinking. Certain kinds of opinions are
so nearly universal in a culture that a sampler can scarcely go wrong
by any method of sampling he uses to determine the nature of these
opinions. Other attitudes and opinions are so diverse, even within a
single family or close knit group of friends, that it is easy to make a
serious mistake in sampling.

The sampler cannot count on being saved from his mistakes by a
friendly dealer, even though the dealer always appeared in the past to
set up the cards so that he could not go wrong very far. This makes
it worth while to learn and use techniques of sampling that may seem
artificial and tedious but that make the sampler independent of the
behavior of the dealer.

D. Other Techniques

The game model can be elaborated to include other techniques of
selection. In the examples of the theater and the library described
in Chapter 3, as well as in other instances, it is possible to observe in
each individual certain indications of his characteristics and even of
his attitudes and opinions. The problem here is not whether to use
the external indications of variables that are hidden from observation.

It is rather how to use them. We want to avoid mistakes of method and false clues. The psychology of stereotypes has shown that too much reliance can be put on outward appearances as a guide to the personal qualities of individuals. Attempts to use them to improve the sampling could be more harmful than helpful. Samples made up of people, each of whom *looks* typical or average, may turn out to be quite atypical, whereas a sample that includes every variety of appearance in approximately its correct proportion may prove to be quite representative of the population.

E. Keeping Score: The Distribution of Results or Errors

Whatever the particular form of the game, we may be sure that if the players can be induced to continue playing it over and over again they will settle down to playing it by some regular method. That is although their methods of play may include certain intentional variations of their behavior from one play of the game to another, they will be using the same system or strategy in deciding what they will do. In a sense they will repeat themselves. The game will be steady or stable, even though the outcome of any one play may be unpredictable.

As we watch the game, we can keep score. We find that the sampler overestimates the count in the population sometimes and underestimates it at other times. He is close at times and far from the correct count at others. Although there are short periods with a preponderance of close results and periods with a scarcity of close results, the succession of results does not seem to follow any consistent pattern. If it did, the players would probably observe the fact and vary their play in an attempt to take advantage of what they thought they had discovered.

As the scores accumulate, we find that they are spread over a range of numbers and that the amounts by which the sampler has over- or underestimated the count in the four piles at each play of the game is also spread. We can call these amounts his *discrepancies* or *errors* in estimating the count. They may be designated as *positive* errors if he overestimates and *negative* errors if he underestimates. If the total of the positive errors persistently exceeds the total of the negative errors, we say he has a positive *bias* or tendency to overestimate, unless of course it looks reasonable that the two totals differ only by chance. The location of all his errors as they spread over various possible amounts is called the *distribution* of errors. In further plays we expect his errors to occur in an unpredictable order but spread in much the same way as this distribution and with much the same degree of bias. The model thus gives us some knowledge about the game and the players, though not as simple and definite as the information provided by a model of the regular motion of the planets in their orbits. Every technique of selection and every selection process that can be represented by such a model may be expected to have a bias, possibly close to zero, and a distribution of errors.

F. The Situation as a Dealer

We could proceed with the game model to further modifications and enlargements of the game, and we should then find very interesting questions about the best way to play it. That would divert us, however, into problems of game theory beyond the present stage of our discussion of sampling. The lesson we draw from the model is that *though the elementary techniques of selection can be assembled into more complex systems, they are always used in a game-like situation.* In a sense, we can regard the situation as the dealer in a game. The selection system we use is our way of playing the game with the situation. The population is set up by the situation, possibly with no concern about us or what we are doing. Whether by accident or by design, whether to thwart us or help us, whether shrewdly or carelessly, the situation arranges the location of different kinds of people in the population from which we choose our sample. The effectiveness of our method of selection is like the effectiveness of a method of play. It depends on how the population is arranged. It depends on how much we utilize the knowledge we have about the situation. Sometimes it even depends on whether the way we play the game affects the situation and the way the population is "set up."

4.4 MODELS OF THE MEASUREMENT PROCESS

A full discussion of measurement is well beyond the province of this volume, but certain essential matters must be considered systematically because they are intimately involved in survey and sampling processes. The problem of measurement will be exhibited by several simple models in this section. They will be discussed further in Chapters 5 and 17.

The importance of the measurement process is not generally recognized. Many theoretical and practical discussions of sampling proceed on the implicit assumption that measurement is not a problem, in fact that it is nearly perfect. In the field of opinion, attitudes, and consumer wants this is a most naive assumption. Greater attention is now being given to measurement problems and "response errors"

in factual surveys as well as surveys of attitude and opinion. A broad
classification by Deming (6, Chap. 2) of the kinds of errors that af
fect surveys and an extended discussion of a mathematical model fo
response errors in surveys by Hansen, Hurwitz, Marks, and Mauldi
(20) are two outstanding examples of the recognition of the impor
tance of measurement error by statisticians whose natural inclinatio
is to emphasize the sampling problem. Hyman's study of interview
ing (21), Parry and Crossley's experimental survey in Denver (22)
and Campbell and Katona's discussion of the reliability and validity
of survey results (23, pp. 41–51) are examples of the concern of psy
chologists and survey directors with the problems of measuremen
error. Even in highly developed scientific technology, measuremen
error is studied and controlled (see, for example, 24).

We shall not review here the evidence about the nature and magnitude of response errors but merely point out several general problems
that can be represented by an appropriate model. It is not unusua
to have even a simple question of fact or opinion answered as indicated in Tables 4.1 and 4.2 in two successive interviews with the
same 100 individuals.

TABLE 4.1

RESPONSES BY 100 PERSONS TO THE SAME QUESTION ASKED IN TWO SURVEY

Type of Response	Answer in the First Survey	Answer in the Second Survey	Persons Who Answered This Way
A. Consistent	Yes	Yes	55
B. Inconsistent	Yes	No	9
C. Inconsistent	No	Yes	7
D. Consistent	No	No	29
			——
			100
Totals for One Survey			
A or B	Yes	Either	64
C or D	No	Either	36 .
			——
			100
A or C	Either	Yes	62
B or D	Either	No	38
			——
			100

When a research analyst examines data such as these, he is inclined
to be more cautious than a casual reader in drawing conclusions. Both

TABLE 4.2

COMPARISON OF TOTALS

Responses	First Survey	Second Survey	Differences
Yes	64	62	2
No	36	38	−2
	100	100	0

ould agree that either there was a considerable change between the
wo surveys or that the responses were not all consistent. If the facts
ere such that they could only change slowly or not at all, then the
nly conclusion is inconsistency.

The totals agree fairly well. Most reports present only totals, such
s those in Table 4.2. The casual reader is likely to assume that the
greement between the total responses at the two surveys gives him
n accurate measurement of the division of opinion or fact. This
ollows the common belief that "errors tend to cancel out."

The analyst may be more skeptical. He has seen results such as
nese tested later by accurate determinations which sometimes show
ost of the errors to have been of one kind. Thus he knows that all
he inconsistent responses could be instances in which the answer
hould be "Yes," but at one of the surveys the respondent erroneously
aid "No"; or the interviewer may have made a mistake in recording
he response, checking "No" instead of "Yes." Recordings of actual
nterviews have revealed just such recording errors (25). The prob-
em of the relation between recorded data and the "true facts" is
ommonly called the question of *validity*. Related to it is the prob-
em of the agreement between repeated measurements of the same
act. This is the question of *reliability*.

The analyst usually is interested in validity as well as reliability.
Ie concludes that the true answers might total only 55 "yeses" or
night total 71 "yeses," depending on the nature of the inconsistencies.
Moreover, he knows that if a proper "yes" can be reported incorrectly
s a "no" in one of the two surveys, it can happen for the same per-
on in both surveys. He needs more evidence, or more support for
heoretical assumptions, to justify drawing definite conclusions about
he group of 100 persons. The same problems confront him in draw-
ng conclusions from the sample of 100 persons concerning some popu-
ation it represents. The problems of sampling error and measure-
nent error are interlocked. So, too, in all serious survey work prob-
ems of validity and reliability, as well as those of formal consistency,

deserve the fullest consideration. They are treated in a number o
excellent references (26, pp. 41–51; 27, pp. 338–341; 28, Chap. 14; 29
Chap. 4; 30, Chaps. 15 and 16; 31, pp. 203–205; 32; 33, Chaps. 2,
and 15) and some discussion of these matters will be found in Chapte.
17. Our attention here will be given to the problems of measuremen
that are most pertinent to sampling.

A. Specification, Definition, and Numerical Components of Data

There are certain assumptions that are implicit in all measurement
A measurement is not merely a number. It is a statement abou
something, usually a certain characteristic of an object or a person
but possibly a relationship or some quite abstract aspect of though
and experience. It need not even include a number, for it may b
just an arrangement of objects in order of increasing or decreasing
degree with reference to some variable. Thus we can tell which of tw
foods is the saltier, but not in numerical terms. We make up many
lists in order of preference. Most measurements of attitudes are o
this kind.

It is essential that the number or ordering or other form of measure
ment include a description or statement adequate to identify what i
measured. This is "specifying" the persons or objects to which the
information contained in the measurements applies. Closely related t
this problem of *specification* are the problems of "concepts" and "definitions" common in statistical work. The characteristics or variable
being measured must be *defined* adequately to make clear just wha
the measurements mean. Many inquiries about attitudes confuse th
meaning of measurements by putting the data under labels somewha
different from the apparent meanings of the questions asked of the
respondents in the sample.

This lack of precision of reference in the identification of the objects on which the measurements are made and in the definition o
the variables measured is quite as much a part of measurement erro
as are inaccuracies in the numbers. Thus a report of an interview
could be in error in several ways, such as those illustrated in Table 4.3
Clearly errors such as these reduce the accuracy of any sampling o
data, no matter how carefully it is selected.

B. Assumptions about Conditions

There are many matters about measurement that are assumed to
be commonly understood. For example, it is assumed that there are
no extraordinary conditions at the time of measurement to interfere

TABLE 4.3

ᴇxᴀᴍᴘʟᴇꜱ ᴏꜰ Eʀʀᴏʀꜱ ᴏꜰ Sᴘᴇᴄɪꜰɪᴄᴀᴛɪᴏɴ, Dᴇꜰɪɴɪᴛɪᴏɴ, ᴀɴᴅ Nᴜᴍᴇʀɪᴄᴀʟ Dᴀᴛᴀ

Report	Comment
Respondent interviewed at 17 Madison St.	Inadequate specification of the respondent because the respondent was present in the sample household as a visitor from another community and should be excluded from the sample in the analysis.
Voted for Eisenhower.	Inadequate definition; the respondent voted for Eisenhower (actually for delegates pledged to him) in the 1956 primaries but did not vote in the election.
Voted in three of the four previous Congressional elections.	Incorrect number; respondent actually voted in only two of the four elections, 1952 and 1948 but not 1954 or 1950.

with obtaining a "normal" measurement. There may be certain unusual conditions whose effects are to be measured; and rather arbitrary and unusual conditions may be specified to "standardize" the environment as well as the state of the object and the measuring instruments; but these are then "normal" for the purposes of the measurement. We should not attempt to measure attitudes toward political issues when the respondent is in severe pain, in a fit of anger, drunken, or in some other greatly disturbed state (for an example in another field, see 34). In surveys it is difficult to avoid some of the milder kinds of disturbance.

C. True Value, Average Value, Expected Value

It is commonly assumed that there is a *true value* that we are trying to measure, a definite quality or fact that we are trying to ascertain. The measurements are approximations to that true value. Actually there is reason to assume that no exact true value exists but only a small range of values that represent the target or goal of the measurement process. However, it is useful to set up models in which there is one exact and true value, all other values of the variable being in error to the extent to which they differ from it. Much of the theory of measurement is concerned with the reduction of the amount of such *measurement error*.

We should not confuse the true value with the average value of a large number of measurements of the same object on the same variable. Under suitable conditions the measurements are closely concen-

trated about this average value. It is seldom possible, however, eve
in the most highly advanced scientific work, to make this averag
coincide *exactly* with the true value.

There is still another value, the *expected value,* which is not a mat
ter of actual measurement but a calculated value for some part of
conceptual model. It may correspond to the average value of a serie
of measurements made on the same object or person, i.e., to the mea
of the distribution of results that is conceived to be generated b
innumerable repetitions of the measurement process with the sam
situation, system of measurement, and state of the object at the tim
of the measurement, insofar as they are represented in the model. I
may or may not be exactly equal to the true value.

D. The Distribution of Measurements and Measurement Error

We can find many physical models for the measurement process
For example, we can ask a number of people who are gathered to
gether in one room to look at their watches and write down the time
Then we can make up a list of the reported times. They will be dif
ferent, perhaps as much as 5 minutes or more in some instances. *A*
hypothetical list of such observed times and their distribution is give
in Tables 4.4 and 4.5. The average will not necessarily be the tru

TABLE 4.4

List of Times Observed

2:17:35 p.m.	2:17:10	2:19:57	2:20:15
2:19:02	2:18:50	2:15:40	2:18:25
2:18:00	2:12:30	2:18:10	2:12:50
2:15:50	2:19:45	2:17:35	2:16:05
2:20:15	2:18:23	2:22:00	2:18:40

time. If now we repeat the experiment, but this time we strike a bel
or give some similar signal, everyone will observe the position of th
hands on their watches at approximately the same time. What car
we expect to find as a result of the better definition of the measure
ment process? What would be the effect of asking everyone to rese
his watch to correct it for the difference between his first reading an
the average? Would the average be close to the true time?

These questions lead to a conceptual model of a process of measure
ment that produces a distribution of measurements not unlike th
distributions of samples. One way of conceiving such a model is t
assume that we are taking measurements at random one by one from

TABLE 4.5

DISTRIBUTION OF TIMES OBSERVED

Time Interval	Number of Observations
2:10:00 to 2:11:59 P.M.	0
2:12:00 to 2:13:59	2
2:14:00 to 2:15:59	2
2:16:00 to 2:17:59	4
2:18:00 to 2:19:59	9
2:20:00 to 2:21:59	2
2:22:00 to 2:23:59	1
2:24:00 to 2:25:59	0
	20

a limitless supply of *possible* measurements on the same object or individual.

In this example, the difference between a distribution of measurements and one of errors is that the measurements are given in terms of *instants* of time (time of the day) and the errors in terms of time *intervals* (seconds, minutes, or hours) between the true time and the observed time. The distribution of errors for the data of Table 4.5 is given in Table 4.6. A distribution of errors can be obtained (at

TABLE 4.6

DISTRIBUTION OF ERRORS

(True time 2:19:33)

Error in Minutes	Number of Errors
8 to 6 minutes slow	2
6 to 4	0
4 to 2	3
2 to 0	9
0 to 2 minutes fast	4
2 to 4	1
	20

least approximately) from the distribution of times and the true value. Likewise the distribution of times can be obtained (approximately) from the distribution of errors and the true value. Hence either can be used in place of the other.

In the preceding model, a group of people generated a distribution of actual measurements of the time of day by looking at their watches. Any interviewer who enters in her record the time of the interview makes such a measurement. Ordinarily it need not be very accurate. Now we can look at other examples. The theatrical producer could measure the audience's response to a play by the total length of time they applauded. Alternatively, he could measure it by determining how loud and vigorous the applause was. Similarly, he could set up a scale of statements and ask members of the audience to indicate which they would use in expressing their opinion of the play. Finally he could use more subtle and presumably more accurate measures. Any method of measurement he uses will be subject to measurement error and will produce a distribution of measurements if it is repeated under suitable conditions on the same persons.

The letters to the editor can also be measured, in inches of space, in strength of support for one side or the other, and in other ways. Again they are affected by measurement error.

The attitudes ascertained in national surveys and in experiments are likewise subject to measurement error.

The theory of mathematical statistics provides many functions and models that can be used to represent the distribution of errors that would be generated by a measurement process if it could be repeated indefinitely. Sometimes they "fit" a set of data quite well, but sometimes it is difficult to find a convenient mathematical function to serve as the model of an actual process for the purpose of representing the distribution of results produced by the process. Very often there are several methods that can be used to make a model represent the data closely or represent the process closely. We must leave the problems of fitting the model to the standard statistical treatises; it suffices that such a conceptual model will be useful if one can be found that fits the process to a fair degree of approximation, and if it is not used beyond the limits set by its imperfect representation of the actual distributions.

E. Systematic and Random Components of Error

Sometimes we can learn about the errors of measurement by examining the instruments that are used or by testing them against more accurate instruments that can be employed as a standard. A cloth measuring tape may stretch and produce measurements that are consistently too short by a constant percentage. A steel tape stretches very little but expands as the temperature rises. A one-foot ruler that is worn down at the left end measures all lengths less than 12 inches with an error of the same amount, not the same percentage. A clock that is set too fast measures all times too late by the same amount. A revised form of a test or attitude scale may produce scores that are consistently lower than the form it replaces. This part of error, found to be the same from one measurement to another in absolute amount, percentage, or some other systematic way, is generally called *systematic error*. The remainder of the error differs from one measurement to another in an unaccountable way and is called *random error*.

The separation of the error into two parts is somewhat superficial and arbitrary because there are some parts of error that are neither constant nor systematic but vary in a way that can be attributed to definite causes which impinge on the measurement system from without. Moreover the random errors may be correlated or may show other relationships in any sequence of measurements. For purely pragmatic reasons we shall keep to the simple concepts of a constant amount of error or *bias* in measurements of attitudes and wants and let the remaining error carry the designation "random error," even though this may not be a perfectly valid separation of actual errors from the standpoint of the assumptions made about the nature of the random error and the models appropriate to it.

From this standpoint, a measurement process will be conceived as producing a distribution of measurements when applied repeatedly to the same object under essentially equivalent conditions. The distribution may be represented by a theoretical distribution derived from an appropriate model of the process. The theoretical distribution has an average or expected value which differs from the true value of the variable in the population by a constant amount, the *bias*. If the bias is subtracted from each actual error of measurement, the difference can be regarded as a random error.

F. Importance of Including Measurement Error and Bias in Appraising Survey Accuracy

Too often in surveys we know little or nothing of the magnitude of the bias of measurement and the variance of random error. Our ignorance of them is no reason to ignore them or assume that they are zero. Of course, they may have been taken into consideration if they are included in the data that are used to estimate the bias and error variance of sampling. Unless they are adequately represented in this way, however, they should be estimated on the best basis possible and combined with the sampling bias and error so that their effect on the survey results may be approximated. When this is not done, the user of survey results is given a spurious impression of accuracy. The value of the results is diminished and the danger of serious losses is increased through unwise use of the measurements.

4.5 MODELS FOR THE ESTIMATION PROCESS

We have considered several models that represent some of the principal features of sample selection and measurement. Since the measurements made on the sample must be combined in some manner to provide estimates of the characteristics of the population, the models of selection and measurement must be supplemented by models of the estimation process. Often the process of estimation is so simple and obvious that any attempt to understand it by examining a model seems superfluous. In other instances, however, estimation is a complex process and some simpler representation of it is needed.

The formation of estimates from data may proceed in various ways, depending on the nature of the data, the techniques of estimation known to the sampler, his ingenuity in applying them, and the additional data available to him from other sources. As in the case of measurement, a thorough discussion of estimation lies outside the scope of this book. However, since the results of sampling operations are affected by the processes of estimation that are used, we must include some of the essentials of estimation in our study. In fact, the methods of estimation are sometimes so completely dovetailed into the methods of sampling that each is rather meaningless without the other.

We shall use much the same kind of a model for the estimation process that we use for the measurement process. A specified estimation procedure is applied to a body of data resulting from sampling persons and then measuring them. It produces an estimate of the value of a specified variable. If the procedure is repeated, the estimate it produces may not differ from the first estimate. In this case we are only concerned with the amount by which the estimate differs from a standard estimate of the specified variable. This difference is the bias of estimation.

If, on the other hand, the repetition of the estimation process on the same data does not inevitably lead to identical estimates, we have at least potentially an unlimited supply of estimates. They form a distribution of estimates with its average value and its measure of spread, the standard deviation of the distribution. We may be able to establish a reference point corresponding to a "true estimate" from the set of measurements and thereby calculate a *bias of estimate* and *error variance of estimate* based on the estimation process alone.

(These terms are used in statistical texts to indicate the bias and error with reference to the true value of the population. We are concerned with only the part or component that is attributable to the estimating method alone, for a particular set of data.)

The estimates in the distribution may differ for various reasons.

1. The scrutiny, editing, coding, and transcription of the recorded measurements may not yield identical results on repetition.

2. The estimation may involve decisions, computations, and combinations of data that are not completely specified or determined by the procedure.

3. The processes of checking and verification are often not perfect and may reveal at each repetition somewhat different sets of mistakes for correction.

The estimation process is familiar in many forms, and there are many physical models by which it can be represented. For example, when we measure the distance between two points on a map, we actually measure the distance between two dots on a piece of paper. We can use the scale printed on the map to convert this distance into an estimate of the number of miles "as the crow flies." To estimate the distance that a train, automobile, or ship must travel to go from one of the points to the other, we must allow for the fact that the path will not be a straight line but will run along the curved surface of the Earth and follow the winding route of a road or roadbed. Moreover the map is distorted by the map maker in his attempt to put the curved surface on a flat sheet. One inch on the map represents different distances in one part of the map than in another. A more refined method of estimation can correct most if not all the effect of the imperfect correspondence of distances on the paper sheet and distances on the surface of the Earth it represents. This is what estimation procedures do in sampling surveys.

As for the letters to the editor, an estimate of the space devoted to an issue could be made from measurements of the weight of the clippings. By weighing whole sheets of newsprint, we could establish a relation between weight and area. With this ratio we could then convert pounds of clippings into inches of type. If the clippings included portions of adjacent columns, the estimate would be biased. However, the ratio derived from weighing and measuring a random sample of properly trimmed clippings would have little or no bias.

Likewise a relation between a scale of statements and the results of more thorough measurement of reaction to a play might be used to estimate reaction from data drawn only from the scale.

A simple example of estimation may be cited at this point. In several surveys, a sample of households is taken in such a way that every household in the population has the same probability of being selected, at least so far as the model governing the selection process is concerned. Then one adult is selected in each household to be interviewed about his own attitudes. Thus the adult in a small household has a greater probability of being selected than adults in households with greater numbers of adult members. A simple method of estimation is then used to correct the effect that this unequal chance of being selected may have on the accuracy of the results. Each adult's responses are counted in the data as many times as there are adults in his household. In effect, he is used as a sample from his household and he represents each adult in it by his responses to the questions. .

Looked at in another way, this method of estimation consists of "weighting" each response by the number of adults in the household of the respondent. In several surveys it has been found that such weighting made little difference in the final results. However, in some surveys it could make a great difference. Clearly it would do so if the question were "Are you the sole adult member of your household?" or "Do you have other eligible voters in your family with whom you have discussed this issue?"

There are often very important reasons why the interviewing should be limited to one person in a household. Obviously if other members overhear the interview questions and the respondent's answers, they may be reluctant to express opinions that appear to be contradictory. There may be other undesirable effects, such as a reluctance to spend time repeating in detail what another member has already said. The other members may be inclined to say "I think the same as he does about it." If it is known that the method of estimation will correct the bias introduced by taking only one adult per household, then this otherwise more desirable method of selection can be adopted in spite of the bias. This is another way in which estimation and sampling are closely linked.

4.6 THE COMBINED EFFECT OF SAMPLING, MEASUREMENT, AND ESTIMATION

When a method of selection, a measurement process, and a technique of estimation are joined in a single survey operation, they constitute a system for obtaining information from a population about its properties and characteristics. We can think of the information as originating in the population and flowing through the system to merge as a final estimate or set of estimates about the population variables. As it flows along, it successively takes the form of (a) a designation of the population, (b) a selection of some of its members, (c) a set of measurements made on them, and (d) a set of estimates. Some collateral information accompanies it, especially information about how the various processes were operated. At each stage a new distribution of results appears. If we can form from these distributions a sequence of comparable estimates, we find that the bias is modified from one stage to the next and the error variance also changes.

There is no one model that can represent closely all the possible survey operations, but a great many can be represented by a model in which each step or stage adds a further component of bias and its own component of error variance. Since the biases may be positive or negative, the combined bias for the whole survey operation may be either positive or negative. The biases of opposite sign will offset each other to some extent when they are summed. It is conceivable that the *net bias* of the final estimates will be zero, though this is highly unlikely, unless this is assured by the design of the survey. It is usually impossible to design a survey so that the net bias is negligible.

The error variances are all positive. Therefore, they accumulate. Ordinarily we expect each step to make the distribution spread a little more. There are exceptions, however, since the introduction of data from other sources, the correction of bias in estimation, and some other technical procedures may reduce the spread of the distribution and center it more closely about the true value.

In our attempts to understand sampling better and to use the results of sampling surveys more wisely, we find that we can learn something but not a great deal about the combined effect of the several stages in the survey operation, including the sampling process. We can also learn something but not a great deal about each process by itself. Part of what we learn we get by logical or mathematical analysis of appropriate models and part by observation, testing, and accumulation of experience with surveys. The two kinds of knowledge interlock and are to a considerable extent dependent each on the other.

The model and the actual survey never fit completely; the model and a sequence of equivalent survey operations repeated under essentially identical conditions seldom if ever fit perfectly. This lack of perfect fit is probably true of all scientific laws and the phenomena they represent. The lack of perfect correspondence is not an intoler-

able limitation. The value of using models to improve our understanding of actual operations and their results would be jeopardize
if the models had to fit exactly. It would also be jeopardized if th
degree of fit of the model were not taken into account whenever it i
pertinent to the analysis and use of these operations and results.

The technical problems connected with the use of models are bein
studied, and a rapidly growing literature is available (e.g. 4, 5, 6, 7
8, 9). Almost all the attention in these references is devoted to sam
pling that is guided by probability models. Because these books, anc
the articles to which they refer, are available, we have not attemptec
to develop the technical aspects of sampling models; rather, we hav
stressed the unstated assumptions that underlie most uses of model
and have sought to broaden our general understanding of how they
work.

4.7 SUMMARY

In this chapter we have seen something of the value of models i
demonstrating clearly what occurs in the real world and in helping u
to discover additional facts and relationships. They also help us t
make sound decisions and take wise actions. They increase the ef
fectiveness of our efforts to attain valuable objectives and avoic
losses.

Models do not represent all the characteristics of the systems o
objects and relations to which they correspond but only those aspect
they are intended to exhibit. The user should understand wherein they
differ.

For each of the variety of possible sampling operations and for the
other phases of survey processes, it is possible to establish a model
The value of the model depends on how well it fits or conforms to the
real system of events it is intended to represent.

Models that represent the survey as a game between a sampler anc
the "situation" reveal the problems of choosing effective methods anc
of evaluating sampling results. Information about the way in which
the population is "set up" by the "situation" may be very useful in
the choice of methods of playing the survey game.

Models reflecting the processes of measurement and estimation are
also useful, especially when they are combined with models for the
selection of the sample of persons from the population to provide a
model for the whole survey operation.

In analyzing distributions of results derived with the help of a
model, it is found that a useful distinction can be made between systematic errors or bias on the one hand and random error on the other.

The manner in which biases combine to form the net bias of the whole survey process is different from the manner in which the random errors combine. This has important implications for the appraisal of sample results and for the design of survey procedures.

The final distribution of estimates determined by the combined model, when evaluated in accordance with the purposes of the survey, provides the basis for comparing alternative procedures and for judging the dependability of survey results. The development of the techniques of constructing mathematical models and using or applying them is a large subject, treated in the excellent books and articles published in the past decade. Certain general conclusions and interpretations will be presented in the next chapter.

REFERENCES

1. Kenneth J. Arrow, "Mathematical Models in the Social Sciences," Chapter 8 in Daniel Lerner and Harold D. Lasswell, editors, *The Policy Sciences*, Stanford University Press, Stanford, 1951.
2. Irwin D. J. Bross, *Design for Decision*, The Macmillan Company, New York, 1953.
3. Paul F. Lazarsfeld, editor, *Mathematical Thinking in the Social Sciences*, The Free Press, Glencoe, Ill., 1954.
4. Russell L. Ackoff, *The Design of Social Research*, University of Chicago Press, Chicago, 1953.
5. William G. Cochran, *Sampling Techniques*, John Wiley & Sons, New York, 1953.
6. W. Edwards Deming, *Some Theory of Sampling*, John Wiley & Sons, New York, 1950.
7. Morris H. Hansen, William N. Hurwitz, and William G. Madow, *Sample Survey Methods and Theory*, Vols. I and II, John Wiley & Sons, New York, 1953.
8. P. V. Sukhatme, *Sampling Theory of Surveys with Applications*, Iowa State College Press, Ames, 1954.
9. Frank Yates, *Sampling Methods for Censuses and Surveys*, second edition, Charles Griffin & Company, London, 1953.
10. Howard M. Bell, *Youth Tell Their Story*, American Council on Education, 1938.
11. Alfred C. Kinsey, Wardell B. Pomeroy, and Clyde E. Martin, *Sexual Behavior in the Human Male*, W. B. Saunders Company, Philadelphia, 1948.
12. Bernard R. Berelson, Paul F. Lazarsfeld, and William N. McPhee, *Voting*, University of Chicago Press, Chicago, 1954.
13. Elihu Katz and Paul F. Lazarsfeld, *Personal Influence*, The Free Press, Glencoe, Ill., 1955.
14. S. A. Stouffer, "Statistical Induction in Rural Social Research," *Soc. Forces*, 13 (1935), 505–515.
15. Clyde V. Kiser and P. K. Whelpton, "Social and Psychological Factors Af-

fecting Fertility. V. The Sampling Plan, Selection and the Representativeness of Couples in the Inflated Sample," *Milbank Mem. Fd. Quart. Bull.*, 24 (1946), 49–93.

16. C. A. Moser, "Recent Developments in the Sampling of Human Populations in Great Britain," *J. Am. Statist. Assoc.*, 50 (1955), 1195–1214.

17. Philip M. Morse and George E. Kimball, *Methods of Operations Research*, The Technology Press and John Wiley & Sons, New York, 1951.

18. John von Neumann and Oskar Morgenstern, *Theory of Games and Economic Behavior*, Princeton University Press, Princeton, 1944.

19. J. D. Williams, *The Compleat Strategyst*, McGraw-Hill Book Company, New York, 1954.

20. Morris H. Hansen, William N. Hurwitz, Eli S. Marks, and W. Parker Mauldin, "Response Errors in Surveys," *J. Am. Statist. Assoc.*, 46 (1951), 147–190.

21. Herbert Hyman, *Interviewing in Social Research*, University of Chicago Press, Chicago, 1954.

22. Hugh J. Parry and Helen M. Crossley, "Validity of Responses to Survey Questions," *Publ. Op. Quart.*, 14 (1950), 61–80.

23. A. Angus Campbell and George Katona, "The Sample Survey: A Technique for Social Science Research," in Leon Festinger and Daniel Katz, *Research Methods in the Behavioral Sciences*, The Dryden Press, New York, 1953.

24. Carl A. Bennett and Norman L. Franklin, *Statistical Analysis in Chemistry and the Chemical Industry*, John Wiley & Sons, New York, 1954.

25. Stanley L. Payne, "Interviewer Memory Faults," *Publ. Op. Quart.*, 13 (1949), 684–685.

26. Leon Festinger and Daniel Katz, editors, *Research Methods in the Behavioral Sciences*, The Dryden Press, New York, 1953.

27. Bert F. Green, "Attitude Measurement" in Gardner Lindzey, editor, *Handbook of Social Psychology*, Vol. I, Addison-Wesley Publishing Company, Cambridge, Mass., 1954.

28. J. P. Guilford, *Psychometric Methods*, second edition, McGraw-Hill Book Company, New York, 1954.

29. Marie Jahoda, Morton Deutsch, and Stuart W. Cook, *Research Methods in Social Relations*, The Dryden Press, New York, 1951.

30. E. F. Lindquist, editor, *Educational Measurement*, American Council on Education, Washington, D. C., 1951.

31. James H. Lorie and Harry V. Roberts, *Basic Methods of Marketing Research*, McGraw-Hill Book Company, New York, 1951.

32. S. S. Stevens, editor, *Handbook of Experimental Psychology*, John Wiley & Sons, New York, 1951.

33. Samuel A. Stouffer, Louis Guttman, Edward A. Suchman, Paul F. Lazarsfeld, Shirley A. Star, and John A. Clausen, *Measurement and Prediction*, Vol. IV of studies in Social Psychology in World War II, Princeton University Press, Princeton, 1950.

34. M. K. Horwitt, "Fact and Artifact in the Biology of Schizophrenia," *Science*, 124 (1956), 429–430.

CHAPTER 5

Some General Principles
of Sampling and Conclusions
Deduced from Models

5.1 THE BASIS OF DEDUCTIVE RESULTS

A rigorous logical or mathematical approach to sampling would start with a set of axioms that are stated in an unambiguous manner and in a specialized language. It would proceed to reason deductively from these axioms to theorems that are valid so long as the axioms are accepted. We are tempted to develop principles of sampling on this basis. However, this has been done very well in several treatises (1, 2, 3, 4, 5) published in recent years by leading sampling experts. Moreover, to do so would tend to cut us off from a large audience of people who have serious interests in sampling problems. For these reasons we shall attempt to formulate here in a manner closer to common sense some of the results that can be obtained in a more scientific manner. We shall try to reflect in our discussion some of the many conclusions of a deductive nature that are not treated in the modern treatises. We shall also include some of the many reasonable and practical assumptions that are not encompassed by their axioms. Such assumptions and deductive conclusions are especially important for an understanding of sampling as it is conducted in this field of attitudes, opinion, and consumer wants, for they tend to bring to light what is implicit in thinking and practice.

Our starting point is the sampler in a situation performing sampling operations to satisfy certain purposes associated with the whole survey. Directly or indirectly the efforts of the sampler are guided by these purposes, and within this general guidance they are specifically aimed at intermediate objectives defined in terms of accuracy, economy, feasibility, convenience, public acceptability, etc. Any part of the sampling operation may be judged in relation to these specific aims, but essentially the whole survey must be evaluated in terms of the guiding purposes.

Each part of the sampling operation must be appraised in the context of the other parts as they jointly contribute to the accomplishment of the purposes. This makes the model of a card game appropriate. The model exhibits the interrelations of the actions of the players and the way that the outcome is determined by their combined action, not the action of any one player. We see sampling as part of a game played with the situation. The sampler enters the situation and finds the population "set up" for him. His problem is to decide how he should play the game. Anyone who uses the results of his sampling operations will want to know how he played the game. Neither the sampler nor the user of his results can avoid taking risks or escape taking losses once in a while. Both want to find a line of action that is most advantageous to them according to their own standards. Both also want to know as much as they can about the gains and losses that result from their actions, to determine, if for no other reason, whether their operations are working out as should be expected.

From the preceding discussion we formulate the following assumptions, which will be the basis for deductive conclusions.

1. There is a *sampler* and a *situation*.

2. There is an operation that can be performed to obtain information in the situation which will be called the *survey operation* or the *survey*.

3. As an integral part of the survey, the sampler operates a definite *selection procedure* which can be repeated in similar circumstances.

4. There is a definite *population* or set of persons in the situation.

5. It is possible to repeat the selection procedure on this population without any effect or change resulting from previous selections; that is, the population is restored to its previous condition after each selection is made, or else the operation can be repeated with a series of populations equivalent to the first one.

6. The repetition of the selection procedure produces a series of *samples of persons*.

7. There is a *measurement process*. It can be applied to each person in a sample. When so applied, it will produce an element of data, a measurement, except that in some instances the attempt will fail to produce a definitive measurement, and a dummy measurement or report of "not ascertained" will be taken to be the result of the process.

8. There is an *estimation process* by which the set of data produced by the measurement process can be converted into determinations of the value of a variable or several variables pertaining to the population. For each such variable there is a *true value* for the population, and the purpose of the survey includes obtaining an estimate that approximates the true value to a satisfactory degree. "Value" in this sense of magnitude should not be confused with "value" in the sense of importance or desirability, as in assumption 10.

9. There are specific *constraints* on the survey and restricting influences of a practical nature. They confine the effort that may be expended in obtaining the information through restrictions on the time and money to be expended and also by many other more or less specific restraining influences affecting how the effort can be applied. Some of these restrictions can be allocated to particular units of operation by such means as budgeting, scheduling, and personnel assignment.

10. The information produced by the survey is valuable for the purposes it serves. Its *value* can be determined, or an equivalent measurement of value comparable to the measurement of costs can be made. Further, it is possible to determine the *worth* of the information by subtracting the costs from the value, or to make an equivalent classification of the results of the survey in categories that are sufficient for distinguishing results of greater and lesser worth and particularly those of negative, negligible, and positive worth.

11. From the standpoint of their effect on the worth of the information, errors of measurement and errors of estimation are not negligible.

12. Relative to the true values, the results of the survey differ by an amount called the *error*. The worth of the result varies with the error but is not necessarily greatest for results of zero error.

13. The survey is a *complex system of operations*, and it can produce information only if certain essential parts are included in the system—the population as an accessible set of persons, the selection process, the measurement process, and the estimation process. These processes are so interrelated that no one of them can be changed freely without inducing changes in the others.

14. One or more *models* of the survey system can be constructed. The system and its model will have the similar characteristics necessary for the model to be used to represent the system in certain specified analyses. No such model will correspond to the survey completely and exactly, short of having the survey represent itself. Nevertheless, conclusions drawn from a study of the model will be ap-

plicable to the survey and its results, insofar as the conclusions are drawn in accordance with valid principles of deduction from models.

15. The sampler has *prior information* about the situation and the population which he can use to make his operations more effective in the attainment of the objectives set for him. This information is also available for use in assigning magnitudes to some of the unknown constants (parameters) involved in the model whenever the sampler's operations are guided or regulated by conclusions drawn from a model.

16. The errors that result from the repetition of a survey have an average or expected value in the long run that differs from the true value by an amount called the *bias*. If the effects of each of the processes in the whole survey can be distinguished and a distinct component of bias determined for each, then the bias of the survey is the *net bias*, the sum of these components. Ordinarily some of the components are not known and the net bias is not known, at least at the time the survey is in operation. The magnitude of the net bias is important in relation to the average or expected worth of the results of the survey.

17. The results of a survey will frequently be applied or extended to populations other than the population specifically designated for the survey. For such populations the bias will be different and the worth also different, but otherwise there is no reason to hold such extensions invalid or spurious. The same is true of using the results for purposes other than those that guide the survey.

From these initial assumptions, augmented by occasional additions of information or supplementary assumptions, a number of general conclusions may be derived. Some of them may serve as principles of sampling, at least until more specific and closely reasoned guides can be developed for the appraisal of survey results and for the design of new survey operations.

5.2 COMPARATIVE WORTH

The worth of sample results compared with information obtained by any other process is not determined by the size of the sample, the particular techniques employed, the cost of the operation, the error variance of the processes producing the results, or any other aspect, except as these aspects of the survey and corresponding aspects of the alternative source of information affect the distribution of estimates appropriate to each.

The worth of a *sampling method,* as distinct from the worth of a *sample,* depends on its total or average output, not on any single result. If an attempt is made to compare the worth of two methods on the basis of a single result produced by each of them, the comparisons will often lead to false conclusions. The method that has produced the result of greater worth may in fact be a method that would have the lesser average or expected worth in a longer sequence of experience. It is even quite possible that a relatively poor method may occasionally produce results of the highest possible worth or results that are exactly equal to the true value they are intended to approximate. Likewise, a single instance of results worth relatively little may be obtained from time to time by use of the methods that prove to be superior in the long run.

This is the principal reason why the use of models to represent the performance of various kinds of operations in repeated trials is a distinct advance over the appeal to rather limited experience in which the results of a method are tested by a later determination of the true value. For example, both the close approximation of the *Literary Digest* poll to the presidential election vote in 1932 and its large error in 1936 are less informative about the dependability of such mail-balloting than an analysis of the operations of polls of this type with the aid of suitable models, when this can be done.

Although the models may not always fit the actual operations very closely, they provide useful approximations to the distributions of results that would be produced by these operations in many repeated trials. The observed instances of large and small errors provide only a few items of data. The determination of the shape and location of the distribution from a few observed points, in the absence of any other information about them, is an impossible task. Likewise, cost comparisons and other analyses incident to the comparisons are beclouded by peculiar conditions that prevail during these particular instances but will not necessarily appear in many others thereafter.

A number of rules and principles have been used by samplers. Some of them approximate the full comparison of the effect variations in procedure have on the worth of the results expected in the long run. When these principles are carried to the full extent of their logical implications, however, samplers may arrive at conclusions quite contrary to those reached in a direct study of comparative worth. For example, a well-known principle has led to insistence on the use of unbiased methods of selection and estimation. Recently this principle has been qualified to permit acceptance of bias when other advantages make it worth while. Principles that in their extreme forms call for rejection of all methods other than those of "probability sampling" have been moderated by recognition of situations in which "purposive" and other currently unpopular methods seem appropriate. Principles that called for the inclusion in the sample of representative members of each major part of the population in proportional numbers have been relaxed. It is no longer held that each local cluster in a sample must be representative of its locality. Likewise it is no longer held that each element of the population must have the same chance of being drawn into the sample. Even the recent emphasis on obtaining the most accuracy for a given cost or incurring the least cost for a given accuracy, a very real advance in sampling theory, is being revised to conform more closely to the concept of greatest gain or maximum profit within the restrictions set for the operation. The notion that the "most information per dollar" should be sought is plausible, but it can be quite misleading as a principle for the design and comparison of surveys. Determinations of comparative worth must be made according to rules and principles that take adequate account of the complexity of the survey system.

5.3 PREDOMINANT FACTORS

Among the many factors in a survey and situation that affect the worth of the results of the survey, there will be some that have a predominant influence. This predominance is not absolute, however, for the elimination or extreme modification of any one among a large number of factors may suffice to destroy the survey. Predominance of a factor depends on the magnitude of the effect it would have if it were varied through some reasonable range of variation. More specifically, it is the degree to which the sampler can affect the worth of the results by making manipulations of the factor that are feasible for him.

Clearly the sampler has to manage the whole sampling operation. His control of some phases of the survey may be weak or non-existent. The control he has or can get over various factors may depend on steps he has yet to take. He faces a question of how he should attempt to exercise whatever control the situation may permit at the moment. Ordinarily he will concentrate his efforts on control of the predominant factors.

In managing a complex and interrelated system, the various factors must be adjusted in conjunction and not separately and independently. The effort available for the adjustment and regulation of controllable factors should be applied economically, i.e., by an allocation that aims to achieve the greatest increase in worth. If the spread of the distribution of results is attributable more to measurement error than to selection error, only a limited gain can be had from reductions of selection error, even to zero. If a known component of bias can be eliminated, such elimination may actually make the sample worse when the composite bias due to the other components before the change is between -50 and -150 per cent of the bias of this component. The net bias would then be greater in absolute value (disregarding its direction as positive or negative) after the removal of the component of bias than before. This complexity constitutes a difficult problem for even so apparently simple a matter as deciding about the removal of a predominant bias component. The problem is comparable in difficulty to that of finding the magnitude of net bias that will provide the greatest worth, for in addition to the difficulty of measuring costs and values, there is the difficulty of determining the joint effect of changes in the bias and in the variance of the distribution of results.

It may well be more profitable to improve the measurement process or to obtain more data from sources outside the survey to be used in the estimation process, or to modify the cost relationships and the value relationships that go into the determination of worth, than to seek immediate improvements in the sampling. On the other hand poorly designed selection procedures may be the predominant factor limiting the worth of the results; if this is true, the most profitable place for improvement is the selection process itself.

5.4 SIZE OF THE SAMPLE

One of the questions frequently asked about sampling is "How large a sample do I need to make sure it is representative?" On our assumptions there is no answer to this question except "It all depends." A small sample may be representative and a large one may not. Most models will show a decrease in the error variance with an increase in the size of the sample when other factors are not changed. Mere change in size does not change the bias, however. Of course, it may not be feasible to enlarge the sample without simultaneously changing some of the other features of the selection process and even some of the features of the measurement and estimation processes. The combined effect of all these changes could be to increase rather than decrease the random-error variance and to affect adversely the net bias. In the simplest model of random selection, the error variance of most estimates is inversely proportional to the number of elements selected for the sample. As the size of the sample increases indefinitely, the error variance goes to zero. However, it does so in a way that is likely to yield sharply diminishing returns in terms of worth. This is because the spread of the distribution of estimates is reduced as the square root of the reciprocal of the sample size. A 20 per cent increase in size produces only a 10 per cent reduction in spread. The cost is likely to increase by something nearer 20 per cent, at least after fixed costs become a smaller part of the total.

In more complex types of sampling, such as cluster sampling with subsampling, the effect of an increase in sample size will depend on the way in which the increase is obtained. If it is merely an increase in the number of people taken from each cluster, there may be very little reduction of error variance. If it is an increase in the number of clusters, the effect will be like that of simple random sampling. In systematic sampling the effects may be more complicated. The increase in size will change the interval between elements in the sample and thereby alter the relation of the pattern of selection to the systematic variations in the structure of the population.

In general, then, the effect of changing the sample size is like the effect of changing any other feature of the survey procedure. It depends on the interrelations among the parts of the survey and on many other factors in the situation. *A detailed analysis of a suitable model, as well as empirical investigation, is needed to determine ever approximately what the effects may be.* As a rule of thumb, but no more than that, the larger the sample the more accurate and dependable the results, *other things being equal.* Unfortunately they are seldom equal, except in certain cases of adjusting the size of one survey or manipulating a model. Size, by itself, is no basis for comparing two surveys.

5.5 THE POPULATION

There are many characteristics of the population that affect survey operations other than those characteristics about which information is being sought. One of the most important from the standpoint of costs and accessibility is the geographic distribution of the population. It determines the ease or difficulty of reaching individuals by any given means of transportation. The pattern of geographic distribution is produced by a complex and continuous process of geographic specialization in which different kinds of people tend to occupy different areas or districts. These tendencies prevent people from becoming thoroughly mixed, as the air or water are on the face of the

Earth. Human populations are distributed more like the solid portions of the Earth. The location of different kinds of earth and rock provides an opportunity for economic activities such as mining and agriculture that would be impossible if all the solid part of the Earth's surface were thoroughly intermingled. In like manner, the pattern of geographic distribution, the specialization into apartment, suburban, occupational, racial, religious, nationality, and political clusters within cities and rural regions provides an opportunity for the sampler to use this geographic structuring of the population to his own advantage in designing sampling procedures. It is by analogy to geological formations that the use of these divisions of the population are termed "strata," although patterns of subdivision of the population according to characteristics other than those of geographic distribution are more often the basis for stratification in sample design.

Information from the tabulations of the Census of Population by city blocks or other small areas provides measures of economic level and demographic characteristics for use by samplers. Maps and directories have been used in many surveys as the basic material for sample selection. Information about the direction of variation in the population can be used to lay out areas for sampling operations, just as the direction of maximum variation in fertility is used in agricultural field trials. The information need not be in terms of areas since variation along major highways and back roads is quite as important as variation from county to county or town to town.

In addition to the geographic distribution, there are many other kinds of information about the population that can be used to strengthen the selection process and improve the survey. For example, the occupation, education, religion, race, age, sex, migration, national origin, income, marital status, parental status, intelligence, temperament, and health of individuals in the population may be useful bases for classification into groups that can be used to advantage. They may be useful in the sampling, in estimation, and even in measurement.

Lack of information about important characteristics of the population or neglect of it may lead to serious mistakes or losses in surveys. A survey taken in a holiday season will usually differ from one taken in ordinary periods. For example, a study of wedding announcements published in a large metropolitan daily led to an erroneous conclusion that such announcements avoided mention of a synagogue or any other association of a wedding with the Jewish faith (6, 7). Although the sample included more than 400 announcements published in June over a period of ten years, it was entirely unrepresenta-

tive because it was taken from a season of the year in which almos'
all rabbis abstain from performing marriages. A blunder of thi
kind could be quite serious in content analysis of propaganda and
news. Instances of this kind of faulty sampling have been reported
in governmental inquiries into the practices of business corporations
The example illustrates an element of timing that frequently causes
trouble in sampling.

In terms of the card game model, we may think of the population
as being prepared or set up for the sampler by the dealer. If the
dealer sets up the population so that it has a definite and non-random
arrangement or so that it is marked by indicators of some useful variables, then the sampler may incur unnecessary losses if he ignores
information about the structure of the population. He may make
greater gains by obtaining information about the structure of the
population and using it in sampling, estimation, and the rest of the
survey operation.

In summary, there is a great deal that the sampler can learn about
the nature of the population that helps him to sample more effectively
The general conclusion is: In an operation that is like a game and that
depends on the way the population is "set up" by various social and
economic processes, *the sampler can operate more successfully and increase the worth of his results by exploiting what is known in advance
about the population.* It is even profitable to allocate a portion of
the available resources for use in increasing this stock of information
(The same is true of more technical uses of information such as for
determining the best periods and days for interviewing or for finding
a suitable sample of people outside their homes.) Finally, information about the population can be used to improve estimates by means
of regression and other statistical methods.

5.6 SUBDIVISIONS OF THE POPULATION

The population can be divided into parts, and these parts can be
subdivided further, on the basis of the information that is available
about the characteristics of individuals and groups of individuals
This subdividing may or may not have a relation to the sampling
and an effect on the survey. It may be merely incidental to some
of the administrative arrangements, such as the assignment of part of
the work to different persons in the survey personnel. It may be arbitrary and irrelevant to the whole survey. When it is not merely
incidental or irrelevant, it may take any one of several forms.

One of the simplest forms of subdivision is the sorting out of unlike or dissimilar persons, such as occurs in the assignment of children to grades in school according to their educational level; the separation of the sexes in some schools, clubs, occupations, and other social groups or individual roles; the grouping of persons who have similar political, religious, or other interests into distinct organizations; and a great number of classifications of people according to legal and social status.

If, for the variables of concern to the sampler, the subdivision makes a complete sorting of people into classifications and if all the people in any one class are exactly alike in the variables, then his problem of selection is reduced to one of taking the right proportion of persons from each subdivision. This is an extreme case of stratified sampling. It reduces the error variance to zero. It also reduces the bias to zero if the number of people taken from each subdivision is correct for the estimation procedure that is to be used. From this we deduce a general principle that one solution to the sampling problem is a subdivision of the population into parts, each made up of persons who are identical on the variables that are to be estimated, and the selection of a fixed proportion of the persons in each subdivision for a sample. We assume that both measurement and estimation are free of error and bias.

Unfortunately it is not often feasible to make such a perfect sorting. Less than perfect sorting can be effected by the use of variables that are correlated with the survey variables. This is a practical method, although it does not necessarily lead to unbiased estimates or to great reductions of the variance.

Another kind of subdivision makes up sets of people, each set being composed of a mixture or assortment of dissimilar persons. If the first type of subdivision, *sorting*, is like filtering, the second type, *mixture*, is like stirring. At least it is the kind of mixture involved in the preparation of an assortment with a predetermined number of each kind in each of the sets. Common examples are the formation of teams and crews in industrial, military, naval, or athletic organizations, each with a predetermined number of members and usually with specified numbers who are specially selected for particular roles or positions. Other examples are certain types of boards and commissions with specified representation of different interests. Less than perfect assortment occurs in families, orchestras, schools, factories, stores, etc. In these the number of members varies more freely, but a balanced group of people with different roles and duties are still needed to complement each other. It would solve the sampling problem if the population could be subdivided into identical assortments.

Then any one assortment would be a fully representative sample with no bias or variance on the variables to be estimated, provided the estimation and measurement were also free of bias and error. Families tend to be mixed by age and sex but sorted by religion. Public schools tend to be sorted by age and mixed by religion.

Most subdivisions are separations in part and mixtures in part. There is, however, a type of subdividing that is indeterminate, namely any subdivision of a population whose members are all identical in the variables that are being studied. It is a population for which sampling can hardly go wrong. For example, any sample of living human beings would have sufficed for the discovery of the circulation of the blood.

It is relatively easy to show by mathematical reasoning that the foregoing discussion can be developed into a system applicable to all possible subdivisions of the population. We find that a sampling distribution of estimates will have a smaller spread if the samples take some individuals from each of the subdivisions that are set up as sortings and also if they take in any one assortment either all the individuals or none. This is to say, the mixtures should be the units of selection and the sortings the strata. Sampling should be a recombination in controlled proportions of what is separated in filtering. It should preserve mixtures or improve the mixing.

For purposes of estimation, the subdividing should conform to that of the additional data available for use in estimation or it should produce subdivisions that are susceptible of recombination into divisions matching those of the additional data. This poses something of a dilemma. The divisions of the population for which data are available are usually not those that would be most efficient for selection of the sample. For example, data from elections are available for areas such as counties and congressional districts that tend to be mixtures of the population. They would be most useful for sampling if they were sortings. The tendency in the delineation of political districts is to mix people in a way that favors the dominant party by including enough but only enough of its kind of voters in each district to assure success in electing its district candidates. Any extreme form of "gerrymandering" works against the principles that should guide the delineation of strata for sampling.

Another difficulty arises in the smallest districts for which the election returns are counted, the precincts. These areas are changed from time to time, and they do not match other small areas for which population data are collected, such as census enumeration areas and tracts. Because they change, it is well nigh impossible in a consid-

‹rable part of the population to work up the results of previous elec-
‹ions for comparison with the current election on a precinct-by-pre-
‹inct basis. Some very rough approximations may be made. Where
‹he changes in boundaries are a result of population changes, these
‹pproximations may be close to the best that can be expected from
‹ast data, anyway.

5.7 GEOGRAPHIC DISTRIBUTION OF THE SAMPLE

Since people tend to differ on the average from one region to another, subdivisions of the population by geographic location tend to
be sortings on some variables and less than perfect mixtures for other
variables. Samples that do not draw the same proportion of the
population from each of the subdivisions will tend to be biased, and
it would seem that a broad scattering of the sample throughout the
population is necessary to avoid serious bias in the survey results.
However, there may be combinations of less widely scattered areas
for which the biases are small because each combination is itself a
good mixing. Moreover, the bias of an uneven scattering may be
reduced by a suitable estimating procedure, such as weighting up the
results from areas that are underrepresented in the sample and weighting down those that are overrepresented.

If we assume that the measurement and estimation processes are free
of bias and that the sampling bias is correctable by some means outside the sampling operation, then *it is not always true that the sampling procedures which produce the smallest error variances are those
that take the same proportion of the population in each geographic
subdivision.* If in the card game two piles were set up with only red
cards in one and black in the other, whereas the remaining two piles
had mixtures of red and black, then the estimates of the proportion of
black cards in the four piles would be distributed with a smaller
spread if the greater part of the sampling were done in the third and
fourth piles. Even a small amount of evidence that something like
this kind of setup has occurred should lead the sampler to take fewer
cards from the first two piles than the others. This may be the situation in geographic subdivisions of the population. When it is, the
sampler has an opportunity to modify the geographic scattering of the
sample in a way that is to his advantage. There may be such opportunities to use weighting to effect a more advantageous geographic
distribution when there is uncorrectable bias in the sampling and
other processes.

5.8 COSTS OF SAMPLING

Since the worth of the survey results is reduced by an increase of costs, the cost of performing the sampling operation and of making contact with the sample for the purpose of measurement is part of the problem of determining the relative efficiency of two or more procedures and designing surveys. Some reduction of value can be accepted if it is more than offset by a reduction in costs; some increase in costs can be accepted if it is more than offset by an increase in the value of the results.

Here a problem arises, since some of the costs of sampling may involve the use of materials and the application of results of research that are also usable for other surveys. It is a problem of the cost of using capital resources and a problem of allocating joint costs. Hence the cost factors in the determination of worth are not independent of the work that is done in other surveys, or even in other kinds of enterprises. This may mean that the worth of a survey cannot be determined without extensive inquiries. Of course, the same is true of the determination of value, since the results of the survey may be used with the results of other surveys.

The costs of sampling are also linked to the costs of the measurement and estimation processes. The costs of measurement may be greater per person for persons obtained by one method of sampling than for those obtained by another. The same is true for estimation. This is one more reason why the appraisal of a selection procedure must be done in the context of the whole survey operation, and indeed for the whole situation in which the sampler must operate.

If in the card game it costs the sampler the same amount to look at a card without regard to the place of the card in the piles, and there are no other sampling costs, then the cost of sampling has no effect on the survey costs except through the number of cards taken. If on the other hand the cost of taking a card is a fixed amount for each card that must be removed to reach it in the pile, then for the same setup of the population by the dealer the sampler's best method of sampling would take more of the cards from the upper part and fewer from the lower part of the piles. He might still go to the bottom of a pile some of the time in order to prevent losses he would incur if he gave the dealer an opportunity to take advantage of the cost situation by loading the bottom of the piles with one kind of card.

Actual examples can be found in situations in which the most remote parts of the population are quite distinct from other parts in important variables. A survey organization located in the Northeast might well go to Texas and California to make sure that its sample includes these states and is representative of opinion on the question of offshore oil wells or to the West coast if the issue concerned shipping or fisheries.

Relatively complicated cost functions may be required in a survey model. The difficulty of making the model match or fit the actual operations may be greater in this respect than in any other. In this simple discussion of the deductive implications of our formulation of the survey process, we can at least conclude that cost analysis is an essential part of the appraisal of the worth of survey results. Any evaluation of data provided without cost to a user of survey results should be tempered by due regard for the limitations that are imposed on the producer by the cost of obtaining data. If there were no cost limitations (in the sense that includes all the effort and resources devoted to a survey), the size and intensity of the survey operation would be expanded until they reached the limits set by other constraints. Compared to such an expansion, all surveys appear inadequate and of low quality. On a more realistic basis that counts the costs as offsets to the values attained, relatively meager operations may appear to be thoroughly worth while and even the best for the purposes. None of this, of course, alters the fact that the user must take full account of the limitations of the results, especially the degree of bias and error to be expected, in his use of data from samples that are severely curtailed by cost restrictions.

In a sense a survey is an exchange of the costs of remaining in ignorance for the costs of obtaining information. The problem is to determine the point or region in which further exchange would not be profitable.

5.9 OPERATIONAL RELATIONS OF SELECTION, MEASUREMENT, AND ESTIMATION

It is easy to see that the worth of the results of a survey may become negative (less than that of taking no survey at all) if one of the essential processes fails completely. If there is no sample, if there is no measurement, or if the measurements are not converted into estimates for the population, then there is no meaningful outcome from the survey and the value is zero. The sampler still faces the lack of information that prevailed at the start. Subtracting the costs of the survey, we find the worth to be less than zero.

It is not very difficult to see that any of the processes can operate in a defective manner short of complete failure and still degrade the survey process so severely that the other processes make no really effective contribution to the results. For example, the measurement process may operate in a nullifying manner if it produces a large proportion of responses classified as "Don't know" or "Not ascertained." If it does, too little information is transmitted from the population to the estimating process. No technique of estimation can correct this malfunctioning. Likewise, if the estimation procedure fails, no selection or measurement process can restore what is lost.

These remarks apply to extreme cases. In less extreme instances there is a possibility that the effects of defects in one process can be reduced by suitable adjustment of the other processes. In addition, opportunities may be exploited to improve the survey process by methods that introduce deliberate distortion in one step and correct the distortion in later steps. Such distortion must be controlled so that the distortion that occurs in the middle stages of the process actually will be removed in subsequent stages and not damage the final results. Some of the controversy about adjusting survey figures neglects this principle by failing to distinguish biases that can be controlled and removed from those that cannot. What really matters is how the survey process operates in its entirety and within the situation. So long as the basic purposes of the survey are fulfilled and the user of the results can be assured on sufficient grounds that they are, the nature of the intermediate processes is not of great concern to him. It may be for technical reasons, such as the advantages of simplifying or standardizing procedures, or it may be for purposes outside the scope of the survey, but any restrictions that are placed on freedom of action in the design and operation of intermediate processes may unnecessarily exclude more advantageous or more efficient procedures.

This conclusion about avoiding unnecessary restrictions in the design and operation of surveys does not warrant relaxation of essential standards, nor does it support complacency based on ignorance of the damage that is being done in the survey by poor performance and inferior methods. Though it implies that the results of surveys that do not conform to some intermediate technical requirements should not be rejected on that basis alone, it also means that results should be accepted only on the basis of definite evidence of their worth.

5.10 TECHNIQUES OF SELECTION

We may recognize two broad classes of selection techniques: (1) those techniques that are independent of the particular setup of the population in the sense that the selection can be made in advance by a means that is essentially a selection of certain places or positions out of all the possible places an individual may occupy, and (2) those techniques that depend to some extent on individual characteristics of the persons in the population or at least some sorting of individuals as they are actually assigned to or fall into their places in the setup. An example of the first is found by first drawing a sample of dwellings or a sample of jobs so that the drawing is independent of the characteristics of the dwellings or jobs and the persons who occupy them, and then finding out who occupies them. An example of the second is a sample consisting of persons who match in their age distribution the age distribution of another sample. Another example of the second kind is a sample consisting of the first ten housewives found at home as an interviewer proceeds down a predesignated street. We call these two classes of selection techniques *place* selection and *indicator* selection.

It is possible to combine place and indicator selection and also to add restrictions and constraints of various kinds. The designation of the size of the sample may be regarded as one such restriction. Other rules might limit the number of individuals of a certain kind or assure that at least a certain number of them was obtained. A rule that might be used for some purposes is "Keep drawing elements from the population until the number of elements of a certain kind that have been drawn equals a given number." These more complex rules all aim to control the distribution of estimates and the cost of the survey in an advantageous manner.

If the selection technique is that of place selection and the population is not already set up in a convenient order, the preparation of the population for selection of the sample can be accomplished by such devices as numbering individuals in serial order, listing alphabetically by name (with a special rule for persons who have the same name), listing in order of application or encounter, assigning places in a list by lot, etc. Often this has already been done by such systems as the assignment of Social Security numbers, automobile licenses, registration for military service, and directory listings. There are methods of achieving equivalent results that avoid the labor of listing. One method of selection that has been used is to take every person who lives at an address that has a house number ending in a given digit (again with special rules for cases that lack house numbers). One possible safeguard against an adverse setup of the population is to use all patterns of place selection equally often, accomplished by using devices such as games of chance or their equivalent, a table of random numbers. The result is a "random sample."

Actually it is the procedure, not the sample, that is random. The sample may well be the same as a sample produced by some other procedure. A sample need not be random to be valid, as some people believe. The worth of the survey results may be increased on the average by some modification of the selection procedure that yields a more advantageous distribution of samples. The development of probability models to guide the selection procedure is a very important step in this direction.

The setup of the population may be favorable to the sampler whether he knows it or not. At least it may offer him an opportunity to improve the effectiveness of his procedure. The sampler can use every known characteristic of the individuals in the population to decrease the error variance of estimates. Here the best strategy of the sampler will depend on cost considerations, for the use of indicators otherwise will at least do him no harm. (There is an exception with respect to the estimation of the sampling error, but one that is not often serious.) We reach the deductive conclusion, then, that *indicators can be used to advantage, subject to cost considerations, and that where the indicators are not distinct, "place" may be used as if it were an indicator, for purposes of selection.* The manner of using indicators is important. They should usually be used as sorters in the sense described previously. The sample should be made representative of the population with regard to the indicators, if that is possible, and in any event the selection of individuals with the same indicators should be by random selection according to place.

Random selection by place undoes the effect of the place assignments by which the population has been set up. If the processes by which the population has been distributed to the places occupied by its members are known, even in part, the sampler may derive from this information some additional indicators for use in his method of selection. One of the simplest of these is based on the similarity or difference of adjacent place holders. If the sampler knows that neighbors tend to be alike, his sampling will avoid taking neighbors, other things being equal. If he knows that they are dissimilar, he will find it advantageous to take neighbors, other things being equal. This

inciple is developed in more general form as systematic sampling
d cluster sampling.

The full selection process depends, therefore, on how the population
set up and how the sampler makes his selection. If the population
in random order and the selection is independent of individual
aracteristics, he can only draw a random sample; if the sampler
aws a random sample, the setup of the population is of no conse-
ence. Otherwise the joint effect of setup and selection procedure
akes the results something else than random. Just what depends
the nature of the game.

5.11 GAUGING ACCURACY AND THE DISTRIBUTION OF RESULTS

In making any use of sample survey results, we need not only the
sults themselves but information about the manner in which they
ay be used. One of the most important parts of this information
r the user is a judgment of the accuracy of the results. If the
mount and direction of error were known, of course, it would be a
mple matter to correct the estimates so as to eliminate the error.
he situation is seldom that simple. At best we may only be able
say that in repetitions of the survey some definite proportion, say
ineteen out of twenty, of the estimates will be between two limits
at are specified. Ordinarily the limits can only be approximated
om the data furnished by one sample.

We may say that the two limits will encompass the expected value
etween them in nineteen out of twenty trials in the long run. But
hat of the short run and what of the present instance? Clearly we
re still playing the game. Maybe we are right in claiming that our
wo limits have caught the expected value like a butterfly in a net.
Maybe we are wrong. One time we win and another we lose. We
lay the game for what we may win on the average. We hope we
ill not lose everything and be compelled to withdraw from the game
mpty-handed. The fear of ruin may make us find a safer and more
onservative way of playing. We may spread the limits or shift them
n one direction to increase the probability of being right.

If what we win depends on how close the limits are to each other,
e shall not spread them beyond the distance at which we begin to
educe the prize more rapidly than we increase the frequency of win-
ing it. Here is another factor in the worth of the survey. When our
ctions depend not only on the magnitude of the estimate but on our

judgment of its accuracy, and when the value we gain depends o
our actions, then it may be as important to measure the accurac
of an estimate as it is to measure the variables being estimated.

There is a fairly simple way to get limits on the estimates. Sup
pose the survey is repeated nine times, the conditions in the situatio
being the same each time. The nine estimates mark off ten interva'
on the scale of the variable. Now suppose, instead, that ten survey
had been taken and that the nine were chosen at random from th
ten. The one that was left out when the nine were taken was equall
likely to have been the first, second, . . . , tenth in order of the mag
nitude of the estimate. One of the ranks from one to ten is missin
from the set of nine. It might have been between any two estimate
of adjoining rank or below the lowest or above the highest. In othe
words it has a probability of 0.1 of being in any interval and of 0.
of being between the highest and the lowest of the nine. Therefor
we can say that the probability is 0.8 that the next survey estimat
will be within the range of the previous nine. Similarly, the prob
ability is 0.1 that the next survey will provide an estimate highe
than the previous nine and 0.1 that it will provide one lower tha
the previous nine.

This does not tell us how close the estimates come to the tru
value. For this we must find out about the bias. Unfortunately th
internal variation of estimates from similar samples does not tell u
anything about bias. All the estimates have the same bias. The sam
is true of the limits that are derived for a single estimate by well
known methods as set forth in standard statistical textbooks. The
are computed from the differences between individual measurement
and their average. They are based on the same kind of internal com
parison that does not reveal the amount of bias. These method
can be extended by means of some of the relationships observed be
tween the limits and other factors, such as various characteristics o
the variables that are being estimated and the size and distributio
of the samples. Finally, data from other surveys on similar variable
may be used to approximate the components of error and the rela
tions between them that go to make up the error variance of the whol
survey. The error variance leads directly to limits of the kind dis
cussed previously and to their corresponding probabilities.

To gauge accuracy it is necessary to find ways of learning about th
bias. Unfortunately this may never be possible or it may not be
possible until some time after the completion of all uses of the survey
results. There seldom is available any closer approximation to the
true values than the estimates that are being considered. Hence the

amount of bias and its direction can only be surmised. What few direct tests of bias there are will help in formulating the surmises. This necessity of considering the biases and making even blind guesses at their magnitudes constitutes a serious limitation of the usefulness of computations of the standard error of estimate. If there could be some assurance that the biases were definitely smaller than the standard errors, the limitation would not be great; but in the field of opinion, attitudes, and consumer wants the biases of measurement and estimation may be considerable, overshadowing in moderately large samples the error variances. When they are reduced to a much smaller magnitude, the sampling biases and error variances of all the processes will become more important.

5.12 INCREASING THE WORTH OF SURVEY RESULTS

We can see from the foregoing sections that the worth of survey results may be increased by any of several modifications of the survey and of its situation. First the processes can be tuned up to work better. This may bring about a reduction in their error variances. It may bring about a reduction in their contributions to the net bias. It may effect a reduction in costs. These changes may also be accompanied by changes in the other aspects of the survey. The net effect will be beneficial if the beneficial effects outweigh the harmful ones brought about by the change.

To accomplish a good tuning up requires a much more exacting analysis than we have made, based on the particular situation and survey system. Mathematical models or their equivalent are required for at least the major parts of the survey. Information is needed about the magnitudes of various quantities involved in the models. For many kinds of sampling such models and information are not yet available. Even so, crude approximations can be helpful.

One of the best means of tuning up a survey is to test the results against much more accurate measures of the true values. Such tests are infrequent. They do not offer much help in determining which of the constituent processes needs adjustment. This may be obtained better from a series of surveys using some components in common. An experimental design could be made up to permit this calibration of component parts.

Fortunately, the effect of bias and error on the worth of many survey results is relatively small until they reach somewhat extreme magnitudes. Moreover, the more extreme estimates may be partly self-correcting in that they cease to be plausible and come to be questioned or rejected by the prospective users. As the information available prior to the estimate accumulates, the value of estimate subject to substantial error decreases and greater importance must be attached to even relatively small components of bias and error. As the amount of relevant prior information accumulates in the measurement of opinion, attitudes, and consumer wants, the problem of sampling will become more like the sampling problems in relatively exact scientific work.

Since the worth of the results of a survey depends finally on the way in which they are put to use, and this use in turn is dependent on the skill and understanding of the users, it is clear that better education of the users in what the estimates are like and how they should be applied will increase the worth of the results. If such education of the users does not increase the worth of the results, it is because the instruction is ineffective or too costly for the improvements it accomplishes. Part of this education must be specifically on the properties and effects of sampling.

Instruction in technical operations and abstract theory may be out of place for the ordinary users. It should still be possible to give them a reasonably good sense of what is being done and how it affects their use of the results. These users, in turn, may be the only source of good guesses about the worth and value of information. Their support may be essential before a survey is approved. They may also discover important factors to be incorporated into the survey model or to be studied in other operations.

5.13 SUMMARY

In this chapter an attempt was made to work from some plausible assumptions by crude processes of deduction to general conclusions about the dependability, improvement, and operation of sample surveys. These assumptions apply broadly to many kinds of sampling operations not usually considered in the current literature of sampling. The discussion was conducted without recourse to mathematical formulations in order that the conclusions might be understood better by readers who prefer this manner of presenting the problems. Still it has some severe limitations and can carry us only part way toward full understanding. There is no substitute for mathematical analysis in carrying us further.

It was shown that when the relative worth of survey results is the standard for appraisal, very few simple unqualified rules of sampling can be formulated in a generally valid manner. The selection process has an effect on the distribution of resulting estimates, but this effect is not independent of the other parts of the survey. In some degree the effect is a joint effect, and only a relatively complex model can represent it adequately.

Increasing the size of the sample tends to improve the results, but not always. Much depends on the nature of the population and the manner in which it is distributed, geographically and otherwise.

The selection of persons for the sample is the combined effect of the setup of the population and the selection technique that is used. Usually, but not always, a wide scattering of the sample is beneficial, especially if it is similar to that of the population itself, but again there are exceptions. Various kinds of subdivisions of the population may be used. The manner of using them depends on the extent of separation and mixing involved in their formation.

Selection techniques of various kinds may be used either to avoid adverse effects of the way in which the population is "set up" or to take advantage of it. The full selection process is a joint resultant of the setup and the technique of drawing persons for the sample.

It is also important to gauge the accuracy of the estimates, since that is essential information for the user and so affects the worth. The worth is also affected by the costs. A cost analysis is important for the appraisal of results in their proper perspective and for the improvement of sampling operations. The worth of results can also be improved by educating the user in the best manner of using them and in the implications of sampling for his interests.

At this point the deductive consideration of sampling must become mathematical. The further development is left to such texts and methodological studies as those referred to in the beginning of the chapter. Another line of studying sampling is the observation of its performance in concrete instances, a more inductive or empirical approach. The next eight chapters will be devoted to this kind of study.

REFERENCES

1. William G. Cochran, *Sampling Techniques,* John Wiley & Sons, New York, 1953.
2. W. Edwards Deming, *Some Theory of Sampling,* John Wiley & Sons, New York, 1950.
3. Morris H. Hansen, William N. Hurwitz, and William G. Madow, *Sample Survey Methods and Theory,* Vols. I and II, John Wiley & Sons, New York, 1953.

4. P. V. Sukhatme, *Sampling Theory of Surveys with Applications*, Iowa State College Press, Ames, 1954.
5. Frank Yates, *Sampling Methods for Censuses and Surveys*, second edition Charles Griffin & Company, London, 1953.
6. David L. Hatch and Mary A. Hatch, "Criteria of Social Status as Derived from Marriage Announcements in the *New York Times*," *Am. Sociol. Rev.* 12 (1947), 396–403.
7. W. J. Cahnman, "A Note on Marriage Announcements in the *New York Times*," *Am. Sociol. Rev.*, 13 (1948), 96–97.

PART II

Empirical
Studies

CHAPTER 6

Introduction to Empirical Studies

6.1 THE NEED FOR EMPIRICAL STUDIES

The chapters in this part will open another approach to understanding, appraising, and designing sample surveys of opinion and related variables. The discussion in the preceding chapters started with a broad survey of the uses and problems of sampling opinion. It moved on to describe and classify sampling procedures and then to develop a number of assumptions and deductions that are applicable to a wide variety of actual sampling operations.

Now we turn to an examination of actual surveys and the evidence they provide about certain aspects of actual sampling operations. Particular facts and features that were ignored or coalesced in previous generalizing may be the center of our interest.

This contrast and shift of attention from the general to the particular may suggest that there is a fundamental conflict between the two approaches, deductive and empirical. Actually they reinforce each other more than they conflict. Without the results of a substantial amount of empirical study, the deductive approach is purely speculative. It does not even make good progress toward a respectable theory of sampling opinion. Without the generalizing and deductive approach there is little in the empirical approach beyond the assemblage of unrelated fragments of fact which cannot be applied in any rational manner to new and different situations.

In conjunction, the two approaches can do more than supplement each other. They can do more than correct each other's deficiencies. They can be combined into a single system of analysis and design that accomplishes what neither could produce with only incidental help from the other. This combination of approaches has been effected only in the past twenty years or so. It has been very fruitful, and it promises to be much more fruitful in the future as it develops and as the results of research are accumulated.

One of the important examples of the combination of approaches arises in cluster sampling (1, pp. 13–21; 2, pp. 349–353; 3, Chaps. 6 and 12D). In order to reduce travel costs in an interview survey, it is convenient to draw a sample composed of a number of "clusters" of persons or households. What effect does this have on the accuracy of the estimates? Empirical studies can answer the question for a particular population and method of dividing it into clusters of a specified size. The clusters can be combined into larger clusters to obtain an answer for certain other sizes. Here the empirical study bogs down. Can the answers to the question be applied to other populations and to other sizes? A suitable model will enable us to extend the empirical results to other sizes of clusters, though with some loss of accuracy in the additional answers. For the extension of the results to other populations, some additional empirical evidence must be obtained for these populations and additional theoretical principles must be invoked. Thus the two approaches intertwine in the understanding of cluster sampling and also in the choice of the type and size of clusters in a survey design.

There are many kinds of data that may be obtained from empirical studies and used to appraise survey results, form estimates, design sample procedures, and operate surveys. For example, the following are commonly needed.

1. Estimates of variation between elements in the population and between various groupings of these elements.

2. Properties of groupings, such as clusters, strata, and systematic selections, commonly expressed as correlations or other coefficients to be used with the variances in 1.

3. Cost factors and analyses, cost relationships.

4. Operational rates and ratios, such as rates at which questionnaires are returned, interviewing rates, degree of success in completing assigned interviews or locating specified classes of elements.

5. Gross comparisons of surveys intended to obtain comparable information.

6. Comparisons of corresponding parts or component processes in two or more surveys.

7. Data of established accuracy for use in testing and correcting ordinary procedures.

8. Data on uses and applications, particularly as they determine or restrict the specifications to be met by survey results.

Most of Part II will be devoted to subjects closely related to examples 2, 4, 5, and 7. The balance of the list will figure more prominently in Part III.

The need for empirical studies and data does not decline with the advance of selection techniques. Indeed it tends to increase. The more advanced techniques are based on mathematical models of a type commonly designated "probability models," since they are essentially applications of probability theory. They involve at least two steps: (a) a probability model is associated with the sampling procedure and with the population which is being sampled, the purpose of this model being to specify the probability that any given set of elements of the population is drawn as a sample; and (b) from the model are derived expressions which describe the sampling distributions of estimates computed from samples drawn through repeated application of the sampling procedure. These results are seldom, if ever, sufficient in and of themselves to allow us to compute numerical values for the distributions or even to choose between two or more alternative probability model sampling procedures. To do so we must have, in addition to these theoretical results, certain empirical data.

Empirical data are needed for the following reasons.

1. The sampling distribution of an estimate usually depends on the values of one or more population parameters (e.g., the variance of some variable or characteristic of the elements of the population). These parameters can usually be estimated once a sample has been selected, but we must know them in advance in order to construct or choose a suitable sample design. Prior experience offers the only solution to this dilemma.

2. The sampling distribution of an estimate also depends on some quantitative facts connected with the procedures used for selection, measurement, and estimation. The actual sizes of clusters, the cost factors of various operations, and the weights to be used in estimation are a few examples. Again, some of these facts will be determined by the survey as it progresses, but fair approximations to them are needed in advance. Prior experience may be useful for this purpose, too.

3. Even though a sample may be designed or chosen according to a specified model, it rarely happens that perfect execution of the sampling plan is achieved. Deviations from design creep in through such channels as mistakes on the part of field workers, inability to contact designated sample elements, and refusals by designated respondents. These and related problems of execution invalidate to some extent the theoretical deductions. The extent to which this is so can only be learned from experience.

4. The model may be too complex to use for certain computations. In this case empirical trials may be the only feasible method of determining how it works in practice.

5. Some of the assumptions on which the model is based may be dubious. Empirical trials are then necessary to determine how the invalidity of the assumptions may affect the dependability of conclusions drawn from the model concerning the survey results.

Not only must we use empirical data in connection with deductive models, but there are also situations in which experience is the *only* guide to performance. Thus many forms of sampling procedure in current use are not specified definitely enough to permit the deductive approach. In a specific situation this may mean that such procedures cannot be considered. However, if they can be considered, any evaluation of their performance must depend heavily on empirical studies. This is true generally of procedures that do not conform to a definite probability model. Quota sampling procedures are a case in point. For example, evidence will be presented in Chapter 10 to the effect that repeated application of a quota sampling procedure gives rise to a series of results that approximate a sampling distribution for certain forms of estimate. However, the variance and bias of the estimate must be obtained from empirical analyses.

As a consequence of these considerations, the Study of Sampling has been concerned with making and examining empirical studies of all phases of the survey process as they affect and are affected by the sampling procedure.

6.2 SOME REFERENCES TO THE USE OF EMPIRICAL DATA IN STUDYING SAMPLING PROCEDURES OUTSIDE THE FIELD OF OPINION RESEARCH

There are many precedents for the use of the empirical approach, both in the field of statistics in general and in the field of sampling in particular. For example, agricultural statisticians have long used uniformity trial data to aid in the development of experimental designs. W. G. Cochran (4) describes this as follows.

In a field uniformity trial, the area under experiment is divided into a number of plots, usually all of the same dimensions; the same variety of the crop is grown and the manurial and cultural operations are carried out on each plot. The yield of each plot is recorded separately at harvest.

Such data are used to determine optimum size and shape of plot, to investigate the relative efficiencies of different types of experimental designs, to increase the efficiency of later`experiments (by covariance analysis), and to check on the applicability to field experiments of the statistical analyses contemplated.

Because of the interest of agricultural statisticians in empirical data provided by uniformity trials, it is not surprising that similar work developed in connection with the use of sampling to obtain estimates of certain types of agricultural information. Representative accounts of such work are given by Yates and Zacopany (5), by Irvin Holmes (6), and by Raymond J. Jessen (7). Jessen puts the case for the use of empirical data very well when he states that his study was undertaken

to investigate the following questions pertinent to the problem of collecting data by the sample survey method.

(a) What is the amount and nature of error in data secured by interview?

(b) What is the best available sampling procedure?

(c) What method of 'expanding' sample data will provide the best estimate of state or subdivision totals?

So far as the sampling of human populations is concerned, the largest amount of empirical research on sampling methods has been carried out in connection with the measurement of unemployment and other labor force characteristics in the United States. This work started under the Work Projects Administration and was then transferred to the Bureau of the Census and converted into a general population survey conducted monthly (8). Accounts of representative portions of this work are given by Frankel and Stock (9), by Hansen and Hurwitz (2, 10), and by Hansen, Hurwitz, Marks, and Mauldin (11). The following quotation from the second paper listed shows the use that was made of empirical analyses:

The analyses summarized below were carried out for the purpose of deciding between alternative sampling procedures in the revision of a monthly national sample for labor force and other characteristics . . .

The applications of the various principles suggested in this paper have been evaluated by estimating 1930 Census labor force characteristics from a sample that was stratified on the basis of 1940 and more recent data.

The final paper is particularly interesting in that its goal is to give an explicit formulation of a mathematical model for response errors. Elsewhere, Tepping (3, p. 588) has written:

Two separate studies are reported in this case study. Both of them present empirical results of studies of variances and covariances and relations between them for different levels and units of sampling. These results can serve as a guide in the design of sample surveys when similar or related types of characteristics are to be sampled.

Earlier work of this kind was done by Dedrick, Stouffer, and Stephan (12) in connection with the use of blocks and block "segments," by Schoenberg, Parten, Brady, Friedman, Wallis, and others (13, 14) in a series of studies of consumer incomes and expenditures, and by Stephan, Deming, and Hansen (15) in designing the sampling procedure of the 1940 Census of Population. In the same decade similar studies combining theory and practice were undertaken in India by Mahalanobis (16) and in Poland by Neyman (17).

Until relatively recently, there has not been a large volume of this type of empirical research done in connection with the sampling of human populations for the measurement of opinions, attitudes, and consumer wants. The earlier work done on survey methods was concentrated more upon such problems as questionnaire construction and interviewing techniques. The empirical research that has been carried out in connection with sampling methods is exemplified by portions of Cantril's book (18), by the comparison between a national quota sample and the Monthly Report on the Labor Force designed and executed by the National Opinion Research Center (see Sec. 8.2A for a description), and by work on callbacks [e.g., Robert Williams (19) and Durbin (20)]. A remarkable series of studies has been conducted by Moser, Stuart, Durbin, and others at the London School of Economics with cooperation from several British survey organizations (21, 22).

Herbert Hyman's study of interviewing (23) and his extraordinary exposition of survey design and analysis (24) are the result of extensive empirical work. Kish, in his analysis of the dangers in uncritical use of binomial variance formulas (25) and in other studies, has worked together empirical data, and sampling theory. The Study of Sampling has assumed as one of its principal lines of study the examination, evaluation, initiation, and presentation of empirical research related to sampling for the measurement of opinions, attitudes, and consumer wants.

6.3 THE STUDY OF SAMPLING AND ITS PROGRAM OF EMPIRICAL STUDIES

A. Collection and Analysis of Existing Data

The first step in the Study of Sampling was to visit a large number of survey agencies * and to obtain from each a description of the

---

* Most of these visits were made during the period 1946–1947. Subsequent visits were made to obtain information about new projects and modifications of methods.

sampling methods in current use and a statement about special studies which had been conducted, either for the purpose of comparing one or more sampling methods or for obtaining information on one or more component parts of survey and sample design. In addition to this original canvass, efforts were made to learn of other available information—either through published reports or through private sources—which would be of use to the project. The more important results of this search for data and studies, and of the use to which the obtained data and studies were put, are listed below:

1. The close contact which the Study of Sampling maintained with survey agencies in opinion polling, market research, the government, and elsewhere resulted in the acquisition of descriptions of a wide variety of sampling methods which had been or were being used in practical survey work. No attempt was made to incorporate these descriptions into this report. However, they were extremely helpful in planning further inquiries as well as in the formulation of general theory. Many of them have been used in the report to illustrate certain specific points.

2. A number of examples were found in which two sampling methods (or two modifications of the same general method) had been used in essentially the same survey situation, thus affording an opportunity to compare their effects on survey performance. The original concept of the Committee on the Measurement of Opinion, Attitudes, and Consumer Wants had been that the entire program of the sampling study could be based on such gross empirical studies. However, the early work of the study, in addition to the recommendations of the Advisory Committee, indicated that the emphasis of the project should be directed toward analytic study of the component parts of the survey operation. The reasons for this shift of emphasis are discussed in Chapter 8.

3. If a sampling method is to be evaluated or compared with another method on the basis of its effects on the performance of a particular survey operation, then it is necessary to take into account all aspects of survey design and operation. Thus we are concerned not only with the design of a sampling method but also with its execution. The project therefore concerned itself with finding existing data which would describe the extent and effects of deviations from design introduced by the use of human agents. Since data of this kind are not ordinarily collected during the course of survey operations, the findings were rather meager. For example, it is of interest to know how quota interviewers go about the task of filling their

quotas and how much effect the specific procedures they use have on survey results. Only one piece of indirect information on this subject was obtained, namely, by plotting the addresses of respondents for a number of city surveys conducted by National Opinion Research Center. This group also made an analysis of the errors committed by field workers during several prelisting operations. The Census Bureau made an elaborate set of checks on the 1950 Censuses and conducted a Post-Enumeration Survey. Most of the detailed results of this survey are yet to be published, but many of the principal results are presented in the introductory chapter to the several Census volumes (26).

4. When we are sampling human populations, the characteristics of the individual population elements may have an appreciable effect on the manner in which the sample design is executed. Consequently, data were collected on families and individuals who refused to be interviewed or were not at home when the interviewer called to determine the effects of these losses and the value of making repeated callbacks. Many studies of this kind were found, but very few of them provided any information with respect to attitudes, opinions, and related variables. This material was assembled by the Study and is presented in Chapter 11. Another aspect of this same problem relates to methods for reducing costs of callbacks, and in this connection some analyses were made of data on the daily cycle of absence from home.

B. Design of Experiments to Provide Additional Information on Various Components of Sample and Survey Design

At the time the Study was initiated by the National Research Council and Social Science Research Council, it was fully recognized that the joint Committee would not be in a position to actually carry out any large-scale studies, primarily because of the cost of such undertakings. Nevertheless, in view of the shortcomings of existing data as outlined in the preceding paragraphs, some attention was focused on the design of experiments to provide needed data. This activity was carried out at three levels: (a) the design of a single large-scale experiment to provide information on all phases of survey and sample design, (b) the design of experiments that could be carried out during the ordinary course of routine operations by service agencies, and (c) the design of experiments that the project might hope to accomplish with the aid of cooperating agencies. In contrast to special tabulations of existing data, such experimentation appeared to be quite expensive. Apart from an indirect relation to the Denver Validity Study (27), the only quantitative results that arose from these considerations were obtained from a national quota sample of NORC in which each interviewer kept a time log of all his activities. These logs provided information on the time schedule of a national quota survey, on the times that were required for the various phases of the operation (interviewing, travel, etc.), on some of the procedures followed by the interviewers, and on the differences that exist between interviewers with respect to these factors. This study is described in Chapter 12. Some interviewing of interviewers was also done to determine how they operate in fulfilling their assignments.

C. Analytic Studies on Particular Topics

A number of analytical studies were made when the needs and available time indicated that they were appropriate. Specifically, some computations were made to show the effects of systematic selection of city areas with respect to attitude and opinion data; computations to show the effects of weighting and adjustment on data having inherent biases were carried out; and variances were obtained for several surveys to illustrate the applicability of the broad concept of sampling theory which this report presents. Many of these studies have been omitted to keep the size of this volume within reasonable limits, but most of them have contributed to the views and conclusions expressed herein.

6.4 PREVIEW OF PART II

The succeeding chapters of Part II will present the results of some of the empirical research of the Study of Sampling and certain related studies published by others. Since the gross comparisons, either between survey results and known population data or between results obtained by two different surveys, illustrate most of the problems that must be faced in evaluating sampling procedures, they will be given first. The detail with which these studies will be presented will depend on their importance, whether or not similar studies have appeared in the literature, and whether or not replication of the studies is needed in the future. Thus the study of the behavior of quota interviewers will be given in some detail, since it is aimed at a topic about which little is known. Other subjects can be discussed in a more cursory fashion.

It should be recognized in advance that the empirical research to be described here will not give definitive solutions to most of the problems which plague the user, designer, and operator of a sample survey. Nevertheless, empirical research does serve to place the sampling problem in its proper perspective in relation to the other phases of survey design. In addition, the Study of Sampling was intended to stimulate greater interchange between individuals and organizations who are conducting, or who could conduct, similar studies. Even at the present time there undoubtedly exists, in addition to the results reported to the Study of Sampling, a large volume of relevant information which has not been published. Those persons and organizations who are willing to have their work and weaknesses exhibited deserve the greatest respect and praise. Their frankness is a sign of the growing maturity of research in this field; it is essential to future progress.

REFERENCES

1. Philip J. McCarthy, *Sampling: Elementary Principles*, Bulletin 15, New York State School of Industrial and Labor Relations, Ithaca, 1951 (reissued 1956).
2. Morris H. Hansen and William N. Hurwitz, "On the Theory of Sampling from Finite Populations," *Ann. Math. Statist.*, 20 (1949), 426–432.
3. Morris H. Hansen, William N. Hurwitz, and William G. Madow, *Sample Survey Methods and Theory*, Vol. I, John Wiley & Sons, New York, 1953.
4. William G. Cochran, "A Catalogue of Uniformity Trial Data," *J. Roy. Statist. Soc. Suppl.*, 4 (1937), 233–253.
5. F. Yates and I. Zacopanay, "The Estimation of the Efficiency of Sampling, with Reference to Sampling for Yield in Cereal Experiments," *J. Agric. Sci.* 25 (1935), 543–577.
6. Irvin Holmes, *Research in Sample Farm Census Methodology: Part I, Comparative Statistical Efficiency of Sampling Units Smaller than the Minor Civil Division for Estimating Year-to-Year Change*, Agricultural Marketing Service, U.S. Dept. of Agriculture, 1939.
7. R. J. Jessen, *Statistical Investigation of a Sample Survey for Obtaining Farm Facts*, Research Bulletin 304, Iowa Agricultural Experiment Station, Ames, 1942.
8. Frederick F. Stephan, "History of the Uses of Modern Sampling Procedures," *J. Am. Statist. Assoc.*, 43 (1948), 12–39.
9. Lester R. Frankel and J. Stevens Stock, "On the Sample Survey of Unemployment," *J. Am. Statist. Assoc.*, 37 (1942), 77–80.
10. Morris H. Hansen and William N. Hurwitz, "Relative Efficiencies of Various Sampling Units in Population Inquiries," *J. Am. Statist. Assoc.*, 37 (1942), 89–94.
11. Morris H. Hansen, William N. Hurwitz, Eli S. Marks, and W. Parker Mauldin, "Response Errors in Surveys," *J. Am. Statist. Assoc.*, 46 (1951), 147–190.
12. Frederick F. Stephan, "Practical Problems of Sampling Procedure," *Am. Sociol. Rev.*, 1 (1936), 569–580.

13. Erika H. Schoenberg and Mildred Parten, "Methods and Problems of Sampling Presented by the Urban Study of Consumer Purchases," *J. Am. Statist. Assoc.*, 32 (1937), 311–322.
14. U.S. Bureau of Labor Statistics, *Study of Consumer Purchases, Urban Series,* Bulletins 642 and 643; and U.S. Bureau of Home Economics, *Family Income and Expenditures,* U.S. Dept. of Agriculture Misc. Publications, 339, 345, 356, 370, etc., U.S. Government Printing Office, Washington, D. C., 1939.
15. Frederick F. Stephan, W. Edwards Deming, and Morris H. Hansen, "On the Sampling Methods in the 1940 Population Census," *J. Am. Statist. Assoc.*, 35 (1940), 615–630.
16. P. C. Mahalanobis, "Recent Experiments in Statistical Sampling in the Indian Statistical Institute," *J. Roy. Statist. Soc.*, 109 (1946), 325–370.
17. Jerzy Neyman, "On the Two Different Aspects of the Representative Method: The Method of Stratified Sampling and the Method of Purposive Selection," *J. Roy. Statist. Soc.*, 97 (1934), 558–606.
18. Hadley Cantril, *Gauging Public Opinion,* Princeton University Press, Princeton, 1944.
19. Robert Williams, "Probability Sampling in the Field: Case Study," *Publ. Op. Quart.*, 14 (1950), 316–330.
20. J. Durbin and A. Stuart, "Callbacks and Clustering in Sample Surveys: An Experimental Study," *J. Roy. Statist. Soc.*, 117 (1954), 388–428.
21. J. Durbin and A. Stuart, "Differences in Response Rates of Experienced and Inexperienced Interviewers," *J. Roy. Statist. Soc.*, 114 (1951), 164–206.
22. C. A. Moser and A. Stuart, "An Experimental Study of Quota Sampling," *J. Roy. Statist. Soc.*, 116 (1953), 349–405.
23. Herbert H. Hyman, *Interviewing in Social Research,* University of Chicago Press, Chicago, 1954.
24. Herbert H. Hyman, *Survey Design and Analysis,* The Free Press, Glencoe, 1955.
25. Leslie Kish, "Confidence Intervals for Clustered Samples," *Am. Sociol. Rev.*, 22 (1957), 154–165.
26. A. Ross Eckler and Leon Pritzker, "Measuring the Accuracy of Enumerative Surveys," *Bull. Int. Statist. Inst.*, 33 (1951), pt. 4, 7–25.
27. Hugh J. Parry and Helen M. Crossley, "Validity of Responses to Survey Questions," *Publ. Op. Quart.*, 14 (1950), 61–80.

CHAPTER 7

Comparison of Survey
Estimates with Known
Population Characteristics

7.1 DEFINITION OF CHECK DATA AND
GENERAL DISCUSSION OF THEIR USE

A. Definition

Any individual who has ever designed, administered, or made use of sample surveys has almost certainly made some comparisons between survey estimates and data for the population from which the sample was drawn. Natural curiosity would be enough to motivate such comparisons, even if there were no other compelling reasons. It has been customary to refer to the population characteristics used for this purpose as *check factors* and to the numerical descriptions of these factors as *check data*. Obviously, characteristics for which the details of the sample design force agreement between sample and population will have little value as check factors.

Since it is only sound practice to test a theoretical result empirically, check factors have been and will continue to be used in all applications of the sample survey method. The fields of attitude measurement and public opinion polling have been no exception in this respect, and the type of check factors used by polling organizations is well illustrated by the following excerpt from a paper by Elmo Roper (1):

With these six controls (geographic region, size of place, sex, age, occupation, and economic level) as the yardstick by which we determine our sample, we find we are able then to check on the accuracy of the sample by several devices. If the people in our sample do not report that they own their homes in approximately the same ratio as the United States census figures show Americans generally to own their homes, or if they don't have the right percentage of telephones or electricity meters, or if the percentage f 1936 Ford cars is high or low as compared to national registration figures, e know that our sample is open to the charge of being unrepresentative ⊃ that extent. If, however, with a fair knowledge of economic and geo-ʳaphical variations we have carefully considered all of the yardsticks I have ᵗentioned and if we then find, following the field work, that the sample ᵗeasures up to par on these various items of checking data, I think we are ˟arranted in feeling sure that we have in fact selected for interviewing ᵊ America in microcosm.

ᵗ addition to the check factors noted in this quotation, education ˟as also found extensive application in the field of attitude and opin-ᵊn measurement. Some instances where check data have been cited ˟ay be found in Cantril's book on public opinion (2, Chap. XI), ˢlankenship's book on consumer and opinion research (3, p. 272), a ˟escription of the methods of the survey of consumer finances (4), ᵊnd Katona's book on psychological analysis of economic behavior ⊃, p. 321).

⸰. Reasons for Using Check Data

Over and above the natural motive of curiosity, there are many ᵊnd varied reasons for using check data in connection with sampling ᵖerations. Although most of these will be treated in some detail ᵗhroughout the course of this chapter, a brief summary of the more ᵗnportant reasons will now be presented.

Among those individuals who do, or should, make use of data de-ᵊived from sample surveys, there are many who are distrustful of ᵗhe results obtained because they do not understand the theoretical ⁼rounds upon which the evaluation of such results must rest. One ᵊf the simplest ways of overcoming this distrust and of educating ᵗhese individuals to the practical meaning of sampling variability is ⊃ present estimates from a large number of surveys side by side with ᵃccurate and independent measures of the population quantities being ˢstimated. Very fine illustrations of this type of approach may be ⊃und in the work of the Bureau of the Census which frequently makes ᵈdvance tabulations on the basis of a sample of the returns and pub-ᶫshes complete tabulations at a later date. For example, Hansen and ᶦurwitz (6) present such data for the 1945 Census of Agriculture. On ᵘuly 30, 1946, the Bureau of the Census released national estimates, ᵃased on a sample of returns, for 61 agricultural items. Measures of ᵃampling variability were attached to each estimate. Complete tabula-ᵗion of the returns showed that 5 per cent of the estimates differed from ᵗhe true figure by more than two standard deviations (approximately ᵛhat would be predicted on a probability basis), and that none differed ᵊy more than three standard deviations. Accounts of this kind, where

there can ordinarily be little doubt concerning the sampling metho
and the applicability of theory, are used primarily as an education
device and not necessarily to justify any particular method
sampling. Indeed, it must always be recognized that there are man
traps in the way of anyone who argues for the accuracy of one metho
by appealing to the example of some other method that has been teste
by check data in a different situation.

Survey operations are complicated processes which involve not onl
sampling but also other components such as interviewing, coding
questionnaires or schedules, and processing of data. The final est
mate is conditioned by all these factors, and so it is natural to us
check data for investigating the effects of the various component
or at least for investigating the differences in their effects betwee
the survey and the enumeration leading to the population values.

As a final broad motivation for the use of check data, we ma
distinguish those instances in which agreement between sample an
population is interpreted as meaning that the sample actually draw
is in fact "representative" (7). It is almost as if some such state
ment as the following were being made: "The sampling method use
in this survey is subject to attack on logical grounds. However, se
the good agreement between sample estimates and check data. Th
should convince you that the method worked well this time and th
sample results can be trusted for subjects of investigation for whic
checks cannot be made."

C. Purpose of This Chapter

As outlined earlier, the original emphasis of the Study of Samplin
in evaluating sampling methods for the measurement of opinions, att
tudes, and consumer wants was on the use of check data. This ap
proach did not prove to be as feasible or as useful as had been antic
pated. It is the primary purpose of this chapter to discuss some
the reasons, with illustrative examples, why this approach was no
particularly fruitful. In spite of the limitations, however, an inves
tigation of the uses of check data proves to be very helpful in under
standing the intricacies of the survey process and the role of th
sampling method in this process.

7.2 LIMITATIONS ASSOCIATED WITH THE USE OF CHECK DATA

A. Non-comparability of Measurement Processes

One of the major stumbling blocks underlying the use of check data for the evaluation of sampling methods arises from the fact that the survey results and the corresponding check data may not be directly comparable. This means that an observed difference may be due not only to sampling error and sampling bias but also to differences in the measurement process. Consequently, if it is impossible to determine that the measurement process used in obtaining the sample results and the one used in determining the appropriate values of the check factors are the same or that they produce a known difference, the conclusions drawn from a comparison are severely limited. In particular, it will be impossible to separate the effects of sampling from the effects of the other factors influencing the survey results.

One of the most dramatic illustrations of this type of situation is provided by the "past vote" question. During the period between 1944 and 1948, it was customary for survey organizations to ask some such question as the following:

Do you remember for certain whether or not you voted in the 1944 presidential election?
——— Yes, voted
——— No, did not vote
——— Too young to vote
——— Don't remember

and to ask further of those who said that they voted:

Did you vote for Dewey, Roosevelt, Thomas, or other?
——— Dewey
——— Roosevelt
——— Thomas
——— Other

Some typical results on this question, together with the known behavior of the adult population of the United States in the 1944 presidential elections, are assembled in Tables 7.1, 7.2, and 7.3 on pages 138 and 139.

The values given in the tables show extreme discrepancies between the actual vote cast in the 1944 election and the responses to survey questions. Most of these discrepancies are far beyond what might be expected on the basis of sampling variability, or even on the basis of sampling variability plus a reasonable estimate of sampling bias.

TABLE 7.1

SURVEY RESULTS OBTAINED IN RESPONSE TO A QUESTION ON VOTING
BEHAVIOR IN THE 1944 PRESIDENTIAL ELECTION

Survey and Date	Number of Respondents	Per Cent Saying They Voted	Per Cent Saying They Did Not Vote	Per Cent Who Do Not Remember	Per Cent Too Young and No Answer
OPOR quota, Feb. 1945 [a]	1250	73.2	20.6	0.0	6.2
NORC quota, May 1946 [b]	2589	64.9	35.1		
AIPO quota (non-South), May 1946 [c]	2735	74.8	18.3	0.6	6.3
B and B quota, June, Aug. 1946 [d]	5924	68.5	25.4	0.9	5.2
SRC area, June, Aug. 1946 [e]	1168	60.0	33.8	1.7	4.5

[a] Office of Public Opinion Research, Princeton University, national quota sample.

[b] National Opinion Research Center. This sample is described in Chapter 8 in connection with the NORC-MRLF comparison.

[c] American Institute of Public Opinion, regular national quota sample.

[d] Combination of two national quota samples taken by Benson and Benson. These samples are described in Chapter 8 in connection with the Bikini comparison.

[e] Combination of two area samples taken by the Survey Research Center, University of Michigan. These samples are described in Chapter 8 in connection with the Bikini comparison.

TABLE 7.2

VOTING AND NON-VOTING

(For those who were old enough to vote and definitely stated that they did or did not vote. Actually, 53 per cent cast a vote.[a])

	Per Cent Saying They Voted	Per Cent Saying They Did Not Vote	Deviation from Actual Percentage
OPOR quota	78	22	+25
NORC quota	. . .	. . .	. . .
AIPO quota (non-South)	80	20	+18 [b]
B and B quota	73	27	+20
SRC area	64	36	+11

[a] These figures were obtained from records of votes cast in the 1944 election (including soldier vote) and the July 1, 1944, Census estimate of the population 21 years of age and over (including armed forces). The greater portion of those in the armed forces in 1944 were civilian late in 1946.

[b] The deviation is here taken from 62 per cent, the value corresponding to 53 per cent when the South is excluded.

TABLE 7.3

CANDIDATE FOR WHOM RESPONDENTS VOTED

(Respondents who definitely stated that they did vote)

	Dewey	Roosevelt	Other or No Answer	Deviation of Roosevelt Percentage
	Per Cent Voting for			
Actual vote cast	45.8	53.3	0.9	. . .
OPOR quota	42.5	56.5	1.0	+3.2
NORC quota	32.4	65.2	2.4	+11.9
AIPO quota	42.1	56.4	1.5	+3.1
B and B quota	38.4	59.2	2.4	+5.9
SRC area	36.0	62.6	1.4	+9.3

They would seem to be indicative of exaggeration, i.e., people tend to say they voted when they did not and tend to say they voted for the winning candidate when they did not. In other words, the election results are not comparable with the responses obtained to the voting questions. In situations of this kind, it is important to realize that the sampling procedure that produces the result closest to the population value may not be the best procedure. There may be differentials in exaggeration between different parts of the population, or it may even happen that the measurement differences and the sampling biases compensate each other much less in each of the parts than they do for the whole sample.

We might possibly argue that the data of Tables 7.1, 7.2, and 7.3 represent a biased selection of cases to prove a point (e.g., cases weighted too heavily toward quota sampling methods and cases all based on questions asked concerning the 1944 presidential election). However, many more instances could be cited, all of which would tend to show the same consistent exaggeration. For example:

1. Two OPOR surveys were examined on which a question had been asked concerning voting behavior in the 1940 presidential election and the results given in Table 7.4 on page 140 were obtained. Though they refer to the 1940 presidential election, these data show the same consistent exaggeration that is exhibited in Tables 7.1, 7.2, and 7.3.

2. The Elmira Study (8), described in Chapter 9, obtained a sample of adults from the city of Elmira by probability methods. This survey was made in June of 1948. The 465 individuals who said that

TABLE 7.4

SURVEY RESULTS OBTAINED IN RESPONSE TO A QUESTION ON VOTING
BEHAVIOR IN THE 1940 PRESIDENTIAL ELECTION

(Office of Public Opinion Research)

	Per Cent, April 1944	Per Cent, June 1944	Per Cent Actual Vote Cast
Total distribution of respondents			
Number of respondents	(1229)	(1206)	
Voted	68.0	69.4	
Did not vote	21.2	19.1	
Too young and no answer	10.8	11.5	
Voting and non-voting			
Voted	76.3	78.4	59.2
Did not vote	23.7	21.6	40.8
Candidate			
Dewey	36.3	39.5	44.8
Roosevelt	62.8	59.8	54.7
Other	0.9	0.7	0.5

they had voted for a particular candidate in the 1944 presidential
election distributed themselves between Dewey and Roosevelt as
follows:

	Per Cent for Dewey	Per Cent for Roosevelt
Survey results	44.9	55.1
Actual vote cast	51.7	48.3

Here again the inflated vote for Roosevelt is shown by the survey
data.

3. In its release of July 16, 1952, AIPO reported information on
voting in the 1948 presidential election. The release stated that out
of respondents who were 21 or over in 1948, excluding "people in
various institutions, hospitals, in the Armed Forces, those ill or bed-
ridden and the non-English-speaking groups," 65 per cent said that
they had voted in 1948. The actual proportion of people voting in
the 1948 presidential election was 52 per cent (9, p. 316). The AIPO
figure of 65 per cent was based on a series of surveys of the adult
population which had taken more than a year to complete and "In
these surveys, using an area sample design, a total of 14,696 persons

was asked whether they had or had not voted in the presidential election of 1948."

4. In the spring of 1948, NORC selected a probability model sample in the three states California, Illinois, and New York. Of the 2007 respondents who said they had voted for a particular candidate in the 1944 presidential election, 65.7 per cent said they had voted for Roosevelt, 33.9 per cent for Dewey, and 0.3 per cent for some other candidate. The actual vote cast in these three states was distributed with 53.1 per cent for Roosevelt, 46.4 per cent for Dewey, and 0.4 per cent for other candidates.

5. Further and more conclusive evidence that the differences exhibited in Tables 7.1, 7.2, and 7.3 are at least partially due to measurement difficulties is provided by an experiment conducted in Denver by the Opinion Research Center of the University of Denver and described by Parry and Crossley (10). They present data showing, for example, that 23 per cent of 920 respondents said they voted in the 1944 presidential election when actually they had not voted; actual voting or non-voting was determined from an examination of the election register.

The preceding data have an immediate interpretation for the work of this Study. If questions about past voting behavior are to be checked against election returns for the evaluation of sampling methods (for both bias and sampling variability), we would have to know the answers to such questions as the following: "How much exaggeration is taking place?" "How does this exaggeration depend on the interviewing methods?" and "How does it vary with respect to the type of election being used and with respect to the time that has elapsed since the election?"

This problem of the comparability of measurement procedures, or alternatively, of the validity of responses, may also arise in connection with factual or census-type questions. The previously quoted paper by Parry and Crossley presents information on the accuracy of reports of community chest contributions, on possession of a currently valid library card, on automobile ownership, on possession of a driver's license, on age, on home ownership, and on telephone, as well as on past voting. In general, the responses were most inaccurate for past voting and community chest contributions, and considerably better, though not perfect, for the other items.

Data similar to the above, which do not seem to be well known, were obtained by the Milbank Memorial Fund from their studies in the city of Indianapolis of the factors affecting fertility, and these data are described by Kiser and Whelpton (11). As a first stage, a house-to-house survey with a short list of questions was made in Indianapolis in the summer of 1941. Its object was to locate all couples having certain desired characteristics. These couples were to be interviewed intensively in the succeeding stages of the study. These characteristics were: husband and wife native white; both Protestant; married in 1927, 1928, or 1929; wife under 30 and husband under 40 at marriage; neither previously married; residents of a large city most of the time since marriage; and both elementary school graduates. The canvassers were recent college graduates, carefully trained and supervised, paid by the hour and told that more experienced interviewers would later revisit certain of the households. At the conclusion of this preliminary survey, more detailed interviews were obtained from 1545 couples by highly trained interviewers under circumstances which were conducive to cooperation. Many comparisons have been made between answers obtained on the short schedules and those obtained from the detailed interviews. Table 7.5 presents one small segment of these data by way of illustration.

Here there is an average difference of +0.12 grades of schooling and −2.7 per cent in rental reported *for the same individuals* (or for the same dwelling unit) at the first survey (which was similar to a census) and at the second survey (which was an intensive interviewing). If the first survey had been taken as check data to test the sampling methods used for a second survey, differences of about this same magnitude might have arisen. From this it might then have been concluded that the sampling method was biased to this degree. However, the differences would actually have been due to lack of complete comparability in the measurements of schooling and rental and would not necessarily indicate bias in the sampling procedure or defects in its execution.

Since responses to questions involving factual characteristics of a respondent can vary from time to time as in Table 7.5, or can differ from what is known to be true as in the Denver Validity Study, we must approach the use of even the more common check factors with caution. The discrepancies may not be as large as in the case of past voting or other prestige items such as community chest contributions, but they will still obscure the effects of other components of the survey operation if not taken into account.

The difficulties suggested by the preceding discussion, and frequently many more, will arise whenever we are trying to evaluate or compare sampling procedures by means of check data. In this book we are primarily interested in the performance of sampling procedures. Yet

TABLE 7.5

COMPARISON OF RESPONSES OBTAINED TO SHORT SCHEDULE AND TO DETAILED
INTERVIEWS WITH WIFE ACTING AS INFORMANT

Highest School Grade, Wife

Number of reports	1177
Percentage:	
Correct	80.6
1 grade too low	3.5
2 grades too low	1.2
3 plus grades too low	0.8
1 grade too high	9.0
2 grades too high	3.6
3 plus grades too high	1.3
Average error	+0.12 grades

Rent Paid by Tenants

Number of reports	475
Percentage:	
Agreeing within 4.9 per cent	73.3
Too low by 5–14.9	11.8
Too low by 15–24.9	4.8
Too low by 25 plus	4.4
Too high by 5–14.9	3.4
Too high by 15 plus	2.3
Average per cent error	−2.7

bserved results depend not only on sampling but also on all phases
f the measurement process. It is very difficult, especially when surveys are not designed in advance for this purpose, to separate the
ffects of the two. Yet this must be done if valid conclusions are to
be drawn about the sampling methods. This is well illustrated by
he work of the Social Science Research Council Committee on
Analysis of Pre-election Polls and Forecasts (9). There was clear
vidence that many factors contributed to the failure of the forecasts
—the use of quota samples, the determination of those who were
going to vote in the election, the treatment of undecided respondents,
and last-minute shifts. Nevertheless, after extensive investigation,
he staff was not able to produce a quantitative statement concerning
he amount which each of these contributed to the final error.

B. The Effects of Sampling Variability

In making a comparison between a population value and a samp[
survey result, it is always necessary to take sampling variability in
account. In other words, only discrepancies greater than those th
might be expected on the basis of sampling variability can provi(
evidence that something is operating to produce a systematic error-
either the sampling procedure, or the estimation procedure, or t)
measurement process, or any combination of the three. Stated
statistical language, we must first test the hypothesis that the lon;
run average of the estimates produced by the sample survey procedu
is equal to the known population value. A significant difference the
points to the desirability of an investigation of the components (
the operation.

In practice, two difficulties are encountered in applying this ty]
of analysis. First, an estimate of sampling variability must be avai
able for making the test of significance. Some methods of samplir
do not readily admit the estimation of sampling variability (e.g., t)
quota method), and others require an excessive amount of comput;
tion to estimate the sampling errors. These points will be develope
at some length in Chapters 9 and 10. Second, if the systematic erro:
are small, extremely large samples are required to detect them wit
any practical degree of certainty.

The problem of detecting small systematic errors when samplir
variability is present can be readily illustrated by a few simple con
putations. Let us suppose that samples of size n are being draw
from a population in which 0.50 of the elements possess a certai
characteristic, and that the survey operation is designed so that
performs as though random samples of size n were being drawn fro(
a population in which 0.49 of the elements possess the characteristi(
This type of situation might arise through the measurement proces
or through faulty execution of the sample design. Suppose we no'
wish to detect this bias of 0.01 through the use of a significance tes
such that if the long-run average of the survey estimates were ac
tually 0.50, we would say that it was less than 0.50 only 2.5 per cen
of the time. In other words, a one-sided test is being used at th
2.5 per cent level of significance. The following question is no\
asked. How large a sample is required in order that the hypothesi
of a long-run average of 0.50 will be rejected 90 per cent of the tim
—or alternatively, that the bias of 0.01 will be detected 90 per cen
of the time? The answer is that a sample of approximately 26,00(

ill be needed. Table 7.6 gives approximate sample sizes for other
opulation values and for other amounts of systematic error.

TABLE 7.6

APPROXIMATE SAMPLE SIZE REQUIRED TO DETECT A SYSTEMATIC ERROR
) PER CENT OF THE TIME WHEN A ONE-SIDED TEST IS USED AT THE
2.5 PER CENT LEVEL OF SIGNIFICANCE

Amount of Systematic Error	True Value of p in the Population			
	0.50	0.60 or 0.40	0.80 or 0.20	0.90 or 0.10
0.01	26,000	25,000	17,000	9,400
0.02	6,600	6,300	4,200	2,400
0.05	1,100	1,000	670	380

It can immediately be seen from Table 7.6 that investigations concerned with the measurement of opinions, attitudes, and consumer
wants will seldom be based on sample sizes large enough to allow
the detection of systematic errors of one percentage point with high
probability. We must ordinarily be satisfied detecting only errors
larger than three or four percentage points, with detection of smaller
errors occurring less frequently. For example, if the true value of
p is equal to 0.50 and the sample size is 9600, a bias of one percentage
point will be detected 50 per cent of the time as compared with detection 90 per cent of the time with a sample size of 26,000. This
situation clearly becomes much worse if we are comparing results
produced by two sampling procedures rather than comparing the result of one sampling procedure with a known population value, or if
we are attempting to detect biases arising from a number of different
sources.

C. Scarcity of Check Data for Attitude and Opinion Variables

Direct check data for attitude and opinion variables are seldom if
ever available for the use of research organizations. There are no censuses or large-scale surveys of attitudes which can provide this type
of information for large populations. Even if population data did
exist, we would almost certainly be faced with the problem that attitudes and opinions may change very rapidly with the passage of time
and with the occurrence of certain events. It may well be that election
returns provide the closest approximation to check data for attitudes
and opinions, though there is plenty of room for debate on the role
of prediction in the measurement of attitude and opinions (12, 13).

The use of election returns implies that we are trying to predict wh will vote and how ballots will be cast by those who do vote, thoug it may be argued that opinions and attitudes may be measured fo themselves without regard to their predictive value. This problem discussed briefly by Parry and Crossley (10), and they provide refer ences to the pertinent literature.

It may be noted at this point that there exists one relatively larg source of data which can be used for evaluating the prediction o behavior, a source which has not yet been exploited as far as th general researcher is concerned. At the present time there is a grea deal of market research which attempts to measure consumer want and to predict what consumers will buy. At the same time, companie have records of their sales in particular territories and these coul serve as the basis for valuable tests, not only of sampling methoc but also of measurement processes. Unfortunately, data of this kin are seldom available for publication.

This lack of direct check data on attitudes has meant that sample used in the measurement of opinions and attitudes have been com pared primarily with population values referring to the personal char acteristics of the respondents—age, sex, education, income, possessio of a car, and other similar items. The assumptions have been (a that the survey and population data are comparable, (b) that ther are high correlations or close relationships between certain of these variables and the attitude and opinion variables, and (c) that, be cause of these high correlations, agreement between sample and popu lation on one set implies agreement on the other set. We shall now examine briefly each of these three points.

The first point, namely that the survey and population data ar comparable, has already been discussed in Section 7.2A, and the illus trations given there have shown that factual and census-type infor mation are not necessarily reported correctly, or even consistently This means, of course, that we must exercise caution whenever chec data are used for validating a sample. For example, it might happe that the bias of a poor sampling procedure and the errors of respons would compensate for one another so that the survey estimates an the check data would show reasonable agreement.

The hypothesized existence of high correlations between certai factual characteristics of respondents and the attitude-opinion vari ables under investigation may be of little help in a practical situation In the first place, there is very little precise information about th correlations between factual characteristics on the one hand an opinion variables on the other. At the present time it is almost imossible to find the published results of a sample survey in this field that do not report these correlations as one of their significant findings. This means that only surveys conducted after the correlations have been determined will be of benefit. In the second place, it is necessary to consider the effects on the relationship of errors in reporting the factual characteristics. If the errors of reporting are different for different groups, and there is evidence that this is actually the case, the observed correlations may be spuriously high or low and may change from time to time as the errors of reporting change. This type of situation might arise, for example, with a variable such as education because people tend to exaggerate their educational attainments. Furthermore, the existence of a high correlation tells us nothing about bias since the addition or subtraction of a constant from either or both of two variables has no effect on the correlation coefficient.

As a final point, we may note that it has long been recognized that agreement between sample and population on one variable does not necessarily mean good agreement on other variables. This situation may, of course, arise because of differential errors of reporting, but it may also arise where no errors of measurement are present. This will be illustrated rather profusely in Chapter 8, which deals with comparisons between different surveys, and reference can be made to the famous experience of the Italian Central Statistical Institute, described in some detail by Neyman (14, p. 558) and Stephan (15, p. 389).

D. Extending the Results of a Comparison over Time

The mere fact that a sample survey estimate has been shown to be in satisfactory agreement with the corresponding population value at a particular point in time is no guarantee that the same sampling method will provide satisfactory agreement at another point in time. The performance of certain sampling procedures may depend on the existence of special relationships in the population, and changes in these relationships will then invalidate the sampling operation. An example of this type of situation is furnished by the use of rent as a quota variable during the period of World War II. The spotty relaxation of rent control after World War II destroyed the usefulness of this variable, and continued application of the previous proportions in the various rental categories would have introduced biases into the sample results. An obvious situation of this kind may be detected and corrected. Yet there may be other equally serious changes in the population which are not seen in time to prevent difficulties.

Roper's experiences in predicting the presidential elections in 194 provide another illustration of these effects (completely described i 9). Stated briefly, the situation was as follows. The Roper sampl overweighted the South in proportion to its contribution to the tota national vote and also overweighted that portion of the population wit relatively high educational attainments. These features were by de sign, the assumption being that there would be appropriate compen sation. The accuracy of Roper's predictions in 1944 attests the valid ity of this assumption for that year. However, this procedure wa definitely not subject to any form of rigid control, and there is rea sonably clear evidence that the compensation behaved in a differen manner in 1948 than in 1944. Although Roper had no reason fo combining his results on the basis of regional and educational distri butions of the population, such combinations can give some clues t the manner in which the compensation is operating. Summary fig ures are provided in Table 7.7.

For the present purposes, the most striking feature of these data is that application of the educational and regional distributions makes very little difference in the results for 1944 and a very noticeable dif ference in those of 1948. Thus the proportions for Dewey and Tru man differ by 13.1 percentage points in the raw sample data, and by only 6.4 percentage points where the regional and educational distri butions are applied. These data show only that the compensating feature operated differently in 1948 than in 1944. They do not prove that the inaccuracy of the prediction necessarily resulted from this dif ference.

E. Extending the Results of a Comparison to Similar But Not Identical Sampling Methods

Two sampling methods, although nominally the same, may differ to a considerable extent in the details of their design and execution. For example, one quota sample may make use of quotas based on some measure of socioeconomic status, and another quota sample may attempt to attain the same ends by means of quota assignments for small geographic areas. These differences make it very difficult to carry the results of a comparison based on one variant of a general method to other variations of the method. More detailed remarks on this point will be found in Chapter 8.

TABLE 7.7 [a]

VOTING INTENTIONS WEIGHTED BY CENSUS EDUCATIONAL DISTRIBUTION AND
GEOGRAPHIC QUOTA, ROPER SURVEYS, 1944 AND 1948

(1944—respondents indicating they had decided for whom they would vote,
per cent)

	Number	Dewey	Roosevelt	Other
Total sample	4886	46.3	53.3	0.4
Combination using both educational and geographic weighting [b]		45.9	53.8	0.3
Actual vote		45.9	53.4	0.7

(1948—respondents signifying intent to vote for one of four candidates, per cent)

	Number	Dewey	Truman	Thurmond	Wallace
Total sample	2799	53.9	40.8	2.8	2.5
Combination using both educational and geographic weighting [b]		51.0	44.6	1.7	2.7
Actual vote		45.4	49.8	2.4	2.4

[a] This table represents a condensation of Table VIII-10, appearing on page 214 of reference 9.

[b] The surveys were analyzed according to two geographic regions (South and North plus West) and three educational groups (grammar school, high school and college) within each geographic region. The results were then combined according to the Census educational distribution and the regional distributions of the 1948 and 1944 presidential vote.

7.3 ADVANTAGES ACCRUING FROM THE USE OF CHECK DATA

If the difficulties already set forth can be taken into account, there are many real advantages associated with the use of check data. Since this book is primarily concerned with the evaluation of sampling procedures, the general remarks that follow will assume that the survey and population measurement processes are the same. First of all, it may be that the benefits accrue in the direction of the check data rather than of the sample. The concept of evaluating sample survey results by means of population data may frequently be reversed. That is, the completeness and accuracy of the population results may be evaluated by comparison with the survey results. Provided sampling variability can be determined in an objective manner, there may be reasons for expecting the survey results to be the more accurate of the two.

A published account of such an investigation has been given by Gabriel Chevry (16). On March 10, 1946, a general census of the population was taken in France. Among other things, information was obtained for industrial and commercial enterprises. An examination of the schedules seemed to indicate that a substantial number of enterprises had not been included in the census, and a check was instituted, based on an area-sampling method. The results are summarized in the following quotation:

> Thus it seems possible to conclude that, if the census of 1946 has been inadequate, it has been so for enterprises of all sizes. According to the information gathered by the investigators of the sample, the fact that such a high percentage of enterprises was overlooked in the census was attributable to negligence on the part of the census enumerators.
>
> For these reasons the Institut National de la Statistique has not published the results of the 1946 Census of industrial and commercial enterprises.

The usual assumption in a case of this kind is that the small number of sample survey enumerators can be more highly trained and supervised than can the large number of enumerators required for a census

Another example of the same type is provided by the 1950 Census in the United States. The Monthly Report on the Labor Force is based on a well-designed and executed national sample of individuals fourteen years of age and over. Comparisons made between MRLF estimates and the 1950 Census reveal significant differences over and above those to be expected on the basis of sampling variability. The reasons for such differences appear to be found not so much in the design and execution of the MRLF sample as they are in such factors as differences in the experience, training, and supervision for the MRLF enumerators and the census enumerators (17, p. 52). Other instances in which comparisons of sample and independent check data have led to improvements in the latter have been reported, but our main interest is in the use of such comparisons to improve the sampling itself.

A. Detection of Bias in the Sampling Procedure

In spite of the difficulties set forth in the preceding portions of this chapter, there are many instances in which check data can provide unmistakable evidence of bias arising from a particular sampling procedure. However, there is a great difference between asserting that bias exists and actually determining the magnitude of the bias.

As an example of this type of situation, let us consider a quota sample whose educational distribution is being compared with Census estimates of the educational attainments of the population. It has long been recognized that many national quota samples underrepresent that portion of the population having a relatively low degree of education and relatively small incomes [see, for example, Cantril (2)]. Some illustrative data on this point are presented in Table 7.8.

TABLE 7.8

EDUCATION DISTRIBUTION FOR TWO NATIONAL QUOTA SAMPLES [a]
(Per cent)

	Quota Samples	Census Estimate as of April 1947 [b]
Number of respondents	(5825)	
No schooling	2.8	3
Attended or completed		
Grammar school	34.0	44
High school	42.8	41
College	20.4	12

[a] See d under Table 7.1 for identification of samples.

[b] U.S. Bureau of the Census, *Current Population Reports—Population Characteristics*, Series P-20, No. 6, Educational Attainment of the Population, 1947.

In Table 7.8, there is a ten-percentage-point observed difference in the "Attended or completed grammar school" category. This difference is composed of three components: sampling variability, sampling bias, and response differences. Under such circumstances we may first inquire whether the observed difference is larger than might be expected on the basis of sampling variability alone. Evidence is presented in Chapter 10 to show that, *within a single survey organization*, estimates made from quota samples will sometimes tend to vary from survey to survey somewhat as predicted from the binomial model. In other words, the standard deviation of an estimate may be roughly approximated by $\sqrt{pq/n}$ times a factor that may be close to 1 and often is less than 2. The use of this result assumes nothing about the existence or size of sampling bias. Let us now apply this result to a consideration of the "Attended or completed grammar school" category. In this instance, the standard deviation becomes $\sqrt{(34)(66)/5825}$ = 0.62 of a percentage point. Suppose now that we become conservative, increase this estimate threefold to 2.0, and assume that the actual sampling error does not exceed two standard deviations on either side of the observed sample percentage of 34. Then the population value must be less than 38. This states that the combined effect of sampling bias and response differences is *at least* equal to six percentage points (44–38), and may well be larger.

Although it may be that people tend to exaggerate their educational attainments less to a Census enumerator than to a person with no governmental connections, there is as yet no direct evidence on this. Moreover, even if this were true, it seems unlikely that the entire six plus percentage points of bias could be accounted for solely by this response difference. Accordingly we are led to conclude that there is reasonably definite evidence of sampling bias associated with this specific quota sampling procedure—exactly how much, however, remaining more or less unknown.

The preceding example illustrates very well the problems involved in using check data to detect bias in a sampling procedure. It may be relatively easy to eliminate most of the sampling variability and verify the existence of bias in an estimate, provided the bias is large in relation to sampling variability. However, the quantitative assignment of portions of the bias to the sampling procedure, to differences in the measurement processes, and to other sources will usually be impossible except in very carefully designed experiments.

B. Building Confidence in the Sampling Operation

Even though a sample survey is known to be adequately designed and executed, independent checks with population data would seem to be worth while. If enough of them can be assembled, they will tend to demonstrate the notion of controlled variability on an empirical basis. This type of demonstration will build up confidence in the good methods and will help to develop the proper amount of restraint and caution necessary when dealing with sample data. An example of this has been cited in Section 7.1B.

C. Adjustment by Means of Check Data

There are many ways in which known population characteristics can be used to increase the accuracy of sample estimates. A simple illustration of this use of check data for adjustment arises in situations in which a population is to be stratified with respect to a specified variable but the strata cannot be identified in advance of drawing the sample. For example, suppose a study is to be made of employee attitudes in a plant where it is expected that nationality background will have a pronounced influence on the attitudes held. Although the plant personnel office may be able to state the proportion of employees falling in each of the desired nationality groups, it may not be feasible to make use of this information in drawing the sample, e.g., if the sample is drawn from records that do not contain information on nationality. The following procedure might then be used. A random sample of all workers is selected, and the chosen workers are separated according to nationality group as determined by reference to other records. The number falling in each group will not necessarily be in proportion to that group's representation in the entire population because of sampling variation. Each nationality group is then analyzed separately, and the results are combined in proportion to the *known* representation of the various groups in the population. This is essentially stratification after the fact, and Cochran (18) has pointed out that for reasonably large samples it is almost as accurate as proportional stratified sampling.

The preceding example represents an instance in which known population data are introduced into the estimation procedure in order to increase the accuracy of the estimate. There are many other situations in which related techniques can be used, but these will not be discussed in detail in this book.

7.4 CONCLUSIONS

This chapter has briefly examined the use of check data for the evaluation of sampling procedures. The primary purpose in using known population characteristics under such circumstances is to detect and measure the sampling bias associated with a particular sampling method. The principal difficulties encountered by the Study of Sampling in attacking the problem in this manner have been:

1. An observed difference between a survey estimate and a supposedly corresponding population value will be the net result of combining the effects of a number of components, the main ones being sampling variation, systematic error due to the sampling procedure, and the bias and variance of the measurement processes. The mere existence of the two bias components can only be verified if their combined effect is relatively large in comparison with the two variance components. The isolation of the systematic error of the sampling procedure then requires that a good approximation be obtained for the bias of measurement, e.g., by comparing the survey measurement and an accurate independent measurement for each individual in the sample. This is seldom possible, especially under uncontrolled and non-experimental conditions.

2. Good check data for attitudes and opinions are non-existent at

the present time, and it is therefore necessary to argue from agreement or disagreement on "checked variables" to "non-checked variables," the determination of intercorrelations between the two frequently being one of the aims of the study. In this connection it may be noted that the check data in market research and other studies of consumer wants have not been exploited, at least so that the results are accessible, to the extent that they could be. Companies have records of their sales in particular territories, and these could form the basis for valuable tests, not only of sampling methods but also of measurement processes.

3. Although two sampling methods may be nominally the same, they usually differ in many respects in their details of execution and other particulars. This means that a study of the over-all performance of a sampling procedure as it is used by one organization may only tell certain general facts that are needed to forecast the performance of a similar procedure as applied by another organization. Other pertinent facts must be added by careful study of the circumstances. The same remark may be applied to the extension of the results of a comparison over a period of time.

Because of these points, the Study of Sampling has not been able to make extensive use of check data for evaluating sampling methods. We believe, however, that over-all calibration of actual sampling operations is very important when it can be done effectively. Opportunities are more likely to occur in studies of consumer wants than in opinion and attitude studies.

REFERENCES

1. Elmo Roper, "Sampling Public Opinion," *J. Am. Statist. Assoc.*, 35 (1940), 325–334.
2. Hadley Cantril, *Gauging Public Opinion*, Princeton University Press, Princeton, 1944.
3. Albert B. Blankenship, editor, *How to Conduct Consumer and Opinion Research*, Harper & Brothers, New York, 1946.
4. U.S. Federal Reserve Board, "Methods of the Survey of Consumer Finances," *Fed. Reserve Bull.*, July 1950, 795–809.
5. George Katona, *Psychological Analysis of Economic Behavior*, McGraw-Hill Book Company, New York, 1951.
6. Morris H. Hansen and William N. Hurwitz, "Dependable Samples for Market Surveys," *J. Marketing*, 14 (1949), 363–372.
7. Market Research Techniques Committee, "Design, Size and Validation of Sample for Market Research," *J. Marketing*, 10 (1946), 221–234.

8. Bernard R. Berelson, Paul F. Lazarsfeld, and William N. McPhee, *Voting*, University of Chicago Press, Chicago, 1954.

9. Frederick Mosteller, Herbert Hyman, Philip J. McCarthy, Eli S. Marks, and David B. Truman, *The Pre-election Polls of 1948*, Bulletin 60, Social Science Research Council, New York, 1949.

10. Hugh J. Parry and Helen M. Crossley, "Validity of Responses to Survey Questions," *Publ. Op. Quart.*, 14 (1950), 61–80.

11. Clyde V. Kiser and P. K. Whelpton, "Social and Psychological Factors Affecting Fertility. III. The Completeness and Accuracy of the Household Survey of Indianapolis," *Milbank Mem. Fd. Quart. Bull.* 23 (1945), 254–296.

12. John Dollard, "Under What Conditions Do Opinions Predict Behavior," *Publ. Op. Quart.*, 12 (1948), 623–633.

13. Morris H. Hansen and W. Edwards Deming, "On an Important Limitation to the Use of Data from Samples," *Proceedings of the International Statistical Conferences*, Berne, Switzerland, 1949.

14. Jerzy Neyman, "On the Two Different Aspects of the Representative Method: The Method of Stratified Sampling and the Method of Purposive Selection," *J. Roy. Statist. Soc.*, 97 (1934), 558–606.

15. Frederick F. Stephan, "Sampling in Studies of Opinions, Attitudes, and Consumer Wants," *Proc. Am. Philo. Soc.*, 92 (1948), 387–398.

16. Gabriel Chevry, "Control of a General Census by Means of an Area Sampling Method," *J. Am. Statist. Assoc.*, 44 (1949), 373–379.

17. U.S. Bureau of the Census, *1950 Census of Population*, Vol. II, Part I, Washington, D. C.

18. William G. Cochran, *Sampling Techniques*, John Wiley & Sons, New York, 1953.

CHAPTER 8

Comparison of Estimates
Obtained from Two or More
Sample Surveys

8.1 INTRODUCTION

In view of the importance of evaluating sampling methods used in the measurement of opinion, attitudes, and consumer wants, it is unfortunate that check data are so seldom available for attitude and opinion variables. Recognition of this lack leads us to consider alternatives, especially the possibility of comparing the results obtained by one sample survey on such variables with the results obtained by other sample surveys. Such an approach had been advocated many times. For example, Cantril (1) presented a set of data showing comparisons between results obtained when similar opinion questions were asked of national cross sections by various pairs of the following polling agencies: National Opinion Research Center, American Institute of Public Opinion, Fortune Poll (Elmo Roper), and the Office of Public Opinion Research. Cantril's summary of these comparisons is reproduced in Table 8.1.

On the basis of these data Cantril concludes that the difference between the polls is not unduly high. His conclusion is undoubtedly warranted if we consider how little effect a difference of three percentage points would usually have on the interpretation of the opinion questions. In a study of sampling methods, however, the primary comparison is between the observed difference and what would be expected on the basis of sampling variability alone. This point is examined in some detail in Chapter 10.

In his paper, Cantril pointed out that all his comparisons were made between surveys employing the quota method of sampling. He suggested that similar comparisons between "quota" samples and "area" samples might afford a basis for evaluating the relative merits

TABLE 8.1

COMPARISON OF SURVEY RESULTS

Type of Question and Time Interval	Number of Comparisons	Average Difference [a]
Total	99	3.24
Same 10-day interval	46	3.05
11–70 days	53	3.41
Political	66	3.15
Same 10-day interval	25	2.81
11–70 days	41	3.36
Non-political	33	3.43
Same 10-day interval	21	3.33
11–70 days	12	3.60

[a] The difference in percentage points (with sign ignored) was obtained for each category of each question on which a two-way comparison was possible. The values in this column are the averages of these percentage point differences.

of the two sampling methods. It was originally hoped that the Study of Sampling would be able to find enough data in the files of survey organizations to make some use of this approach. Accordingly, a thorough search was made to find available data that would be useful in assessing the relative reliability of various types of sampling methods. This investigation revealed so much variation in the questions asked, the populations sampled, the dates of the surveys, and even in essential parts of sampling methods which were nominally the same, that there was very little likelihood of finding enough strictly comparable surveys taken by different sampling methods. Moreover, a large number of comparisons would be required. Not only would we need a sufficient number of comparisons to approximate the sampling distribution of the difference between the estimates produced by two methods under a defined set of circumstances, but the same number of comparisons would be needed for each of a wide range of circumstances to show the effects of different interviewers and field workers, of different populations, and of different subjects of investigation. In addition, accurate check data would be needed to test which of two methods was more reliable if a set of comparisons showed a difference in expected values, and as previously noted, these check data would seldom be available.

Even though these early investigations, plus the recommendations of the study's advisory committee, showed that it was not possible for the major emphasis of the Study of Sampling to be placed on the study of comparisons, every effort was made to obtain and study the comparisons that seemed pertinent. An analysis of several of these comparisons will be found in the succeeding sections of this chapter, each comparison being treated as a complete unit. Following the sections on the individual comparisons, a summary is given which attempts to set forth the major points that have or have not been demonstrated.

As a closing to this introduction, it may be well to review briefly the limitations and advantages of these comparisons taken as a group. The principal limitations are:

1. A sampling method must be characterized by a sampling distribution, that is, by its performance in repeated trials. Since a measure of sampling variability could not be accurately computed for many of the samples included in these comparisons, it was not normally possible to state whether or not an observed difference could be due to chance fluctuations alone. However, in certain comparisons differences were so extreme that chance fluctuations were ruled out as the sole cause.

2. Adequate check data were frequently lacking so that it was difficult to assess which of two methods produced results nearer the truth.

3. The two samples contributing to a comparison have been taken at a particular point in time, with a particular set of interviewers and have been used to obtain information on a particular set of topics. We can well argue that under different circumstances (i.e., with respect to time, interviewers, subject of investigation, and the like) the results of the comparison might no longer be applicable. In other words, any single comparison cannot furnish a general and definite proof of the relative merits of two types of sampling procedure.

4. Although costs and administrative considerations are of primary importance, of necessity, they have been omitted from these accounts.

On the positive side it may be said that:

1. These comparisons provide a description, even if only a brief one, of a fairly wide variety of sampling procedures.

2. They supply very clear-cut demonstrations of the differential effects of the sampling method on different subjects of investigation.

3. They may serve to stimulate future comparisons and thus start the accumulation of experience which can be applied widely.

8.2 SPECIFIC COMPARISONS

A. The NORC-MRLF Comparison

Early in 1946, NORC prepared a plan for including in one of its regular surveys a set of questions almost identical with those used by the Census Bureau in its Monthly Report of the Labor Force. The field work for this survey was carried out in May 1946, but NORC was not able to complete the analysis and interpretation of the study. The data were transmitted to the Study of Sampling for analysis, and a detailed account of this analysis has been given in a working paper of the project.

The NORC quota sample was obtained according to regular procedures, the quotas to the interviewers being given in terms of the following overlapping categories:

1. By sex and age
   a. Men under 40 years of age.
   b. Men 40 years of age and older.
   c. Women under 40 years of age.
   d. Women 40 years of age and older.
2. By race and economic level and farming status
   a. Non-farm white respondents in rent group A, the highest 2 per cent of the population.
   b. Non-farm white respondents in rent group B, the next 14 per cent.
   c. Non-farm white respondents in rent group C, the next 52 per cent.
   d. Non-farm white respondents in rent group D, the lower 32 per cent.
   e. Non-farm Negro respondents.
   f. White farm respondents.
   g. Negro farm respondents.
3. By suburban or other location
   a. Respondents located in the suburban portion of the metropolitan districts.

The Survey produced information for 2589 respondents and for the 7365 members of their households (including the respondents).

The Monthly Report of the Labor Force consisted of a national area sample of approximately 25,000 households which were surveyed monthly. Each month a portion of this sample was replaced by a fresh selection of households so that every six months the sample was completely changed and the survey was maintained as a rotating panel of households. The sampling procedure has been described in some detail by Hansen, Hurwitz, and Madow (2, Chap. 12).

This comparison furnishes an example in which there appears to be rather close agreement between the results obtained through the application of two widely differing sampling procedures. The principal source of this comparison was the labor force questions that NORC included on its questionnaire. It might be noted that special MRLF tabulations were needed in order to make the results comparable, since NORC did not use three screening questions ordinarily used by MRLF in obtaining its standard labor force classification. The principal results of this comparison are summarized in Table 8.2.

TABLE 8.2

MAJOR ACTIVITY [a]

(Percentage of population 14 years of age and older of the sex indicated)

	Survey	MRLF	Difference
Both sexes			
At work	47.8	48.4	−0.6
With job, not at work	2.9	1.9	1.0
Unemployed	2.5	2.3	0.2
Other types of activity	46.8	47.4	−0.6
Males			
At work	71.9	73.6	−1.7
With job, not at work	4.9	2.9	2.0
Unemployed	4.2	4.0	0.2
Other types of activity	19.0	19.5	−0.5
Females			
At work	25.0	25.3	−0.3
With job, not at work	1.1	1.0	0.1
Unemployed	0.8	0.8	0.0
Other types of activity	73.0	73.0	0.0

[a] The question was "What was (name of member of household) doing last week?" If more than one type of activity was mentioned, the one to which the greatest number of hours was devoted was recorded as the "major activity." Employed and unemployed workers whose major activity was "home housework," "in school," etc., are included under "other types of activity."

The Survey percentages given in Table 8.2 are based on the NORC sample of 2589 respondents, plus the 4776 other members of their households.

The figures set forth in the Table 8.2 show remarkably close agreement, considering the two sets of data are subject to some differences of procedure and definition as well as to errors of sampling and interviewing. As a matter of fact, the agreement is probably better than we would expect to obtain if the entire comparison had been repeated one month later. Because of this fact, the Study of Sampling made many other comparisons with the belief that major biases, if they existed, would turn up somewhere. If they happened to be balanced out by sampling error or other disturbances in one place, they would be augmented by such disturbances elsewhere. Essentially three kinds of comparisons were made. They were: (*a*) comparisons between the Survey (respondents plus other members of their households) and MRLF on more detailed labor force questions (breakdowns of "other types of major activity," class of worker, etc.), (*b*) comparisons between the Survey and MRLF for various population characteristics (age, sex, household size, and urban-rural distributions), and (*c*) comparisons between the NORC respondents and the other members of their households. The results of this work may be summarized as follows:

1. The gross tabulations of the labor force questions show very close agreement between MRLF and the NORC Survey figures. The more detailed tabulations of these questions show some evidence of bias, but as will be discussed later, the combined effect of random error and procedural differences obscures the sampling biases and prevents us from determining how large they may be as measured from the true values of the percentages being estimated. As we proceed into detailed tabulations, the number of cases available for estimating percentages becomes smaller and consequently the sampling error becomes larger.

2. For population characteristics, small differences are to be found, some definitely due to errors in the quotas or in the execution of the NORC quota sample and others possibly due to chance. A much larger sample would be needed to reveal clearly the biases that may be included in them, but no gross biases in the population variables were revealed.

3. It would appear that in this example of quota sampling, for variables that could be used to compare respondents with other members of their households, the respondents were very representative except for a large deficiency of young working women and moderate deficiencies of younger men and older women. There were excesses of independent workers and employers and deficiencies of employees; and there was also a small distortion of the education distribution.

B. The Bikini Comparison

The Committee on Social Aspects of Atomic Energy of the Social Science Research Council conducted a nationwide survey on public reaction to the atomic bomb and world affairs in the summer of 1946. Though the survey was not designed to provide information on sampling procedures, an area sampling method and a quota sampling method were employed. The results of these surveys have been given in a report [(3); see also (4)]. Moreover, the Study of Sampling has analyzed the results rather extensively for what they might show on the comparability of the sampling methods. Since these surveys were centered around the Bikini test of the atom bomb, this comparison will hereafter be referred to as the Bikini comparison.

The quota sample was designed in the usual manner with controls on geographic section, city size, urban-rural, sex, age (only informal instructions to interviewers), color (Negro assignments were made only in thirteen southern states) and socio-economic status (wealthy, average plus, average, poor, and on relief). The survey was carried out by Benson and Benson. A total of 5984 interviews were obtained, 3090 in June and 2894 in August.

The area sample was carried out by the Survey Research Center, University of Michigan, and its design was similar to the one described by Goodman (5). Approximately 1200 interviews were obtained, 585 in June and 592 in August.

These samples can be regarded as "typical" quota and area samples with regard to sampling technique, but a large difference in size exists, partially because of the relative costs of intensive and extensive interviews. Thus the quota sample used short answer and check list questions (extensive interviews), and the area sample used more free-answer questions calling for detailed probing by the interviewer (intensive interviews). The intensive interviews were longer, required more highly trained interviewers, and needed more care in the coding and analysis of results than did the extensive interviews. Consequently, they were much more costly than the extensive interviews, and they provided more detailed information. As a result of this difference in interviewing costs, as well as certain differences in sampling costs, the area samples were only about one-fifth as large as the quota samples. This fact must be kept in mind in making the comparisons. Actually, for the purpose of comparing the sampling methods, we should like to know how large an area sample could have been obtained for the same cost as that of the quota sample, assuming the same interviewing techniques were used in each. Unfortunately, this cannot be determined in the present instance. We could compare the total costs of the two studies, but this would not tell us very much about the relative costs of the sampling methods because of the already mentioned differences in interviewing, coding, and analysis.

Though the quota sample and the area sample were used to obtain information on the same subject, public attitudes toward atomic energy, there were very few instances in which exactly the same opinion questions were asked. This was by intent because of the difference in purposes and interviewing techniques. There was, however, a large number of factual items on which nearly comparable questions were asked. The Study of Sampling has analyzed these two samples extensively, not only with respect to the factual items but also with respect to the opinion comparisons available. Some selected results from this work will now be presented.

In contrast to the reasonably close agreement shown by the NORC-MRLF comparison, the Bikini comparison presents a situation where there is close agreement for some variables and poor agreement for other variables. Many of the situations where poor agreement exists seem to be subject to the difficulties in interpretation that have been set forth in the preceding chapter and in the introduction to this chapter. The principal results of this comparison may be summarized as follows:

1. The two samples differed rather markedly on a number of factual items (see 3), but neither appeared to be consistently closer to independent estimates of the same facts. This is solely a matter of observation and should not be interpreted as having any additional meaning with respect to the reliability of either method. Direct comparisons were possible for only a small number of opinion and information questions. The agreement in these comparisons was such that the same conclusions concerning public attitudes toward atomic energy would probably have been drawn from the results of either sample. This, of course, does not necessarily hold true for all questions that were asked or might have been asked. Each would have to be examined in its own right. There were a large number of items on which it was impossible to make comparisons, frequently because of inability to separate sampling differences from the effects of differences in the questions and in the interviewing procedures.

2. The two samples agreed closely on race (a quota control variable), on religion, and on the cross tabulations of age by education and religion by education. The agreement on religion and on these

two cross tabulations is particularly noteworthy since it is on such items as these that we might expect large differences between a quota and a random sample. The income-by-education distribution shows larger deviations than the other two and suggests that there might be other variables for which large differences also exist.

3. Fairly large differences showed up with respect to sex (a quota control variable), income (quota control variable for socio-economic status), education, veteran status of males, voting in the 1944 presidential election, and occupation.

In order to show the range of the agreement and disagreement referred to and to illustrate the difficulty of interpretation, we shall now present some of the data obtained in this comparison. These data are given in Tables 8.3–8.7 on pages 165–169.

These tables illustrate very well the difficulties that are encountered in evaluating an empirical comparison between two sampling methods. Consider the income comparison in Table 8.3. The quota sample is concentrated much more in the $1000–$1999 class than is the area sample, with smaller percentages falling in the very low-income and very high-income classes. The differences are substantially larger than we could expect on the basis of any reasonable assumptions about sampling variability. It is impossible, however, to allocate a large portion of the differences to sampling bias because the two income distributions are not strictly comparable. The principal reason for this arises from the phrasing of the income question, the quota sample asking for the average total weekly income of the respondent and his immediate family (in June and August, 1946) and the area sample asking for the total income in 1945 of the respondent and his immediate family. Thus not only is one figure for 1945 and the other for mid-1946, but the figures are derived from two different ways of estimating, one on a weekly and one on a yearly basis.

Another example is provided by the comparison of opinions 2 and 2′ in Table 8.7. The indicated difference of nine percentage points could have arisen, at least in part, from interviewing and question factors. The quota sample merely asked about recent tests of the atomic bomb whereas the area sample specified tests by the Navy made in the preceding month in the Pacific. Without prompting or probing, a person might conceivably think of the New Mexico test as a recent one, and the interviewer would never be the wiser.

Tables 8.4, 8.5, and 8.6 present comparisons between the area and quota sample with respect to the cross tabulations of education by religion, education by age, and education by income. It can be seen from

TABLE 8.3

SELECTED COMPARISONS FROM THE BIKINI SURVEYS FOR
POPULATION CHARACTERISTICS

(Percentage distributions; number of cases are given in parentheses)

	Area	Quota	Difference	Check Data
Education	(1168)	(5825)		
Grammar school (complete or incomplete)	42.8	36.8	6.0	47 [a]
High school incomplete	18.7	21.4	−2.7	18
High school complete	20.3	21.4	−1.1	23
College incomplete	10.5	10.5	0.0⎤	12
College complete	7.7	9.9	−2.2⎦	
Income	(1134)	(5697)		
Less than $500	8.4	5.4	3.0	5.8 [b]
$500–999	9.7	13.7	−4.0	7.7
1000–1999	20.7	32.7	−12.0	20.5
2000–2999	25.3	24.4	0.9	25.5
3000–4999	23.0	14.4	8.6	27.0
5000 and over	12.9	9.4	3.5	13.6
Religion	(1172)	(5850)		
Protestant	72.9	74.8	−1.9	
Catholic	20.9	18.3	2.6	
Jewish	3.6	3.3	0.3	
Other	0.2	0.1	0.1	
No preference	2.4	3.5	−1.1	
Veteran status of males	(511)	(2932)		
Veteran of World War II	19.8	27.4	−7.6	27.3 [c]
Other	80.2	72.6	7.6	72.7

[a] Census estimates of the educational attainment of the civilian population 20 years and over as of April 1947 (*Current Population Reports*, Series P-20, No. 6).

[b] These figures were taken from *Current Population Reports*, Consumer Income, Series P-60, No. 2, March 2, 1948.

[c] These values were obtained from Census estimates of civilian males and veterans, 20 years of age and over, for July 1946.

TABLE 8.4

DISTRIBUTION OF AGE BY EDUCATION
(Per cent)

Age	No High School	Some High School	Completed High School	College	Total
21–29					
Area	5.6	4.4	6.1	4.8	20.9
Quota	3.7	5.7	7.5	4.7	21.6
Census [a]	5.9	5.9	9.7	3.4	24.9
30–59					
Area	25.0	12.3	12.3	11.4	61.0
Quota	22.6	13.0	12.2	13.8	61.6
Census [a]	28.2	10.1	11.7	7.5	57.5
60 and over					
Area	12.2	1.9	1.8	2.2	18.1
Quota	10.4	2.8	1.8	1.8	16.8
Census [a]	12.7	1.6	1.9	1.4	17.6
Total					
Area	42.8	18.6	20.2	18.4	100
Quota	36.7	21.5	21.5	20.3	100
Census [a]	46.8	17.6	23.3	12.3	100

[a] These percentages are estimates of the educational attainment of the civilian population twenty years of age and over as of April 1947 (*Current Population Reports*, Series P-20, No. 6).

an examination of the tables that the largest difference between the two samples is 5.5 percentage points, this difference occurring in the category "Some High School," $1000–1999 income. In other instances the differences are smaller, and the agreement seems to be very good except for the distribution of income by education. The area sample has a greater proportion of individuals who are high in income and low in education than does the quota sample. In the quota sample this is compensated by the larger number of respondents in the $1000–1999 category at all education levels. One of the difficulties in actually assessing these differences arises from the non-comparability of the income questions, as has already been pointed out.

The agreement on the other two cross tabulations is of special interest since we frequently see the statement that agreement on marginals for a quota sample does not prevent serious biases from arising in the cellular distribution. This is quite true, but here we have an

TABLE 8.5

DISTRIBUTION OF RELIGION BY EDUCATION
(Per cent)

Religion	No High School	Some High School	Completed High School	College	Total
Protestant					
Area	31.4	13.5	14.6	13.4	72.9
Quota	27.4	15.8	15.3	16.4	74.9
Catholic					
Area	8.8	4.2	4.6	3.2	20.8
Quota	7.1	4.2	4.6	2.4	18.3
Jewish					
Area	1.1	0.8	0.5	1.2	3.6
Quota	0.6	0.6	1.0	1.1	3.3
Other					
Area	0.0	0.1	0.0	0.1	0.2
Quota	0.0	0.0	0.1	0.0	0.1
No preference					
Area	1.5	0.2	0.5	0.3	2.5
Quota	1.6	0.7	0.5	0.6	3.4
Total					
Area	42.8	18.8	20.2	18.2	100
Quota	36.7	21.3	21.5	20.5	100

instance of cross tabulations agreeing very well where neither of the two variables was controlled (religion and education).

The Washington State Poll

The accounts of the two preceding comparisons were based on extensive analyses actually carried out by the Study of Sampling. In contrast, the details of the remaining comparisons to be described in the chapter were drawn either from published articles or from correspondence with individuals intimately connected with each comparison. For this reason, some of the succeeding comparisons are not described as fully as might be desired.

The Washington Public Opinion Laboratory is operated jointly by the University of Washington and Washington State College for the purpose of making statewide polls of opinion and conducting other social research projects. Prior to the presidential election of 1948 it had established a polling organization and completed polls on foreign and domestic problems, old-age pensions, and adult education. Its

TABLE 8.6

DISTRIBUTION OF INCOME BY EDUCATION
(Per cent)

Income [a]	No High School	Some High School	Completed High School	College	Total
Less than $1000					
Area	14.2	2.2	1.1	0.6	18.1
Quota	13.7	3.2	1.7	1.0	19.6
$1000–1999					
Area	11.6	3.6	3.2	2.2	20.6
Quota	13.7	9.1	6.7	3.5	33.0
$2000–2999					
Area	9.5	6.3	6.2	3.3	25.3
Quota	6.4	5.7	6.8	5.2	24.1
$3000 and over					
Area	7.4	6.3	9.9	12.4	36.0
Quota	2.9	3.4	6.2	10.8	23.3
Total					
Area	42.7	18.4	20.4	18.5	100
Quota	36.7	21.4	21.4	20.5	100

[a] Not strictly comparable. See p. 164.

fifth poll was taken between October 26 and November 1, 1948, to determine the voting intentions of a sample of the population and facto
related to voting behavior. The relationship of its results to th
failure of the polls in the 1948 elections has been examined (6).

For the purposes of the Study of Sampling, the important featu
of this experience arises from the fact that the pre-election surve
was conducted as two polls, one by an "area" sampling method an
the other by a "quota" sampling method, to permit comparison of th
two survey designs. This was the first use of quota sampling by th
Laboratory; the four previous polls were taken by area samplir
methods. The number of interviews assigned to interviewers was 52
for the quota sample and 516 for the area sample. The objectiv
was 500 completed interviews; the assignments included an allowan
for anticipated losses due to sickness or other causes that might pr
vent one or more interviewers from completing their assignment
Only six days were allowed for the completion of the assignments.

The area sample was based on a division of the state into five re
gions which were subdivided into 28 strata by size of communit

TABLE 8.7

SELECTED COMPARISONS FROM THE BIKINI SURVEYS FOR OPINION VARIABLES

Percentage Distributions [a]

Quota pre-Bikini:

1. "Do you happen to know whether there is any plan to test the atomic bomb in the near future?"

Yes	75
No and DK	25

Area pre-Bikini:

1'. "Have you heard of the test the Navy plans to make of the atomic bomb?"

Yes	76
No and DK	24

Quota post-Bikini:

2. "Do you happen to know whether there has been any test of the atomic bomb recently?"

Yes	89
No and DK	11

Area post-Bikini:

2'. "Have you heard anything about the atomic bomb tests the Navy made last month in the Pacific?"

Yes	80
No and DK	20

Threat of bombs

Quota:

3. "Do you think there is a real danger that atomic bombs will ever be used against the United States?"

	Pre	Post
Yes	64	63
DK	14	13
No	22	24

Area:

3'. "Do you think there is a real danger that atomic bombs will ever be used against the United States?"

	Pre	Post
Yes	17	16
Yes, but	46	43
DK	11	11
No, but	10	12
No	14	14
Not ascertained	1	2
DK atomic bomb	1	2

[a] DK indicates that respondent said he did not know.

Within these strata, a sample of 53 incorporated places and rural precincts was drawn by a method combining simple random sampling of the smaller places and sampling with probability proportionate to size for the larger places. In nine places samples of households were drawn from available lists, and in the others they were drawn by prelisting a sample of blocks and rural precincts and sampling from the resulting lists. For each household in the sample, the member to be interviewed was selected from a list of all adults in a manner that gave each an equal chance of being selected. If that member of the household could not be interviewed after three attempts, the interviewer was permitted to substitute a member of a nearby household, subject to certain restrictions.

The quota sample was assigned to the same interviewers and taken concurrently with the area sample. Quotas of respondents to be interviewed were assigned by sex, three age groups, and four economic levels. The interviewers also were provided with a handbook (7) describing quota methods.

Neither poll was executed in a way that conformed completely to the sample design. About 8 per cent of the assigned interviews were not completed in the quota sample. This is rather unusual in quota interviewing because the interviewer is free to seek any respondent who meets the quota requirements, but apparently the period allowed for the interviewing was too short. About four-fifths of the unfilled quota assignments were within the lowest economic class. About 11 per cent of the area sample assignment was not completed. This also was unusual in view of the opportunity to make substitutions after three unsuccessful calls. Interviews were completed with 485 respondents in the quota sample and 469 in the area sample. Of the latter, about one-fifth were interviews with substitutions.

Each sample was analyzed in the same manner, the final estimate of percentage of voters favoring each candidate being based only on the responses of those who said they were registered to vote. The "don't knows" were divided proportionately in each instance. The resulting percentages were as follows:

Candidate	Actual Washington State Vote	Area Sample	Quota Sample	Difference
Dewey	42.7	46.0	52.0	−6.0
Truman	52.6	50.5	45.3	5.2
Wallace	3.5	2.9	2.5	0.4
Others	1.2	0.6	0.2	0.4

At first glance, the six-percentage-point difference shown by the Dewey figures would seem to indicate that there is definitely a difference in performance between the area sample and the quota sample. Moreover, the area sample percentage is closer to the recorded Washington state vote. Actually, however, this difference is not statistically significant if we assume binominal sampling variability, and the true sampling variability is very likely to be larger than this.

If we could demonstrate conclusively that the area sample was superior to the quota sample in this instance (i.e., possessing smaller sampling bias), we would still face the problem of generalizing the result to other situations. Unfortunately this would be a difficult if not impossible task. The principal reason why the results cannot be generalized is that neither sample was executed completely according to design, and the samples provide no means of evaluating the effects of the deviations from design. Thus the quota sample had a substantial number of unfilled assignments, particularly in the lower economic levels, and not all the respondents designated for the probability sample were interviewed; some were missed entirely and substitutions were made for others. Moreover, this was the first time that any of these interviewers had been assigned quotas. It is well known that much care and instruction are needed in order to insure adequate coverage of the lower economic classes by quota interviewers, and any improvement in the coverage of the quota sample in these categories would probably increase the proportion of Truman's vote. In view of these considerations, we can only state that under the conditions of this comparison the probability sample results are definitely superior to those of the quota sample, but there is no way of generalizing to other situations in which either type of sample was used. Even if the undetermined effects of deviations from design did not preclude the possibility of generalization, it would still be necessary to examine the detailed design of each sample before any general conclusions could be drawn.

D. McCall Evaluation Study

During April and May, 1946, McCall Corporation sponsored a nationwide study of the magazine-reading interests of women. The study was carried out by Alfred Politz Research under the direction of the Third Qualitative Study Committee and is described in a report (8). The sample was selected on a quota basis and contained only women who lived, for the most part, in urban places, who were at home when the interviewer called, and who showed the interviewer any copy of one or more of the 22 magazines on the list. Because of the manner in which this sample was selected, the Third Committee recognized that it was not adequate to provide quantitative measures of readership or circulation (nor was it necessary that it do so). However, there remained the problem of whether it was adequate to measure readership interests. In order to answer this question, an experiment was designed and carried out in Milwaukee, Wisconsin. This experiment was called the Evaluation Study. Since this Study involved the comparison of two different sampling procedures (one of which was a quota method), a brief account will now be given of its results and their interpretation.

Two independent samples of homes were selected in Milwaukee County. For the first sample, a crew of experienced quota interviewers were assigned quotas on the basis of four socio-economic levels, the proportion of interviews assigned to each category being the same as in the nationwide survey. In addition, controls were set up to scatter the interviews throughout the interviewing area. Ten interviewers completed 1041 interviews and encountered 51 refusals in the course of interviewing.

The second sample was chosen by specially trained interviewers who had never worked on quota surveys. These interviewers started from points distributed throughout the area according to the density of population and attempted to obtain an interview in every seventh dwelling unit along a rigidly prescribed route. Up to five callbacks were made in order to obtain interviews in all the designated dwelling units. The 21 interviewers attempted interviews in 2078 homes and obtained 1734 interviews (138 homes in which no one was ever found at home; 159 refusals; and 47 not completed for miscellaneous reasons). In the Evaluation Study this sample was referred to as a "precision sample." This terminology will be used here in order to avoid confusion on the part of those who may wish to refer to the original report. The same questionnaire for eliciting information on the possession and readership of magazines was used in both samples. Moreover, the two crews were trained together on the treatment of the questionnaire.

Analysis showed the most striking difference between the two samples to be in respect to socio-economic status. This was a controlled variable for the quota sample (A, B, C, D—the description of the categories and the assigned proportion in each being given in the report). Though the selection of the precision sample did not depend on socio-economic status, the interviewers were given the same definitions as were the quota interviewers and were instructed to classify the households in which interviews were obtained. The two distributions are:

Socio-economic Level	Interviewed in Precision Sample, per cent	Assigned for Quota Sample, per cent
A	1.9	10
B	31.7	28
C	62.3	38
D	4.1	24
Number of cases	1734	

Since socio-economic status is such a subjective factor, we may suspect that a classification difference in the two samples led to these extreme discrepancies. The two crews were working with the same definitions, however, and these differences held up when a more objective measure of socio-economic status was used, namely rental. The two distributions of monthly rental value are given in Table 8.8.

TABLE 8.8

MONTHLY RENTAL VALUES FOR EVALUATION STUDY

(Contract rent for rented homes and estimated rent for owned homes)

Rental Class	Precision Sample, 1734	Quota Sample, 1036
Under $9.50	0.5	0.0
9.50–14.49	0.7	1.4
14.50–19.49	1.2	4.0
19.50–24.49	2.1	7.5
24.50–29.49	8.5	14.1
29.50–39.49	32.4	19.9
39.50–49.49	27.4	13.5
49.50–59.49	14.3	13.9
59.50–74.49	7.8	12.1
74.50–99.49	3.3	6.9
99.50 and over	1.8	6.7
Total	100.0	100.0

Thus the quota sample had more low rentals and more high rentals than did the precision sample, just as for socio-economic status. Note that the median monthly rentals were $41.20 and $41.80, respectively, in fairly close agreement, in spite of the extreme differences in the actual distributions.

The agreement between the two samples was somewhat better for other factual characteristics given in the published report. A comparison for owned homes and number of families with children is given in Table 8.9 on page 174.

As might be expected from the observed differences in the socio-economic and rental distributions, the two samples did not agree at all well on quantitative measures of magazine readership. Respondents were asked to produce copies of a single issue of *Good Housekeeping, Ladies' Home Journal, McCall's,* and *Woman's Home Companion.* In addition, the interviewers paid the respondent the full copy price for

TABLE 8.9

OWNED HOMES AND FAMILIES WITH CHILDREN

Family Characteristic	Precision Sample, 1734	Quota Sample, 1036
Percentage of owned homes	53.4	54.9
Percentage of families having:		
Children under 2	10.9	14.4
Children 2–12	35.8	37.1
Children 13–20	26.5	27.1
No children 20 or under	26.8	21.4

permission to cut out of the table of contents a corner that would identify the issue, the two crews of interviewers being trained together in this operation. Table 8.10 stems from this objective meas-

TABLE 8.10

POSSESSION OF WOMEN'S SERVICE MAGAZINES, APRIL 1946 ISSUE PER 100 HOMES

Magazine	Precision Sample, 1734	Quota Sample, 1041	Circulation Based on Total Homes in Area [a]
Good Housekeeping	10.8	14.7	8.8
Ladies' Home Journal	14.6	20.3	13.3
McCall's	11.5	16.9	11.8
Woman's Home Companion	10.4	14.2	11.9
Total	47.3	66.1	45.8

[a] Based on ABC figures (circulation in Milwaukee County).

ure. Thus the quota sample found many more magazines than did the precision sample. These differences in magazine possession can, to some extent, be traced to the differences in rental levels. For example, if we compare magazine possession in the two samples for those homes paying under $40 per month, there is essentially no difference. The upper rental groups still show a difference, but in each instance the quota sample has a higher proportion of individuals in the upper portion of the rental classes than does the precision sample.

When we turn to questions that are more closely related to opinion variables, the differences between the two samples are much less extreme. For example, respondents were asked the question, "With

ιat degree of interest do you look forward to receiving each of these
ιgazines?" A comparison of their responses is given in Table 8.11.

TABLE 8.11

COPIES RECEIVED WITH KEEN INTEREST PER 100 HOMES

Magazine	Precision Sample, 1734	Quota Sample, 1036
Good Housekeeping	8.1	8.8
Ladies' Home Journal	10.3	10.7
McCall's	8.0	9.9
Woman's Home Companion	7.0	7.3

ιe relatively close agreement is somewhat surprising here since the
port indicates that "interest" is related to rental value in approxiately the same degree as is "possession." Similar close agreement
as found in response to the question, "What four things are of most
iterest to you in the magazines you read?"

The published report presents the following interpretation of the
valuation Study:

The averages of the rental values of the homes in the two samples are
proximately the same. In many cases the quota sample under-represents
ιe lower rental brackets which is often mistakenly considered to be the
ιuse for over-stating the number of copies found. In the Milwaukee ex-
eriment the quota sample did not under-represent the lower rental values
nd yet found substantially more copy-owning homes. This may serve as an
lustration of the point that a sample controlled on known characteristics
e.g., rent) can be highly inaccurate on the unknown characteristic under
ιudy (number of copies or readers) even though the known and unknown
ιe highly correlated.

This points to the possibility that samples of the quota type may not
ecessarily be satisfactory for the measurement of extensive properties, such
s the number of copy-owning households or readers of a periodical. On the
ther hand, in reference to readers' interest, the two types of samples are
ι much better agreement. The interest in all reading subjects combined
ppears higher in the quota sample than in the precision sample but the
elative position of the various reading subjects is highly similar in both
ιmples. The rank order correlation is .82. This may indicate that if
nown or unknown biases are present in a sample of the quota type, these
ιiases tend to affect quantitative findings much more than they do findings
f a qualitative nature.

Assuming that there were no *major* sources of bias existent in the
precision sampling procedure, and as far as the Study of Sampling
ιas been able to determine from an examination of the method this

assumption is reasonable, the statement quoted may be made mc
precise. In particular

1. The original quotas for socio-economic status were badly
error. This only illustrates a well-known fact, namely, that the u
of loosely defined controls for socio-economic status is very diffic
to justify on any objective basis. This is why so many organizatio
have substituted geographic location as well as rent in place of soc
economic levels as a control variable.

2. Errors in the determination of quotas will have a substant
effect on the sample results if the variable under investigation is high
correlated with the control variable. This holds true for "possessio
in the present study, and it would seem to be only a matter of lu
that it did not hold for "interest," "interest" also being related
rental level (and also, therefore, to socio-economic status).

3. Even though samples (or samples and populations) may agree
respect to averages on certain characteristics, the actual distributio
may vary widely. These differences in distributions may have
marked effect on sample results.

E. University of Washington Opinion Survey

During the first week of June 1947, a cross section of the adu
population of Seattle was polled on attitudes toward the use of an
mals in medical research. This survey was undertaken as a cla
project by students registered in Public Opinion Analysis, a cla
taught by Professor Allen L. Edwards. Two samples were used
this survey, one a quota sample and the other a block sample.

The block sample was chosen by taking a systematic sample fro
a list of Seattle blocks and then taking a systematic sample of dwel
ing units on the chosen blocks. One individual was interviewed fro
each of these dwelling units. The choice of the individual to be i
terviewed was left up to the discretion of the students, with the sing
exception that equal numbers of men and women were to be obtaine
No information is available for callback procedures or for refus
rates. The sample was relatively small—246 completed interviev
were obtained.

The quota sample was selected on the basis of age, sex, and soci
economic status. The same students chose this sample, and it was n
independent of the block sample in that the interviews were take
insofar as possible, in the areas or districts in which the block inte
views were obtained. The number of interviews completed for th

uota sample was 328, somewhat greater than the number obtained
or the block sample but still a relatively small sample.

As far as the attitude portion of this survey was concerned, the
ifferences between the quota sample results and the block sample
esults were not very large. A few typical comparisons are given in
'able 8.12.

TABLE 8.12

COMPARISON OF BLOCK AND QUOTA SAMPLES WITH RESPECT TO
OPINION ITEMS

Question	Block, 246	Quota, 328	Differ- ence
". . . read or heard about the use of animals in experiments?"			
Yes	85.0	80.8	4.2
No	15.0	18.3	−3.3
Do not know	0.0	0.9	−0.9
Pound animals [a]			
Painlessly killed	31.3	29.6	1.7
Medical research	66.3	62.8	3.5
Do not know	2.4	7.6	−5.2
Use of live dogs and cats for cancer research [b]			
Approve	80.9	79.3	1.6
Disapprove	13.4	17.7	−4.3
Do not know	5.7	3.0	2.7

[a] "Which of these two things do you think should be done with animals that
emain unclaimed at the city pound and for which no homes are available?—
'hey should be painlessly killed by the city authorities *or* they should be turned
ver to medical schools for research purposes."

[b] "Would you approve or disapprove of using live dogs and cats for research
pon the problem of cancer?"

The agreement between the two samples in respect to the factual
haracteristics of the respondents was not nearly as close as for the
pinion questions. Data on a number of these items are given in
'able 8.13. This table shows fairly extreme differences between the
amples on socio-economic status, home ownership, and type of house
n which the respondent lived. In addition to the items shown, rea-
onably close agreement arose for age, sex, religion, race, and news-
aper readership.

TABLE 8.13

COMPARISON OF BLOCK AND QUOTA SAMPLES FOR FACTUAL CHARACTERISTICS
OF RESPONDENTS

Characteristic	Block, 246	Quota, 328	Difference
Socio-economic status			
Wealthy	2.4	1.8	0.6
Upper-middle	18.7	14.9	3.8
Middle	61.4	50.9	10.5
Poor	17.5	32.3	−14.8
Home ownership			
Own	73.2	60.1	13.1
Rent	26.8	39.9	−13.1
Education			
Grammar school (completed or incomplete)	20.7	20.1	0.6
High school incomplete	16.3	22.3	−6.0
High school complete	34.2	24.7	9.5
College incomplete	17.5	21.0	−3.5
College complete	11.4	11.9	−0.5
Vote in 1944 presidential election			
Yes	71.5	68.9	2.6
No	28.5	31.1	−2.6
Type of house			
One family	80.5	68.3	12.2
Two family	3.7	7.3	−3.6
Three or more family or apartment	15.9	24.4	−8.5

In many respects this comparison is very similar to the McCall Evaluation Study. Socio-economic status was used as a quota control in both instances, and both show extreme differences on this variable. For the study discussed here, the definitions of socio-economic status and the proportions falling in each category were taken from NORC's handbook (7). Variables closely related to socio-economic status also show disagreement. However, these discrepancies do not seem to carry over to the attitude questions on the use of animals in medical research.

F. The Broadcast Measurement Bureau Evaluation Study

In the spring of 1946, the Broadcast Measurement Bureau (hereafter referred to as BMB) conducted a nationwide survey designed to measure radio station audiences. This survey was made on the basis of a sample of families selected by a quota method, and the information on station listening was obtained through the use of a mail questionnaire. After the completion of this survey, an evaluation study was carried out by Alfred Politz Research to test for the existence of bias in the original sample of families and to test the effects of non-response to the mail questionnaire. This evaluation study is described in a report submitted to BMB by Alfred Politz Research (9).

The original BMB sample was distributed among 3091 counties or subcounty areas. Within each county, the local representatives of several research agencies selected a quota sample of families on the basis of geographic location, city size (including rural regions), and local living conditions (three levels of socio-economic status). These families were sent a questionnaire relating to their frequency of listening to radio stations. A gift was included to elicit greater cooperation. Follow-up letters were sent until at least 50 per cent of the families had responded in any area for which a report was to be made. This entire operation is described in the January 1946 issue of *To Date*, a publication of BMB.

For the evaluation study, Alfred Politz Research selected the area consisting of the Bronx in New York City and the southern portion of Westchester County. Three groups were considered in this area: the total "list" used in the regular BMB survey, the "answerers" to the BMB survey, and a "control" sample selected according to a probability model. Techniques identical with those of the BMB survey were used to elicit information on radio listening, and in addition, a short personal interview questionnaire was administered for the purpose of obtaining data on factual characteristics of the families. The control sample was chosen by first selecting a sample of blocks (with probability proportionate to 1940 population in part of the area and with equal probability in the remainder) and by then subsampling these blocks with an appropriate sampling rate.

A complete description of the entire experiment is contained in the report *BMB Evaluation Study*, and we shall here present only a few representative findings. These are summarized in Table 8.14.

This table shows that there were substantial differences between the four groups, both in respect to listening and in respect to factual characteristics. The "answerer" sample shows a consistent overestimate of listening (i.e., in comparison with the "control" sample), however, even though the relative standings of the stations are not changed to any appreciable extent. These relationships also held for night-time listening.

TABLE 8.14

Comparison of "List," "Answerer," "Non-answerer," and "Contro
Samples on Radio Listening and Factual Characteristics

(Percent of base)

Characteristic	"List" [a]	"Answerer"	"Non-answerer"	"Contro
Number of Cases	786 [b]	427 [b]	359 [b]	784 [b]
in Base	842 [c]	439 [c]	403 [c]	833 [c]
Daytime weekly station audience indices				
WCBS	77.6	78.7	76.0	76.8
WNBC	74.3	78.2	68.4	71.4
WJZ	60.0	60.8	58.9	57.0
WOR	57.4	59.3	54.7	55.3
WNEW	33.8	34.4	32.7	29.6
WHN	31.3	30.8	31.9	27.7
WMCA	21.0	22.5	18.9	18.3
WQXR	13.3	17.2	7.5	11.7
WINS	12.8	14.6	10.1	9.9
WOV	8.0	8.7	6.8	6.0
Size of household				
1 person	2.6	2.9	2.1	4.7
2 persons	14.8	13.5	16.6	22.3
3 persons	26.0	29.1	21.6	27.8
4 persons	28.5	26.2	32.0	25.2
5 persons	15.4	16.2	14.2	10.0
6 or more	12.7	12.1	13.5	10.1
Rental value				
Less than $29.50	2.4	2.3	2.5	6.1
29.50–39.49	19.9	18.8	21.6	27.3
39.50–49.49	27.6	25.1	31.4	29.1
49.50–59.49	18.9	20.6	16.5	18.7
59.50–74.49	15.1	15.9	13.9	11.3
74.50–99.49	7.9	9.8	5.0	4.6
99.50 and over	8.2	7.5	9.1	2.9
Average under $100.00	$51.25	$52.72	$49.11	$47.14
Houses with phone	57.6	57.7	57.5	45.4
Houses with a car	38.3	39.2	37.0	25.6
Owned houses	23.4	23.6	23.2	12.0
Color				
White	85.8	89.3	80.7	91.1
Not white	14.2	10.7	19.3	8.9

[a] Figures obtained by appropriate weighting of "Answerer" and "No
answerer" groups, since these were not sampled with equal rates.
[b] Base for radio listening.
[c] Approximate base for factual characteristics.

The Broadcast Measurement Bureau concluded from these results that the quota sample—mail questionnaire approach—gave satisfactory results for their purposes. However, it should be noted that this experiment was carried out in a small, urbanized region, and we must therefore make some far-reaching generalizations in order to extend these conclusions to other sections of the United States (e.g., to rural regions).

Life Magazine Experiment in Syracuse

The Magazine Audience Group released its eighth report on a *Continuing Study of Magazine Audiences,* August 15, 1946. Field work for this study was carried out between August 1945 and April 1946. Prior to the actual conduct of the study, a large amount of experimentation relating to sampling and to the measurement of readership was performed. One of these experiments was made in the city of Syracuse and involved the comparison of two different methods of intracity sampling. This experiment is described in the report, and a brief summary of its results will now be given.

The first of these samples was selected by taking every twelfth block and instructing the interviewers to visit every sixth dwelling unit in the chosen blocks. An alphabetic listing of the persons living in these dwelling units was then prepared, and every third person was drawn into the sample, the interval of three being carried over from one dwelling unit to the next. Four attempts were made to interview a designated individual, after which a person with similar characteristics was substituted.

Simultaneously, the city was sampled with a modified "quota control" system, using individual quotas for age and sex but substituting geographic quotas for households instead of the usual standard of living groups. Geographic allocation of quotas was accomplished by grouping census tracts according to population and rental (1940 census plus up-to-date local housing information) and social characteristics (commercial, industrial, new and old residential areas). Essentially the same intracity sampling plan was later used for the national study, census tracts being replaced by enumeration districts and an additional control being instituted (for type of dwelling). Seven hundred interviews were obtained by each sampling technique.

The published account of this experiment states that the two samples were in close agreement on family income level of respondents, education, home ownership, rental value of the home, and in percentages of people having autos, telephones, washing machines, and the like. The quota control sample produced too many people living in

apartment houses, however. This discrepancy explains the later add
tion of the already noted control for type of dwelling unit (in lar₉
cities).

Data were presented for family income level and education and a₉
reproduced in Table 8.15.

TABLE 8.15

COMPARISON OF BLOCK AND QUOTA SAMPLES FROM *Life's*
SYRACUSE EXPERIMENT

(Percentages)

	Block System, 700	Quota Control System, 700	Difference
Families having income			
Less than $1250	6.3	6.2	0.1
1250–4000	65.5	64.2	1.3
4000 and up	28.2	29.6	−1.4
Respondents' education			
Grade school or less	39.9	37.2	2.7
High school	41.5	42.7	−1.2
College or better	12.2	13.4	−1.2
No answer	6.4	6.7	−0.3

On the basis of this experiment, the Magazine Audience Group de
cided to use the quota control system of sampling, enumeration dis
tricts being used for the geographic allocation of quotas and a con
trol being instituted for type of dwelling. The major consideration₉
in this decision were the relatively close agreement of the two sam
ples and the greater practicability of the quota control system—th
callback problem being eliminated. It may be noted that this par
ticular form of block sampling spreads the interviews very thinly ove₉
the city and thereby intensifies the problem of callbacks (i.e., tim
and cost required).

H. University of Iowa Test of Sampling Techniques

A laboratory experiment designed to compare various aspects of ₉
quota sample and an area-type sample has been described by Meier
Burke, and Banks (10, 11).

The authors started from the records of a house-to-house survey i₉
Iowa City which yielded records on income distribution, owner-rente₉
status, and occupation. These records were arranged by street an₉
number so that samples could be drawn on the basis of geographic lo₉

cation of the population elements. The "area" samples (so designated by the authors) were selected by dividing the total area of Iowa City into 370 subareas (the method of subdivision was not stated), taking a systematic sample of these areas (with a sampling ratio of 0.0135), and choosing a random subsample from each of the designated areas (a constant sampling ratio being used). Five samples, of sizes 15, 20, 25, 50, and 100, were drawn, and estimates were made from each of the five samples for income levels, home ownership, and occupation.

A complete description of the "quota" procedure is not possible from the published report, but evidently something of the following was done. A local real estate man divided the city into a number of broad areas on the basis of rent and general income level. Quotas were assigned to these broad areas on the basis of known population figures, and "interview streets" were obtained from within the areas by an unstated procedure. The sample was chosen from the selected streets by one of four methods: (a) every household along a given street, (b) every third household, (c) any five households per block, and (d) every household along intersecting streets. Five samples (of size 15, 20, 25, 50, and 100 respectively) were chosen by each of the four methods. Estimates were prepared as for the "area" sample. In the subsequent analysis, the samples chosen by these four methods were grouped together as the "quota" sample since there did not appear to be any significant difference between them.

Though the original data from the experiment were not published, the authors report the results obtained from three different types of analyses, namely,

1. The percentage of owned homes and of persons in the C and D income levels was compared with the *known* population proportions in terms of the standard deviation of *random sampling* for the quota sample (all individual samples thrown together) and for the area sample (all individual samples thrown together). Significance was obtained for home ownership on the quota sample (assuming the computed ratio had a normal distribution), but not for the other items. Though this is labeled as a "test of bias," it may better be interpreted as a combined test for bias and for the applicability of the variance formula for random sampling to the non-random quota type of sampling actually used. As is well known, the variance formula for random sampling does not necessarily apply to other situations.

2. The preceding analysis was extended to the individual sample sizes, the value of the "critical ratio" now being squared and summed over the individual sample sizes in order to obtain a quantity dis-

tributed like χ^2—*provided the variance formula for random sampling applies and there is no bias.* A significant value for χ^2 would indicate that there was bias and/or that the random sampling variance was not applicable. Here again significance was obtained only for home ownership, both the "area" and the "quota" samples giving approximately the same results.

3. The ratios of corresponding χ^2's for the "area" and "quota" samples were taken as F quantities and their significance noted. Moreover, the relative sizes of these values were taken as evidence of the superiority of one method over the other. It is impossible to see exactly what this F ratio is supposed to indicate because of the observed limitations on each χ^2.

As noted in the comments of Banks, this experiment throws little or no light on the relative merits of "area" and "quota" sampling as they are actually used. The main point of contention, the subjective selection of respondents by quota interviewers, is not present in this comparison. In their rejoinder to Banks, the authors acknowledge this and stress the fact that they are evaluating only the selection of specific areas within which interviews are to be obtained. In spite of this, the article never describes *how* their quota areas were selected. It seems most reasonable that the paper has merely demonstrated that for their situation a stratified sample of areas has a somewhat smaller variance than does a random sample of areas, both of these variances tending to be larger than for a random sample of individual households because of the "grouping" effect.

8.3 SUMMARY AND CONCLUSIONS

A. General Observations

Section 8.2 has presented an account of a number of comparisons of sampling methods, most of them comparisons between results obtained by a quota sampling procedure and results obtained by methods that take a large portion of the respondent selection from the hands of the interviewer. There undoubtedly exist many more such comparisons which did not come to the attention of the Study of Sampling. It is to be expected that they would add some strength to the conclusions to be drawn from those included here. They might also modify the conclusions somewhat and add new general results. Other comparisons, involving two or more quota samples, will be treated in Chapter 10. Of those comparisons discussed, only the NORC-MRLF and the Bikini comparisons were actually analyzed by the project. Descriptions and data for the others have been obtained from published accounts or from personal correspondence.

The cited comparisons show a wide range of agreement and disagreement, the most striking fact being that any single comparison produces at the same time both good agreement for some variables and poor agreement for others. The only major exception to this would seem to be the NORC-MRLF comparison where the quota sample results are quite consistently close to those produced by the Census area sample. In general, the largest differences occur with respect to such factual variables as income, socio-economic status, education, and the like. Opinion and attitude variables do not show this extreme disagreement, even though the samples do not agree on the factual characteristics. Especially striking illustrations of this are shown by the Bikini comparison, the McCall Evaluation Study, and the University of Washington Opinion Survey.

In general, the cases of reasonably close agreement have occurred in instances where the quota sample has used some form of rent quotas or small geographic area control in place of socio-economic status. The quota sample of the NORC-MRLF comparison used rent quotas, and the *Life* magazine Syracuse experiment quota sample used small-area geographic quotas. The rent quotas give the interviewer a more objective criterion than socio-economic status to use in selecting respondents, and the geographic controls restrict the area within which he may choose respondents. Actually, the results obtained from the McCall Evaluation Study and from the University of Washington Opinion Survey suggests that quotas for socio-economic status may be seriously in error. The notion of socio-economic status is somewhat indefinite and difficult to apply to individual respondents. Moreover, in both of these examples national quotas for socio-economic status were applied to one single city, a procedure which is not likely to work well.

Further evidence relating to the advantages of using small geographic area controls is provided by an experiment described by Hochstim and Smith (12).

B. The Evaluation of an Observed Difference

In the analysis of relatively undesigned comparisons, such as have been discussed in this chapter, the allocation of an observed difference among the various components which might cause the difference (sampling variability, sampling bias, measurement differences, and the like) is a more or less impossible task. In this problem all the limitations discussed under the topic of check data are still applicable, and in addition it is necessary to take into account the sampling variability arising from both of the surveys under consideration.

C. Difficulty of Generalization

Once a comparison has been found in which the evidence is reasonably conclusive that a real difference exists between the results produced by two different sampling methods, there is a strong temptation to attempt to generalize the results of the two broad classes of sampling methods to which the two specific ones belong. The extension of the differences found in the preceding comparisons to all surveys that employ similar types of sampling would be unwarranted and hazardous. We shall now set forth a few of the reasons why this is so, with illustrations.

The two samples contributing to a comparison have been taken at a particular point in time, with a particular set of interviewers, and have been used to obtain information on a particular set of topics. We can well expect that under certain changes in the circumstances (i.e., time, interviewers, subject of investigation) the results of the comparison will no longer be applicable. This is particularly true where the performance of one sampling method depends to a great extent on the existence of certain degrees of correlation between the personal characteristics of individuals in the population and the subject of investigation. These correlations may change in time or may be different for different opinion variables, thus leading to changes in the performance of the two methods.

The point can be illustrated to some extent by the differential effects that exist within a single comparison. For example, the area and quota samples in the Bikini comparison agreed quite well on religion but showed substantial differences in educational distributions. Consequently, we would expect them to produce similar results for attitude and opinion variables highly correlated with religion and possibly different results for variables highly associated with education. The same effect is shown in a striking manner by the McCall Evaluation Study. There were extreme differences between the two samples in rental distributions, rather large differences in the proportion of homes possessing various magazines, but rather small differences in the more qualitative features of magazine readership (e.g., "What features interest you most?"). Even these relationships will change over a period of time.

A second major difficulty that arises in attempting to generalize the results of a particular comparison concerns the extent to which we are willing to regard a sample as a typical quota sample or as a ypical area sample. In effect, we are asking the question, "What is he complete class or universe of quota samples?" or "What is the omplete class or universe of area samples?" For example, we can nquire whether the quota sample used in the NORC-MRLF comarison was typical of most other quota surveys. An examination of ts details clearly shows that it was not fully representative in a umber of respects. The interviewers were instructed to do almost ll their interviewing by calling on respondents in their homes. Some uota samples are taken in offices and stores and on the street as well s in the home. The amount of interviewing done in the evening and ver week ends when working people may be found at home varies reatly among survey organizations. The type of training given to nterviewers may also affect greatly the procedures they use to fill their uotas. Since quota sampling leaves much of the selection procedure o the discretion of the interviewers, quota samples done by one surey organization are not strictly comparable to those done by anther. Hence the extension of conclusions from this test to other uota sampling operations requires careful examination of the comarability of the conditions and procedures in each quota survey.

In the NORC survey the interviewers were asked to make a special ffort to attain accuracy. They knew their results would be comared with those of the Census. This surely improved their faithfuless in following instructions. It is doubtful, however, that any efort at conscientious performance of their work would reduce biases nherent in the quota procedure; and striving to obtain a very accuate sample might even introduce biases unwittingly as a result of ncorrect conceptions of what an accurate sample would be like. The uota assignments were checked carefully in the central office in the ight of factual data from previous surveys, but here again the oportunity to introduce a spuriously high degree of accuracy was negliible. (It is important to note that a substantial adjustment was nade in the MRLF results for preceding months *after* this survey vas taken, and any attempt to force agreement with previous MRLF urveys, except for the determination of farm quotas, would have esulted in discrepancies in relation to the MRLF after it was revised. The principal survey results were presented at a national conference efore the MRLF revisions were available.) Hence, some account nust be taken of the unusual care given to the performance of the urvey, and this may make it differ from some surveys less carefully erformed.

This point is amply illustrated by a consideration of the socio-economic control, or its replacement, for the quota samples used in the cited comparisons. The general details are:

1. Bikini Comparison—socio-economic classification with five level

2. NORC-MRLF Comparison—quotas were set on the basis c rentals, four classes being used.

3. BMB Evaluation Study—socio-economic classification with thre levels.

4. *Life* Syracuse Comparison—geographic allocation of quotas b census tracts.

5. McCall Corporation Evaluation Study—socio-economic classifi cation with four levels.

6. Washington State Comparison—socio-economic classification with four levels.

In addition to these differences in the way the quotas were set, ther were many other differences relating to the design and execution c these samples. Consequently, the conclusions derived in one instanc must be applied very carefully in a new situation. There can be n hope of valid generalization unless conditions and procedures ar comparable.

D. Lack of Cost Data

One of the important points that should be taken into account i making a comparison between two sampling methods is their relativ costs. Unfortunately, the Study of Sampling has been unable to fin adequate cost data for any of the comparisons it has examined. Ther is a general belief, however, that it is relatively more expensive t obtain a sample of given size by area methods than by quota meth ods, especially if the sampling operations are not of a repetitive na ture. This belief has been expressed specifically with respect to th initial preparation of lists and maps and with respect to callbacks t find designated respondents not found on previous calls.

An example of the difficulties encountered in the matter of costs i provided by the Bikini comparison. The area samples were onl about one-fifth the size of the quota samples. This does not mea however, that the area sampling procedure was five times as costl as the quota sampling procedure since other aspects of the survey that affected sampling costs were also different.

A comparison of the total costs of the two studies would not tel very much about the relative costs of the sampling methods becaus of the differences in interviewing, coding, and analysis. The intensiv interviewing techniques used with the area sample made the inter

iews longer, required more highly trained interviewers, and needed
more care in the coding and analysis of results than did the extensive
interviews. Actually, for the purpose of comparing the sampling
methods, we should like to know how large an area sample could be
obtained for the cost of the quota sample, with the same interview-
ng techniques used in each.

. Future Experimentation

The contents of this chapter have illustrated the shortcomings of
ross empirical comparisons between sample surveys when they are
sed to evaluate the relative merits of two alternative sampling pro-
edures. The major difficulty is simply that relatively unplanned
omparisons confound the effects of so many factors that it is imossible to separate the effects of sampling from those of other factors
hat produced an observed difference. It is also impossible to make
ure that comparisons in which different methods seem to be in close
greement are not actually quite dissimilar, their differences being
idden by compensating effects from other factors.

This does not mean that comparisons should be ignored in future
esearch. It does mean, however, that surveys conducted to provide
uch comparisons should be most carefully designed and executed in
rder that the analysis of the results will unequivocally allocate the
bserved differences among the possible factors contributing to the
ifferences. Further comments on this problem will be found in Chap-
er 13.

In conclusion, we should like to refer to one instance where a care-
ully designed comparison between quota samples and probability
amples has been carried out and analyzed. Moser and Stuart (13)
escribe an experiment wherein both probability and quota samples
ere selected in each of three English cities. Though they report on
he effects of different types of quota controls and on the magnitudes
f sampling errors for the quota samples (these results will be referred
 in Chapters 10 and 12), their major purpose was the direct compari-
on of the two sampling procedures. The questionnaire used in this
xperiment contained the ordinary demographic questions and, in ad-
ition, a large number of items broadly relating to leisure. Moser and
tuart say of their experiment:

We have seen, first from the national survey quoted in part I and then
om the experiment, that, on some fundamental questions, the quota sam-
les were biased. Occupation and education were the most conspicuous cases.
n most of the other questions, differences between quota results and check
ata or random sample results were relatively slight and we conclude that

they do not afford evidence of other systematic biases in the quota sample
There are indeed differences, but—leaving aside questions of purely statistic;
significance—they are rarely of the order that would worry the commerci;
survey practitioner. The two biases noted are serious and should be cor
rected by additional control, if possible, and at any rate given more atter
tion in quota sampling practice. . . .

The experiment has revealed relatively few major differences in results c
the quota and random samples. This finding should be correctly interpreted
it does not mean that quota sampling is theoretically sound. We have, s
to speak, one more "observation" to show that, in the hands of practitioner
of long experience, quota sampling can give fairly accurate overall estimate;
. . . But this practical success of quota sampling does not make it theoreti
cally sound. For this reason alone, we do not think that it is a metho
suitable for surveys in which it is important that the results are derive(
(and known to be derived) from theoretically safe sampling methods. I;
saying this, we are not forgetting the non-response problem.

Thus even with this carefully designed experiment, the fundamenta
status of quota sampling, as far as evidence derived from direct com
parisons is concerned, is left more or less as outlined in the precedin
portions of this chapter. Instances of serious bias can be found
close agreement with check data or with probability sample result
exists for many items; the sources of serious bias are frequently re
lated to the socio-economic control; the actual allocation of bia
among possible sources is extremely difficult; and it seems impossibl
to place quota sampling on a sound theoretical basis.

REFERENCES

1. Hadley Cantril, "Do Different Polls Get the Same Results?" *Publ. Op
   Quart.*, 9 (1945), 61–69.
2. Morris H. Hansen, William N. Hurwitz, and William G. Madow, *Sampl*
   Survey Methods and Theory, Vol. I, John Wiley & Sons, New York, 1953
3. Social Science Research Council Committee on Social Aspects of Atomi
   Energy, *Public Reaction to the Atomic Bomb and World Affairs,* mimeo
   graphed, Ithaca, 1947.
4. Leonard S. Cottrell, Jr., and Sylvia Eberhardt, *American Opinion on Worl*
   Affairs in the Atomic Age, Princeton University Press, Princeton, 1948.
5. Roe Goodman, "Sampling for the 1947 Survey of Consumer Finances,"
   J. Am. Statist. Assoc., 42 (1947), 439–448.
6. Frederick Mosteller, Herbert Hyman, Philip J. McCarthy, Eli S. Marks
   and David B. Truman, *The Pre-election Polls of 1948,* Bulletin 60, Socia
   Science Research Council, New York, 1949.
7. National Opinion Research Center, *Interviewing for NORC,* University o:
   Denver, Denver, 1945.
8. McCall Corporation, *A Qualitative Study of Magazines: Who Reads Then*
   and Why: The Third in a Series of Continuing Studies, New York, 1946.

9. Alfred Politz, *BMB Evaluation Study,* a report submitted to the Broadcast Measurement Bureau, New York, 1947.

10. Norman C. Meier and Cletus J. Burke, "Laboratory Tests of Sampling Techniques," *Publ. Op. Quart.,* 11 (1947), 586–593.

11. Norman C. Meier, Cletus J. Burke, and Seymour Banks, "Laboratory Tests of Sampling Techniques: Comments and Rejoinders," *Publ. Op. Quart.,* 12 (1948), 316–324.

12. J. R. Hochstim and D. M. K. Smith, "Area Sampling or Quota Control? Three Sampling Experiments," *Publ. Op. Quart.,* 12 (1948), 73–80.

13. C. A. Moser and A. Stuart, "An Experimental Study of Quota Sampling," *J. Roy. Statist. Soc.,* 116 (1953), 349–405.

CHAPTER 9

Estimation of Variances for Selected Probability Model Sampling Procedures

9.1 INTRODUCTION

The accuracy of an estimate made from a sample selected according to a probability model sampling procedure is ordinarily assessed by determining the variance of the estimate (or the mean square error in the event that the estimate has a bias component) and using this variance to determine confidence limits for the quantity being estimated. This process is carried through in the following steps:

1. From the probability model we derive, on a purely theoretical basis, an expression for the variance of the estimate. The expression so obtained is necessarily written in terms of certain unknown population parameters. For example, if simple random sampling is used to estimate the proportion, p, of individuals in a population who have a specified characteristic, the sample proportion of individuals possessing the characteristic is used to estimate the value of p. The variance of this estimate is equal to $p(1 - p)/n$, where n is the sample size. This expression ignores the finite population correction factor and the effect of the measurement process, but these could be incorporated into the expression for sampling error if so desired.

2. As the sample design becomes more and more complex, the expressions for the variances contain the values of more and more unknown population parameters. Ordinarily we do not have advance information concerning the values of the population parameters involved in the variance expression for a particular sampling procedure and form of estimate. Hence these parameters must be estimated from

192

the sample. In other words, the sample must provide an estimate of its own accuracy.

There are two major problems associated with this procedure. The first of these has to do with the selection of a sampling procedure for a particular survey. In order to choose between two or more competing designs, or in order to fix the details of a specific sampling design, it would be advantageous to compare, among other things, the relative accuracy of the estimates that would be produced by the designs. Since this type of decision must be made in advance of selecting and analyzing the sample, it is necessary to know or to have reasonably good estimates of the parameters entering the variance expressions, except in certain special cases. This information can come only from past experience and, in particular, only from past experience that has been analyzed with this use in mind. The second problem arises from the theoretical and computational burdens involved in estimating the variance of a sample estimate for many of the more complex sampling procedures. As a matter of fact, these burdens become so heavy that usually no attempt is made to obtain the variance of an estimate, except when it can be approximated by projections from comparable surveys in which variances were computed or when there seems to be justification for substituting a greatly simplified model such as that of simple random selection.

These points have long been recognized, for example, in relation to the sampling work of the Bureau of the Census [see Hansen and Hurwitz (1)] and in relation to sampling for agricultural items. Some work on the latter topic which is illustrative of the general approach has been described by Jebe (2). He states that the principal objects of investigation of his study were:

1. What effect upon various procedures for sampling an agricultural population will be introduced by selection of the primary sampling units with
   a. Equal probability and
   b. Probability proportional to some measure of size?
2. Of what magnitude are the bias contributions to the "mean square error" for the biased estimates that were chosen for examination?
3. What is the magnitude of the "within county" component of the variance relative to the "between county" component? Will the magnitude of the "within county" component affect the choice of the primary units for different stratifications and methods of estimation?
4. What sample survey designs are to be recommended in terms of the preceding considerations?

In spite of the large amount of investigation carried out along these lines, there has, thus far, been little or no work of a similar nature done in connection with sampling for the measurement of opinion, attitudes, and consumer wants. The purpose of this chapter is to present some results that are indicative of what can be accomplished in this direction. It is hoped that this discussion will stimulate further analyses on related problems and that, in time, a body of experience will develop to which we can refer for help and guidance on any specific sampling problem.

9.2 CLUSTER SAMPLING WITH EQUAL-SIZED CLUSTERS *

A. Description of the Survey and Sample

In June of 1948 a sample of adults was selected from the city and environs of Elmira, New York. This sample was drawn for the 1948 Voting Study (3), a study which repeats and supplements similar undertakings carried out in Erie County, Ohio; these earlier studies have been described by Lazarsfeld, Berelson, and Gaudet (4). In addition, the sample was used for the preliminary phases of a project on intergroup relations under the direction of the Sociology Department of Cornell University.

The sample was drawn in the following manner: up-to-date maps were procured for the population area, and these maps were used to divide the area into small clusters of dwelling units. In general, these clusters were individual city blocks. In order to keep the clusters of somewhat the same size, however, this condition had to be relaxed whenever a block appeared to be too large or too small (structures were indicated on the maps) and convenient combinations or subdivisions could be made. The clusters were numbered in a serpentine fashion, going back and forth across the maps, and after choosing a random starting point, every third cluster was drawn into the sample. A listing was next prepared of all dwelling units in these sample clusters, and after choosing a random starting point, approximately every fifth address was taken into the sample. The clusters were kept in their proper order, and the interval of five was carried over from one cluster to the next. The interviewers were then sent to these fixed addresses. One adult was chosen from each sample dwelling unit by means of a procedure similar to that described by Leslie Kish (7).

The field work for the survey closed with the following results:

* Portions of this material are similar to those previously presented by McCarthy (5, 6).

	Number	Per Cent
Total attempts	1267	100.0
Completed interviews	1029	81.2
Mortality	238	18.8
Refusals	92	7.3
Ill	33	2.6
Out of town	3	0.2
Unobtainable (not at home after several calls, etc.)	87	6.7
Did not speak English	12	0.9
Errors (wrong person, under age, etc.)	11	0.9

The interviewers were instructed to make at least two planned callbacks (i.e., on the basis of appointment or other information) in order to obtain an interview with the designated respondent. In some instances as many as eight calls were made. Also, a refusal was accepted only after it had been made to three different interviewers.

Estimates were obtained from this sample by analyzing it as a whole. In other words, the proportion of adults in the population who would express a preference for a particular political candidate was estimated by the corresponding sample proportion, and similarly for all other items of analysis.

B. Description of Theory Applicable to Sample Design

Although this is a relatively simple sample design for use in city sampling, it has a number of features which complicate computation and theory when we attempt to evaluate the sampling errors (i.e., variance plus bias squared) of estimates made from the sample. These features are:

1. The primary sampling unit is the cluster of adults living on what is approximately a city block. Since adults living in such close geographic proximity will tend to have "like" educational attainments and "like" economic status, among other characteristics, they will also tend to have "like" opinions and attitudes. This fact is frequently expressed by saying that there exists a positive *intraclass correlation* between individuals living on the same block with respect to such characteristics. Thus the sampling of such clusters will in general be less efficient than the sampling of individuals, and the survey results must be analyzed block by block in order to evaluate the degree of this loss of efficiency.

2. Even after the clusters were chosen, the sampling proceeded through two more stages. Dwelling units were sampled from these clusters, and only *one* adult was interviewed from each of the selected dwelling units. These two stages must also be taken into account if exactly applicable theory is to be obtained.

3. The clusters and dwelling units within the clusters were chosen not by simple random sampling but by systematic sampling. It is extremely difficult to take care of systematic procedures from a theory point of view and impossible to estimate variances from a single systematic sample. For a discussion of these points, see Lillian Madow (8).

4. The blocks or clusters used as primary sampling units varied widely with respect to the number of adults living on them. This means that the estimates made from the sample have a bias component and that, even if the clusters had been selected in a simple random fashion, only approximations can be obtained for their variance. This point is discussed more fully in Section 9.3.

In view of these difficulties, it was decided to analyze the survey results in a manner most illustrative of the effects of cluster sampling in relation to the measurement of opinions and attitudes, and to dispense with any attempt to estimate actual sampling variances for this particular survey. The following steps and assumptions were involved.

1. In order to simplify the theory and computations, only those blocks were used for analysis from which between five to ten interviews were obtained, the large and small blocks being ignored. There were 79 blocks which satisfied these conditions, the average number of adults for these blocks being 75. Accordingly, we shall proceed as though we were dealing with city blocks as primary sampling units, each block having 75 adults who live on it. These 79 blocks were spread over the entire population area and were not concentrated in any one section of the city.

2. The systematic features of the design were ignored. That is, it was assumed that the clusters were drawn at random and that the dwelling units within a block were drawn at random.

3. The fact that dwelling units were selected within a block and one adult was chosen from each dwelling unit was ignored. In other words, it is assumed that within a block a random sample of individuals was obtained.

4. The effects of the sample mortality were ignored.

Under these assumptions, we are dealing with the following type of situation. We have a population of adults which is divided into N clusters, each cluster having M adults living on it. A sample of n clusters is chosen at random, and m individuals are chosen at random from within each cluster. The total sample size is therefore mn. From this sample we wish to estimate the proportion of individuals, p, in the population who possess a certain characteristic (e.g., who would give a favorable answer to a specific question). As an estimate of this proportion, we use the sample proportion which will be designated by $\hat{p}$. That is

$$\hat{p} = \frac{x}{nm},$$

where x is the number in the sample possessing the characteristic. From theoretical considerations it can be shown that $\hat{p}$ is an unbiased estimate of p and that the variance of $\hat{p}$, $V(\hat{p})$, is given by

$$V(\hat{p}) = \frac{M - m}{M - 1} \frac{1}{nm} V_w + \frac{N - n}{Nn} V_b,$$

where

$$V_w = \frac{\Sigma p_i (1 - p_i)}{N} \qquad (N)$$

and

$$V_b = \frac{\Sigma (p_i - p)^2}{N - 1}, \qquad (N)$$

p_i being the true proportion of individuals possessing the characteristic in the ith cluster. The within-cluster variance is V_w, and V_b is the between-cluster variance. The symbol in parentheses to the right of an expression indicates the range of the summation. For example, in these equations the sums are taken from $i = 1$ to $i = N$.

If we knew the values of the within-cluster variance and the between-cluster variance in advance of designing a sample, then it would be possible to choose n and m to give any required degree of accuracy. It is the purpose here to give values for these two quantities as derived from the Elmira survey under the conditions previously described, and to show the effects of various choices of n and m on the accuracy of estimates. It should be noted that the effect of using only blocks of approximately the same size will be to understate the actual size of the between-cluster variance. Actually, these quantities cannot be given exactly since only a sample is available. They can be estimated from the sample by the following formulas:

$$\hat{V}_w = \frac{M-1}{M} \frac{m}{m-1} \frac{\Sigma \hat{p}_i(1-\hat{p}_i)}{n} \qquad (n)$$

$$\hat{V}_b = \frac{\Sigma(\hat{p}_i - \hat{p})^2}{n-1} - \frac{M-m}{M} \frac{1}{m-1} \frac{\Sigma \hat{p}_i(1-\hat{p}_i)}{n}$$

In these formulas, $\hat{p}_i$ is the sample proportion of individuals possessing the characteristic in the ith cluster. The estimates so obtained are unbiased estimates of the within and between variances.

C. Computation of Variances

Five items were selected from the questionnaire for analysis in line with the foregoing theory, and it was assumed that the following proportions were to be estimated.

Item	Proportion Being Estimated
1	Proportion of adult males
2	Proportion of adults having graduated from high school
3	Proportion of adults who endorsed a statement saying that Catholics in the U.S. were getting too much power for the good of the country
4	Proportion of adults who endorsed a statement saying that labor unions were doing a "fine job" in the U.S.
5	Proportion of adults who felt that our relations with Russia should be firmer

For these five items, the results given in Table 9.1 were obtained.

TABLE 9.1

ESTIMATED VALUES OF THE WITHIN- AND BETWEEN-CLUSTER VARIANCES
FOR CITY BLOCKS

(Each block contained approximately 75 adults)

Item	Estimated Value of Proportion	Within Variance	Between Variance
1	0.442	0.2467	0.0008
2	0.451	0.2272	0.0211
3	0.060	0.0541	0.0025
4	0.154	0.1337	0.0 [a]
5	0.758	0.1811	0.0028

[a] The estimate of the between variance was negative in this instance and was therefore taken to be zero. This does not mean that the true between variance is zero, but only that it is probably small and that the unbiased estimating procedure used will sometimes produce small positive values and sometimes small negative values, depending on the particular sample being studied.

The values given in Table 9.1 are not too meaningful in themselves, but they can be used to illustrate the effects of sampling and subsampling clusters of adults in the measurement of opinions and attitudes. For example, assume that a sample of 525 adults is to be drawn from a population of 23,625 individuals and that this population is grouped in clusters of 75 (having the characteristics exhibited in Table 9.1). This sample can be drawn in any one of the following ways:

Case I. One-third of the clusters are chosen at random and one-fifteenth of the individuals within each cluster are then selected at random.

Case II. One-fifth of the clusters and one-ninth of the individuals within each cluster are selected. It is, of course, not possible to take exactly one-ninth of 75 individuals but we could approximate this by taking eight for some clusters and nine for the remainder.

Case III. One-fifteenth of the clusters and one-third of the individuals within each cluster are selected.

Case IV. One-forty-fifth of the clusters and *all* the individuals are selected.

A comparison of these four cases is contained in Table 9.2.

TABLE 9.2 [a]

A COMPARISON OF VARIANCES FOR DIFFERENT WAYS OF SAMPLING AND
SUBSAMPLING CLUSTERS OF 75 ADULTS

(Total population size, 23,625; total sample size, 525)

Item	Estimated Value of Proportion	Random Sample Variance	Variance as a Percentage of Random Sample Variance for Case			
			I	II	III	IV
1	0.442	0.000470	94	91	75	25
2	0.451	0.000472	114	139	261	626
3	0.060	0.000107	104	115	168	328
4	0.154	0.000248	96	91	68	0 [b]
5	0.758	0.000349	97	98	101	111

[a] The values given in this table differ somewhat from those published in the earlier cited references. The difference is due to the omission in the earlier work of a factor necessary to obtain unbiased estimates of the within-cluster variance. The figures given in this table are very sensitive to even a small change in the within-cluster variance, particularly when a large proportion of each cluster is being drawn into the sample.

[b] See footnote to Table 9.1.

The simplest way to interpret the figures in Table 9.2 is in terms of sample size. Thus the values indicate the number of adults who must be chosen by the appropriate sampling scheme in order to give the same accuracy as 100 adults chosen at random. These values should be interpreted as indicative of what may happen in practice rather than in absolute terms. They are themselves based on sample data. Moreover, the situation would undoubtedly change if the cluster size were increased or decreased, if different variables were studied, or even if different populations (e.g., different cities) were under investigation. It may be noted that item 1, proportion of males, and item 2, proportion of individuals who have graduated from high school, behave in the manner to be expected. City blocks would tend to have the same sex distribution as the entire population (because of marriage and the sex ratio at birth), and so the efficiency of the estimate increases as the proportion of individuals taken from a block increases. On the other hand, individuals living on a block will tend to have like educational attainments, and so this efficiency decreases as the size of sample from each block increases. The other items are opinion variables and are more difficult to interpret in general terms. Thus there seems to be no clear explanation for the behavior of item 4, the proportion of adults who endorsed a statement saying that labor unions were doing a "fine job" in the United States.

9.3 STRATIFIED CLUSTER SAMPLING WITH UNEQUAL-SIZED CLUSTERS

A. Description of the Survey and Sample

In August of 1951 the Extension Service and Experiment Station of the New York State College of Agriculture and the Bureau of Agricultural Education of the New York State Education Department cooperated in conducting a survey among the farmers of New York State. The purpose of this survey was to secure information about the opinions of farmers in New York State on the agricultural programs carried out by the United States Department of Agriculture and cooperating state agencies. For the purposes of this survey, farmer was defined as an individual who received at least half his income from operating a farm.

The sample for this survey was chosen in a relatively straightforward fashion. The materials of the Master Sample of Agriculture [see King and Jessen (9)] formed the basis for the selection. An estimate was made of the number of Master Sample segments that would be required to give the desired sample size, and these segments were al

cated proportionately among the 56 counties to be covered in the
rvey (i.e., in proportion to the total number of segments in each
unty). Within each county, the sample segments were chosen by
mple random sampling. The interviewers were given maps show-
g the selected segments and visited every dwelling unit in each seg-
ent. Interviews were to be obtained with each individual who satis-
d the foregoing definition of a farmer. The field work of the survey
osed with the following results.

Category	Number of Cases
Completed interviews	1530
Eligible farmers not interviewed	179
Individuals who were "Census farmers" and not "Survey farmers"	855
Non-farmers	4168
Not otherwise classified	66

The quantities to be estimated from this survey were all of the
rm "the proportion (or percentage) of survey farmers in New York
ate who would give a specified response to a stated question." In
dition, analyses were to be made of the interrelations between two
 more opinion or informational questions and between such vari-
les and a variety of factual characteristics of the farmers (e.g.,
;e, size of farm, and type of farm). Each of these estimates was
epared by analyzing the sample in its entirety. That is, the pro-
rtion of farmers who favor a certain governmental program was
timated by the sample proportion of farmers favoring the program.

Description of Theory Applicable to Sample Design

Although the sample design set forth in the preceding subsection is
mple to describe, was easy to execute, and led to results that were
raightforward to analyze, it is not a design that is particularly sim-
e when we attempt to obtain exact estimates of the sampling errors
.e., variance plus bias squared) of the survey results. The difficul-
es, both theoretical and computational, follow.

1. The design contains a stratification since a random sample of
gments was chosen within each of the 56 counties (i.e., each county
a stratum). This fact did not complicate the estimation process,
it it does complicate the computation of variances. As noted in
hapter 20, the variance of an estimate made from a stratified sample
pends on the within-stratum variances, and consequently the re-
lts for each of the 56 counties must be analyzed separately in esti-
ating such a variance.

2. Within each county a random sample of Master Sample segmen
was chosen, and every survey farmer living within the chosen seg
ments was to be interviewed. This introduced the clustering effe
described in Chapter 20. That is, farmers living in close geograph
proximity will tend to have somewhat like opinions and attitudes, an
a sample drawn in this way may be less accurate than a simple ran
dom sample of the same size, depending on the degree of "likeness
within segments. In order to evaluate this effect, we must analyz
the survey results segment by segment within each of the 56 countie

3. The Master Sample segments vary in respect to size. That i
some segments will have one survey farmer, some will have two sun
vey farmers, and so on. This fact has two implications for the est
mate described in the preceding subsection. The estimate is biase
and only an approximation can be obtained from the probability mod
for its variance. Both of these results follow from the fact that th
estimate is in the form of a ratio of two random variables. Not onl
would the number of farmers favoring a certain program vary fron
sample to sample (the numerator of the estimate) but the total num
ber of farmers in the sample would vary from sample to sample (th
denominator of the estimate). This latter fact follows from the var
ation in size of Master Sample segments.

Since the variability in size of Master Sample segments adds to th
difficulty of analyzing the sample results, we are presenting a descrip
tion of this variability in Table 9.3.

TABLE 9.3

DISTRIBUTION OF MASTER SAMPLE SEGMENTS BY NUMBER OF
COMPLETED INTERVIEWS

(Interviews with survey farmers; average number of interviews per segment 2.1

Number of Completed Interviews per Segment	Number of Segments	Percentage Distribution
0	121	17.6
1	169	24.5
2	146	21.2
3	123	17.9
4	74	10.8
5	31	4.5
6	15	2.2
7 and over	9	1.3
Total	688 [a]	100.0

[a] A few of the segments actually covered in the survey are missing from th
distribution because of incomplete information (those from 2 of the 56 counties

It might be noted that the existence of a large number of zero segments does not inconvenience the theoretical considerations of this sample design. From a practical point of view, however, this information is necessary to estimate the number of sample segments that must be chosen to give approximately the required number of interviews in the sample. The zero segments in this sample came primarily from the areas adjacent to incorporated places in New York State.

The theory necessary to estimate the variance of estimates obtained from this sample design has been discussed by Hansen, Hurwitz, and Gurney (10) and by Cochran (11), although this theory must be modified to some extent since we are here concerned with the estimation of proportions (or percentages) rather than with the estimation of the population total for a quantitative variable. Starting with Cochran's formula 6.22 (11) and making the appropriate modifications, we obtain the following results. Let

$$n = 1530 = \text{the total number of interviews}$$

and

$$\hat{p} = \frac{x}{n}$$

where x is the number of favorable responses in the sample and $\hat{p}$ is the sample estimate of the proportion of survey farmers in the entire state who would give a favorable response to a specific question. The problem is to obtain an estimate of the variance of $\hat{p}$ which we shall designate by $\hat{V}_{Rc}(\hat{p})$. The subscript Rc follows Cochran (11) and is an abbreviation for "Ratio estimate combined over strata." Then

$$\hat{V}_{Rc}(\hat{p}) \doteq r(1 - r)\frac{\hat{p}^2}{n^2}\Sigma N_i\left(s_{yi}^2 + \frac{s_{xi}^2}{\hat{p}^2} - 2\frac{\hat{\rho}_i s_{yi} s_{xi}}{\hat{p}}\right), \qquad (k)$$

where $\doteq$ means approximately equal to,

$r = 0.0280 = $ the proportion of segments drawn into the sample,
$k = 56 = $ the number of strata or counties,
$N_i = $ the number of Master Sample segments in the ith county (total number and not sample number),
$y_{ij} = $ the number of interviews obtained in the jth segment of the ith county,
$x_{ij} = $ the number of favorable responses among the y_{ij} interviews in the ijth segment,
$s_{yi}^2 = $ the sample variance of the y_{ij} within the ith stratum,
$s_{xi}^2 = $ the sample variance of the x_{ij} within the ith stratum,
$\hat{\rho}_i = $ the sample correlation between the y_{ij} and the x_{ij} within the ith stratum.

There are a number of observations which should be made before applying the foregoing theory. First, as already noted, the expression for $\hat{V}_{Rc}(\hat{p})$ is only an approximation to the variance of $\hat{p}$. Moreover, no attempt has been made to determine the accuracy of this approximation. Second, $\hat{p}$ is a biased estimate of the population quantity and the amount of this bias is unknown, although other investigations and experience would lead us to believe that it is very small compared to the variance. Third, the possible effects of the non-interviewed eligible farmers have been ignored.

C. Computation of Variances *

For purposes of analysis, fourteen questions were selected from the survey questionnaire. Each of these questions was asked of all 1530 respondents, and each had to do with knowledge of or opinion toward agricultural programs carried on by the United States Department of Agriculture and cooperating state agencies. A typical question is the following:

Do you feel the Department of Agriculture should make partial payments to farmers for using *soil-building* practices such as the use of lime and superphosphate, and for *soil-conserving* practices such as contouring or strip cropping or building farm ponds?

           _____ Yes
           _____ No
           _____ Undecided—don't know

All fourteen questions had response categories of this kind, and in each instance it was assumed that the proportion of farmers who would give "Yes" or "Favorable" response among the population of all farmers (i.e., including the "Undecided—don't know" category) was being estimated.

The results of performing variance analyses on these fourteen questions are summarized in Table 9.4.

An examination of Table 9.4 shows that for twelve of the fourteen questions $\hat{V}_{Rc}(\hat{p})$ is larger than the estimated binomial variance (i.e., assuming simple random sampling), and that the average ratio of $\hat{V}_{Rc}(\hat{p})$ to the binomial variance is 1.144. As pointed out earlier, the estimate $\hat{p}$ also has a bias component. If we were able to take this into account and make the comparison between the mean square error of $\hat{p}$ and the binomial variance, the factor 1.144 would undoubtedly be some-

---

* The analyses described in this subsection were initiated and supervised by Mr. D. S. Robson, a staff member of the Biometrics Unit, New York State College of Agriculture, Cornell University.

TABLE 9.4

COMPARISON OF VARIANCE ESTIMATES OBTAINED BY ASSUMING SIMPLE RANDOM
SAMPLING AND BY USING THE THEORY OUTLINED IN SECTION 9.3B

(All variances are multiplied by 1,000,000)

	1	2	3	4
	Estimated Propor-	Estimated		
	tion Favorable,	Binomial		Column 3
Question	$\hat{p}$	Variance [a]	$\hat{V}_{Rc}(\hat{p})$	Column 2
1	0.7078	135	151	1.119
2	0.2941	136	173	1.272
3	0.4451	161	177	1.099
4	0.8327	91	128	1.407
5	0.6340	152	182	1.197
6	0.5876	158	193	1.222
7	0.9523	30	35	1.167
8	0.3373	146	163	1.116
9	0.7033	136	170	1.250
10	0.2935	136	165	1.213
11	0.8542	81	86	1.062
12	0.9673	21	15	0.714
13	0.6196	154	197	1.279
14	0.6046	156	141	0.904

[a] This is $\hat{p}(1 - \hat{p})/n$, which would be the appropriate way to estimate the variance if a simple random sample of 1530 farmers had been drawn from the population of all farmers in New York State.

what increased. It is not felt, however, that the effect of this bias would be large enough to change seriously any of the observations that can be made on the basis of the present data. These results are of interest in two ways. First, they indicate rather clearly that a sample selected in this manner will be less efficient than a random sample of the same size, and second, they give a general idea of the magnitude of this loss of efficiency.

Actually, there are two counteracting forces at work in the present sample design. The county stratification would tend to make it more efficient than simple random sampling, but the clustering of interviews would tend to make it less efficient. However, the effect of the county stratification on the variance is likely to be very small, as will be illustrated in the following paragraphs, and so the values in Table 9.4 indicate primarily the effects of clustering. The individual questions were examined to see whether there was any obvious reason why this clustering effect differed from question to question (for other than sampling reasons), but no such reasons were apparent.

There are many situations in which the technique of stratification leads to very substantial gains in efficiency. This is particularly true when the elements of a population differ widely in respect to size (e.g., manufacturing plants) and where some population characteristic for a quantitative variable closely related to the measure of size is being estimated. It does not seem to have been generally recognized that these very substantial gains are not usually available when we are dealing with the estimation of proportions, as in the measurement of opinions and attitudes. This does not mean that gains cannot be obtained but only that they will be small.

To illustrate these statements, we have chosen some data from this New York State survey. All farmers were asked the question:

The Department of Agriculture through P. & M. A. has set up a marketing order on milk sold on the New York Market. The State Department of Agriculture has set up marketing orders for the Buffalo and Rochester Markets. As a farmer, do you think the marketing orders by which the price of milk is established are a good idea or do you feel they should be dropped?

_____ A good idea
_____ Should be dropped
_____ Undecided—don't know

Having an opinion on this question should be closely related to whether or not a farmer operates a dairy, and this is borne out by Table 9.5.

TABLE 9.5

PROPORTION OF FARMERS "UNDECIDED" ON MILK-MARKETING ORDERS AS RELATED TO FARMING ENTERPRISE

Farming Enterprise	Number of Farmers	Proportion "Undecided"
Dairy only	391	0.212
Dairy principal, but other enterprise secondary	755	0.205
Other enterprise principal, but dairy secondary	60	0.367
No dairy	294	0.565

The differences between these four groups with respect to proportion "undecided" are fairly marked for attitude and opinion work, and it is these differences that make stratification pay off. Suppose now that our sole purpose in making this survey were to estimate the popula-

)n proportion of "undecided" and that we could stratify by these
ur farming enterprise groups—proportional allocation being used to
termine the size of the simple random sample within each group.
mple computations show that, for samples of the same size, the ratio
the variance of the estimate from the stratified sample to the vari-
ce of the random sample is 0.90—certainly not a spectacular gain
accuracy. Of course stratification has uses other than that of sim-
y increasing the accuracy of a single population estimate, and a
scussion of these uses will be found in Chapter 20.

9.4 ALTERNATIVE WAYS OF
ESTIMATING SAMPLING ERROR

The preceding two sections have illustrated well the theoretical
d computational difficulties encountered in trying to estimate the
riance of estimates derived from samples drawn according to many
the commonly used sampling procedures. Because of these difficul-
es it frequently happens that no attempt is made to obtain the
riance of an estimate, except perhaps by proceeding on the assump-
n that the sample was obtained by simple random selection. Ac-
ally, *easily computed* but *rough* estimates of variance can often be
tained by merely introducing rather minor changes into the origi-
l design.

This procedure can best be illustrated with reference to systematic
mpling. Suppose that we wish to draw a systematic sample from an
dered list of population elements and that a sampling interval of one
twenty will provide the required sample size. That is, a random num-
r is selected between one and twenty and, starting with the corre-
onding element, every twentieth element is drawn into the sample.
sample drawn in this manner does not provide an estimate of variance
less it is assumed that the ordering of the population elements was a
ndom one. However, this difficulty can be overcome by selecting a
mber of small independent systematic samples rather than one large
stematic sample. For example, ten independent systematic samples
ould be obtained in the following manner: the sampling interval
ould be taken as one in two hundred instead of one in twenty, and ten
ndom starting points between one and two hundred would be chosen.
he resulting ten samples would still have the same total size as the
iginal sample, but now the variance of a proportion obtained from the
tal sample can be estimated in a very simple manner. Let the total
mple size be n, the size of each small sample $n/10$, the estimate of the
oportion from the ith small sample $\hat{p}_i$, and the estimate for the com-

bined samples $\bar{p}$. With this notation, an estimate of $V(\bar{p})$ is given b

$$\hat{V}(\bar{p}) = \frac{1}{10} \frac{\Sigma(\hat{p}_i - \bar{p})^2}{9} \tag{1}$$

If t independent small samples are used, this formula becomes

$$\hat{V}(\bar{p}) = \frac{1}{t} \frac{\Sigma(\hat{p}_i - \bar{p})^2}{(t - 1)}$$

This same procedure has been discussed by Deming (12, p. 96), wh
has designated it the Tukey plan. It might be observed that est
mates made from "ten random starts" systematic samples may ha
somewhat different variances than those obtained from "single ra
dom start" samples and that in some instances their variances wi
be closer to those of a simple random sampling procedure than to th
variances of the single systematic sampling. Still further variatio
of this type of approach have been suggested by Deming (13).

This process has an added advantage when it is impossible
foresee how large a sample can be taken in an allotted time or wi
a fixed amount of money. The interviewing or measurement can
stopped at the end of any number of small samples, and the combin
tion of these will result in a "well-designed" sample from the enti
population. Ordinarily, failure to secure information from a lar
portion of a single sample will mean the introduction of biases in
the sample results—for example, if the missed portion comes entire
from one particular geographic area of a city.

Once a sample that was not designed for this type of analysis ha
been drawn, it may be somewhat more difficult to obtain simple pr
cedures for estimating variances. However, some approximations a
usually possible. In order to illustrate this, we have applied it to th
sample survey of New York State farmers described in the precedir
section. The 56 counties used in the survey were arranged in alph
betic order, and within each county the segments were arranged in th
random order in which they were drawn. Using this as a continuo
listing, ten systematic subsamples were defined, the estimate of
particular proportion was obtained from each, and the variance ana
ysis described previously was applied to these ten estimates. Th
process was used on the first four of the items appearing in Table 9.
and the results are given in Table 9.6. In this instance, the subsamp
estimate of variance takes into account the clustering effect, the effe
of unequal-sized clusters, and at least a portion of the county stratif
cation effect. It does not fully represent the county stratification be
cause the number of sample segments in the various counties were n

ultiples of ten. This means that some counties would not be repre-
nted in some subsamples and that in general the proportional allo-
tion features of the original design were not maintained exactly.

TABLE 9.6

OMPARISON OF VARIANCE ESTIMATES FOR NEW YORK SURVEY OF FARMERS

(All variances are multiplied by 1,000,000)

Question	Estimated Proportion Favorable, p	Estimated Binomial Variance	Estimated Variance, 9.3B Theory	Estimated Variance from Subsamples
1	0.7078	135	151	223
2	0.2941	136	173	244
3	0.4451	161	177	97
4	0.8327	91	128	108

It is difficult to interpret the values given in Table 9.6 since each
them is also subject to sampling variability. However, it may be
oted that the subsample variances bear the same relationship to the
nomial variance as does the theory variance (with respect to
greater than" or "less than") with the exception of question 3. The
ork required to obtain these estimates is least for the binomial vari-
ice, not much greater for the subsample variance (if this analysis
planned when the original punched cards are prepared), and con-
derably greater than either of these for the theory variance. The
absample procedure is one that should be seriously considered in any
ractical situation. It is discussed further in Chapter 10 and by Dem-
ag (13). Kish (14) has cogent criticisms of the indiscriminate use
f binomial variances for other than simple random samples.

9.5 SUGGESTIONS FOR FURTHER RESEARCH

The contents of this chapter have illustrated the application of the-
ry to the problem of evaluating the accuracy of estimates derived
om several sample designs that have been used in the measurement
f opinions and attitudes. Empirical work of this nature has lagged
ar behind similar work in other fields where sampling techniques are
sed. If research workers are going to want to specify sampling ac-
uracy in advance of drawing a sample, or to assess the sampling ac-
uracy of results after the sample has been drawn, then more such
ork needs to be done. The conclusions drawn in dealing with the
stimation of quantities other than proportions (which form the basis
f most of the work in the measurement of opinions and attitudes) do

not necessarily carry over. For example, we cannot ordinarily e
pect to get very substantial gains in accuracy in the estimation of
population proportion through the use of stratification; improv
methods of estimation that help greatly in dealing with agricultur
statistics are not likely to be of much benefit; and if the size of clu
ter is kept small or if the clusters are sampled at a low rate, the var
ances of estimates derived from cluster sampling are not likely to I
much greater than would be obtained for simple random samples. A
these observations are extremely tentative and need to be substant
ated or negated by further empirical research.

REFERENCES

1. Morris H. Hansen and William·N. Hurwitz, "On the Theory of Samplix
   from Finite Populations," *Ann. Math. Statist.*, 20 (1949), 426–432.
2. Emil H. Jebe, "Estimation for Sub-sampling Designs Employing the Coun
   as a Primary Sampling Unit," *J. Am. Statist. Assoc.*, 47 (1952), 49–70.
3. Bernard R. Berelson, Paul F. Lazarsfeld, and William N. McPhee, *Votin*
   University of Chicago Press, Chicago, 1954.
4. Paul F. Lazarsfeld, Bernard R. Berelson, and Hazel Gaudet, *The Peopl
   Choice,* second edition, Columbia University Press, New York, 1948.
5. Philip J. McCarthy, "Sample Design," Chap. 20 in Marie Jahoda, Mortc
   Deutsch, and Stuart W. Cook, *Research Methods in Social Relations,* TI
   Dryden Press, New York, 1951.
6. Philip J. McCarthy, *Sampling: Elementary Principles,* Bulletin 15, N. `
   State School of Industrial and Labor Relations, Ithaca, 1951 (reissued 1956
7. Leslie Kish, "A Procedure for Objective Respondent Selection within tk
   Household," *J. Am. Statist. Assoc.*, 44 (1949), 380–387.
8. Lillian H. Madow, "Systematic Sampling and Its Relation to Other Samplir
   Designs," *J. Am. Statist. Assoc.*, 41 (1946), 204–217.
9. A. J. King and R. J. Jessen, "The Master Sample of Agriculture," *J. Ar
   Statist. Assoc.*, 40 (1945), 38–56.
10. Morris H. Hansen, William N. Hurwitz, and Margaret Gurney, "Problen
    and Methods of the Sample Survey of Business," *J. Am. Statist. Assoc.*, 4
    (1946), 173–189.
11. William G. Cochran, *Sampling Techniques,* John Wiley & Sons, New Yor
    1953.
12. W. Edwards Deming, *Some Theory of Sampling,* John Wiley & Sons, Ne
    York, 1950.
13. W. Edwards Deming, "On Simplifications of Sampling Design through Repl
    cation with Equal Probabilities and without Stages," *J. Am. Statist. Assoc
    51 (1956), 24–53.
14. Leslie Kish, "Confidence Intervals for Clustered Samples," *Am. Sociol. Rev
    22 (1957), 154–165.

HAPTER 10

Sampling Variability
of Quota Sampling Procedures

10.1 INTRODUCTION

The preceding chapter has illustrated the approach that is used
1en we wish to evaluate the sampling variability of estimates derived
om the application of probability model sampling procedures. Al-
ough this approach may lead to theoretically and computationally
fficult problems, it is, nevertheless, a well-defined one. Such a clear
proach does not exist where quota sampling procedures are involved.
his makes a thorough comparison of quota procedures and probabil-
y model procedures difficult or impossible, as has been amply dem-
strated in Chapters 7 and 8. However, more can be done to ana-
ze the performance of quota procedures than the critics admit or
e advocates of quota methods have attempted. It is the purpose of
is chapter to present some tools of analysis with which we can esti-
ate the sampling variability resulting from repeated applications
a quota sampling procedure and to apply these tools in a number
specific situations.

The contents of this chapter are not to be interpreted as an attempt
put the estimation of sampling variability for quota samples on
e same sound theoretical footing as now exists for well-designed and
ecuted probability model samples. Nor will any attempt be made
obtain an evaluation of the systematic error (or bias) which may
ise from a particular quota sampling procedure. The principal de-
rent to these efforts lies in the freedom the interviewers exercise in
eir final selection of individuals for the sample, a freedom that can-
ot be described in terms of precise probability models. However,
nder certain reasonable assumptions, which will be described in the
ext section, it seems clear that the repeated application of a specified
uota sampling procedure will generate an empirical distribution for
n estimate and that we can then attempt to estimate the variance

211

of this distribution from one or more samples drawn according to t
procedure. The systematic error associated with the specified quo
sampling procedure and estimate cannot be obtained in this mann
Its evaluation must depend on the use of check data, as well as on t
use of the sample data, and the difficulties associated with this a
proach have already been described in Chapter 7.

10.2 QUOTA PROCEDURES AND THEIR SAMPLING DISTRIBUTIONS

To talk about the sampling variability of a quota sample, we mu
first justify in some sense or other our assumption of the existen
of a sampling distribution of an estimate made from the application
the procedure. When we are talking about probability model san
ples, the existence of a sampling distribution is derived from the pro
ability model itself, even though empirical experiments are frequent
carried out to demonstrate the applicability of the theoretical resul
In the case of quota sampling, this approach cannot be used since it
not possible, in advance of drawing the sample, to specify the pro
ability that any given sample will be the one chosen. Interviewe
are free to select any persons they please as respondents, just so lo
as the persons selected conform to preassigned quotas. Consequentl
we are forced to look at quota sampling almost entirely on an empir
cal basis.

The empirical justification of the existence of a sampling distrib
tion for a specified quota procedure would proceed as follows. A san
ple is actually selected according to the defined procedure, the mea
urement process is applied, and an estimate of a population paramet
is prepared. The value of the estimate is noted, the sample is r
turned to the population, and the entire process is repeated. It is a
sumed that the population, the sampling method, the measureme
procedure, and the estimation process remain unchanged from samp
to sample. Variation from sample to sample, and this might possib
be zero, would then arise primarily from the fact that interviewe
sometimes select different individuals from sample to sample, thoug
other sources of variability might be incorporated into the sample d
sign, such as varying interviewers from sample to sample or varyin
sampling points from sample to sample. On these repeated applica
tions we should not allow such items as instructions to interviewer
definition of quota categories, size of the quotas, and the like to var
Changes in these and similar factors would constitute a change i
the sample design.

As this outlined repetitive experiment unfolds, the following type record is kept. Each time a sample is drawn, the value of the estimate is entered in a table. Ultimately this table will contain every value of the estimate that it is possible to obtain. In addition, we recapitulate at frequent intervals the results of the past sampling experience by computing the proportion of times that each possible value of the estimate arises. Alternatively, we may compute the proportion of times that an estimate of less than or equal to a stated value was obtained. We now say that a sampling distribution exists under this procedure and for this estimate if the following condition holds: after many samples have been drawn, the proportion of times any given value of the estimate occurs (or alternatively, the proportion of times that a value less than or equal to the specified value occurs) is nearly equal to a constant, say p, and this proportion is usually nearer to p when more samples are drawn. The collection of all possible values of the estimate and their corresponding long-run proportionate occurrences is then called the empirical sampling distribution of the estimate. Sampling models that are consistent with this evidence of the performance of the method can then be used deductively.

As previously noted, the variability in repeated application of a defined quota sampling procedure results primarily from the net effect of the variability of each interviewer. There may also be some effect from changing interviewers and sampling points from sample to sample. It has long been observed in empirical studies covering many fields of investigation that such net effects will actually give rise to a sampling distribution in the sense already defined. If the net effect did not conform at least approximately to some sampling distribution, each interviewer would have to follow a very complicated systematic pattern in selecting his respondents from sample to sample. Provided the repetitive drawings did not extend over a long period of time, it is not reasonable to expect that this would take place, and so we shall feel free to assume that the repeated application of a quota sampling procedure will generate a sampling distribution for estimates prepared from the samples. Note that this assumption tells us nothing about the variance or expected value of this distribution, and for certain types of distribution they would not even exist. This latter possibility will be ignored in the subsequent work.

10.3 METHODS OF ESTIMATING SAMPLING VARIABILITY

A. Repeated Application of the Procedure

Since sampling variability is by definition the amount of variabilit
that arises through repeated application of a given sampling pro
cedure, the most straightforward way to estimate it is to use situation
where the same variable has been measured on a number of successiv
surveys. Suppose that a given quota sampling procedure is repeate
k times and that a stated population proportion or percentage is est
mated on each repetition. It will be assumed that these repetition
occur under essentially similar circumstances. In general this mean
that the population, the measuring instrument, the instructions to th
interviewers (including the definition of the quota categories and th
quotas assigned to each), the training of the interviewers, the size c
the sample, and the like do not change from repetition to repetition
Whether or not we wish to allow the sampling points and the inter
viewers to vary from repetition to repetition will depend on the prac
tices of the organization whose work is being studied. It is only nec
essary to observe that the estimated sampling variability will contai
a component for each factor allowed to change.

From each of the k repetitions, we obtain a value of the estimat
Let the individual values of these estimates be $p_1, p_2, \ldots, p_k$ and l
their average value be $\bar{p}$. We assume that these k values constitute
simple random sample from the population of all possible values of th
estimate, a population which is described by the sampling distributio
described in the previous section. Under these circumstances, the var
ance of this sampling distribution can then be estimated by

$$\hat{V}(p_i) = \frac{\Sigma(p_i - \bar{p})^2}{k - 1}$$

Whenever we are dealing with a problem that involves the estima
tion of a proportion, it is natural to inquire how the sampling vari
ability compares with what would be obtained by simple random sam
pling. This interest has already been exemplified by the work of th
preceding chapter in which the simple random sampling variance wa
used as a yardstick to compare various probability model samplin
procedures. Such a comparison is of particular interest here since i
may give some indication of the behavior of interviewers as selectin
devices under a quota procedure. For example,

1. Suppose each interviewer has a fixed panel of respondents whom
he interviews each time an assignment is sent from the home offic

If now a series of repetitive surveys is examined on which the same interviewers have been used, there will be no variability in the persons interviewed from survey to survey. However, if a small proportion of the interviewers are changed from survey to survey, the estimate of variance will not necessarily be zero, but it will be small in comparison with that predicted on the basis of simple random sampling.

2. On the other hand, we might be randomly using two sets of interviewers, one set of which is biased toward selecting, for example, Republicans, and the other set is biased toward selecting Democrats. A series of repetitive surveys might now give, for a political question, an estimate of variance greater than what would be predicted on the basis of simple random sampling.

In order to carry out this comparison between the variance of a quota sample estimate and the variance of a simple random sample of the same size, it is necessary to make use of the normal approximation to the binomial distribution. That is, if repeated random samples of size n are drawn from a population for the purpose of estimating a proportion, p, the sample estimate of this proportion will be approximately normally distributed with variance $p(1 - p)/n$, assuming that n is "large." From this follows the result:

If k quota samples, each of size n, are drawn from a population, if the expected value of the quota estimate is p (not necessarily the true value in the population since the quota estimate may be biased), and if the distribution of the quota estimate is the same as that of a simple random sample estimate (the random sample, of size n, being drawn from a population in which the true value of the proportion is p), the quantity

$$\frac{n\Sigma(p_i - \bar{p})^2}{p(1 - p)} \qquad (k)$$

will be distributed approximately as χ^2 with $(k - 1)$ degrees of freedom. $p_1, p_2, \ldots, p_k$ and $\bar{p}$ are as previously defined.

Since the value of p is unknown, this χ^2 expression must be approximated by

$$\frac{n\Sigma(p_i - \bar{p})^2}{\bar{p}(1 - \bar{p})} \qquad (k)$$

Excessively large values of this quantity indicate quota variability larger than simple random sampling variability, and excessively small values indicate quota variability smaller than simple random sampling variability.

An example will serve to illustrate the foregoing procedure. During the period between September 1945 and July 1949, the National Opinion Research Center conducted a series of 29 comparable national surveys. All the samples were drawn by the same quota procedure, and each was approximately of the same size, the average sample size being 1280. Since this series extended over such a long period of time, there was undoubtedly considerable variability in the interviewing staff, at least as between the early surveys and the later surveys. The proportion examined in each of these surveys was the proportion of respondents who had graduated from high school (with no further schooling) or who had had some high school education. There was no direct quota control for education, but, of course, some indirect control was exercised through the fixing of sex, age, and economic status quotas. These data are given in Table 10.1.

TABLE 10.1

DATA FROM TEMPORALLY ORDERED NORC REPETITIVE NATIONAL SURVEYS
SEPTEMBER 1945 TO JULY 1949

(Proportion of respondents with high school or some high school education)

Survey	Estimated Proportion	Survey	Estimated Proportion	Survey	Estimated Proportion
1	0.38	11	0.46	21	0.44
2	0.41	12	0.44	22	0.44
3	0.42	13	0.42	23	0.45
4	0.42	14	0.42	24	0.47
5	0.39	15	0.46	25	0.47
6	0.42	16	0.44	26	0.44
7	0.41	17	0.45	27	0.45
8	0.43	18	0.45	28	0.44
9	0.46	19	0.46	29	0.43
10	0.43	20	0.43		

The previously described analysis was performed on these data with the following results:

$$\bar{p} = 0.436$$

$$\bar{p}(1 - \bar{p}) = 0.2459$$

$$\hat{V}(p_i) = 0.000483$$

$$\chi^2 = n(k - 1)\hat{V}(p_i)/\bar{p}(1 - \bar{p}) = 70.4$$

$$\Pr(\chi^2 \geq 70.4) < 0.001$$

It will be observed that the value of chi-square is highly significant, the probability being less than one in a thousand of obtaining a value this large or larger by chance, under the stated hypothesis. The variance estimated from the repetitions is 2.52 times as large as the simple random sampling variance.

Although our first inclination might be to regard this highly significant value of chi-square as an indication of greater than simple random sampling variability for these 29 surveys, there is another factor that may be increasing the observed variability. During the period from 1945 through 1949 the proportion of individuals in the higher educational categories was increasing, in large part because of the use of GI Benefits by veterans of World War II. This essentially means that the population characteristics with respect to education were changing in the period under study, and this was not taken into account in the previous analysis. Accordingly, a least-squares trend line was fitted to the data of Table 10.1 with the following results:

$$
\begin{array}{ll}
\text{Estimated variance about} & \\
\quad \text{the trend line} & = 0.000310 \\
\chi^2 \ (27 \ df) & = 43.57 \\
\Pr \ (\chi^2 \geq 43.57) & = 0.023
\end{array}
$$

The removal of trend reduced the variation in the observed results, but the survey variability is still significantly higher than the predicted binomial variability at the 5 per cent level of significance. The variance about the trend line is 1.61 times as large as the simple random sampling variance.

The analysis illustrated by this example has been applied in a number of other situations, and the results are summarized in Tables 10.2, 10.3, and 10.4. Before discussing the specific contents of these tables, a few general comments should be made concerning the surveys and variables to which the tables refer. These are:

1. The variables used in the Benson and Benson and in the Roper surveys were not quota control variables, and there is no particular reason for expecting them to be any more or less highly correlated with the quota control variables than any other subjects that we might investigate with a quota sample. They all relate to the personal characteristics of the respondents.

2. As noted in the footnotes to Table 10.2, the variables used in NORC 1 represent subgroups of the population defined by the cross classification of sex with other variables. Since sex is a quota con-

trol variable for these surveys, there might be some tendency for the random variation to be reduced, but this effect should be slight. The same remarks apply to NORC 2.

3. The chi-square values shown for any one group of surveys are not completely independent since they were derived from the same surveys and since chi-squares were computed from each subcategory of a specific variable (e.g., religion: Protestant, Catholic, etc.).

4. National Opinion Research Center 1 and 2 are not independent since they are derived from the same six surveys.

TABLE 10.2

SURVEYS AND VARIABLES USED FOR COMPARING VARIANCES ESTIMATED FROM REPETITIONS OF NATIONAL QUOTA SAMPLING PROCEDURES WITH VALUES ESTIMATED FROM BINOMIAL THEORY

Survey Organization	Number of Surveys	Number of Comparisons
B and B (Bikini surveys, see Chapter 8)	4	17 [a]
Roper [1948 election surveys, see (1)]	6	8 [b]
NORC 1	6	34 [c]
2	6	67 [d]

[a] The analyses were carried out for the various subcategories of education, occupation, religion, union membership, and veteran status.

[b] The analyses were carried out for the various subcategories of education, union membership, and veteran status.

[c] The analyses were carried out for the various subcategories of the two-way classification of sex by education, sex by occupation, and sex by religion.

[d] The analyses were carried out for the various subcategories of the three-way classifications of sex by education by age, sex by education by economic level, and sex by education by religion. No chi-squares were computed where the percentage falling in a cell was less than 3 per cent of the total sample.

There are a number of facts which are more or less obvious from an examination of these three tables. The quartered distributions of Table 10.3 seem to indicate that the chi-square values are fairly well spread over their possible range as predicted by the theoretical distribution, though there appears to be some tendency for the Benson and Benson and Roper surveys to be concentrated in the upper quarters. These observations are substantiated by the summary data given in Table 10.4 where the combined chi-squares for the Benson and Benson and Roper surveys are significant at the 1 per cent level. However, it now appears that the NORC 2 comparisons show the observed variation to be somewhat smaller than we would expect on

TABLE 10.3

QUARTERED DISTRIBUTION [a] OF CHI-SQUARE VALUES OBTAINED FROM COMPARISONS

Frequencies Falling in Quarters
of Chi-Square Distribution

Survey	First	Second	Third	Fourth	Total
B and B	4	1	3	9	17
Roper	0	2	2	4	8
NORC 1	6	10	5	13	34
2	17	22	17	11	67
Total	27	35	27	37	126

[a] From a table of the chi-square distribution, for each stated number of degrees of freedom determine three values, $\chi^2(0.75)$, $\chi^2(0.50)$, and $\chi^2(0.25)$ such that the probability is 0.25 of obtaining a value of chi-square less than the first, is 0.25 of obtaining one between the first and second, is 0.25 of obtaining one greater than the third. These three values determine the quarters of the distribution and were used in preparing this table.

the basis of simple random sampling theory. It might be noted that one further set of analyses was made for the NORC surveys in which cross classifications of quota control variables were used (sex by age, sex by city size, and sex by economic level). The chi-square values obtained in this instance were markedly concentrated in the lower quarter of the distribution, as we would expect for controlled or partially controlled variables.

The conclusions that seem to be warranted from the discussion and data of this subsection can be stated as follows. If we are provided with a series of quota samples that can be regarded as independent

TABLE 10.4

SIGNIFICANCE OF VALUES

Number Significant at

Survey	Number of Comparisons	Upper 0.05 Level	Upper 0.01 Level	Lower 0.05 Level	Lower 0.01 Level	Combined Chi-Square: Probability of Obtaining One as Large or Larger than Observed
B and B	17	4	1	1	1	0.01
Roper	8	2	1	0	0	0.01
NORC 1	34	1	0	1	0	0.32
2	67	0	0	3	1	0.93

repetitions of a well-defined procedure under similar conditions, the
it is possible to estimate the sampling variability of the sample result
This process gives absolutely no information concerning the system
atic error or bias of the procedure and estimate. Moreover, it is nec
essary to examine the surveys carefully to see that no extraneous fac
tors are contributing to the observed variability, factors such a
changing population characteristics. As a result of applying thi
technique, it has been possible to make some comparisons betwee
empirical evaluations of this quota sampling variability and the vari
ability that we would expect on the basis of simple random sampling
These comparisons have certainly neither proved nor disproved tha
the sampling variability of the examined quota samples follows tha
of the binomial model. Nevertheless, the results do suggest that th
binomial model may be used with a suitable multiplier to obtain use
ful approximations to the true sampling variability after the perform
ance of a particular survey has been analyzed and while there ar
no changes that might disturb the stability of the survey operation.

B. An Alternative Use of Repeated Application of a Sampling Procedure

The form of analysis used in the preceding section requires a rea
sonably large number of repetitive applications of a sampling metho
in order to provide an estimate of sampling variability which is itsel
not subject to large sampling errors. Moreover, the larger the numbe
of repetitions that we use, the longer will be the time span betwee
the first and last of the series and the greater will be the chance tha
the estimate will contain the effects of such factors as changes in th
population. It may sometimes happen that a small number of repe
titions (two, for example) can be analyzed to increase the accuracy o
the estimate of sampling variability. This will also decrease th
chance that time changes are entering into the estimate of variability
The procedure being suggested can perhaps best be described in term
of an actual sample.

During World War II, the Office of Public Opinion Research o
Princeton University conducted a number of national surveys based
on a quota sampling procedure. Each survey asked questions of approximately 1200 respondents. For purposes of this analysis, a number of opinion questions were chosen, questions that had been repeated on two successive surveys and that did not appear to have
been affected by trends or by the news. One such question was:
"After this war is over, do you think every young man should be required to serve one year in the Army or Navy?" The number of re-

spondents and the proportion who replied "Yes" to this question, for each of eight geographic regions and for each of two successive surveys, is given in Table 10.5.

TABLE 10.5

PROPORTION OF RESPONDENTS REPLYING "YES" TO QUESTION ON POSTWAR SERVICE IN THE ARMED FORCES

Geographic Region	Survey 1		Survey 2	
	Number of Respondents	Proportion	Number of Respondents	Proportion
1	82	0.793	74	0.770
2	307	0.739	308	0.679
3	251	0.717	226	0.633
4	164	0.598	156	0.583
5	156	0.673	118	0.644
6	44	0.614	44	0.659
7	94	0.596	94	0.649
8	114	0.737	180	0.667
Total	1212	0.695	1200	0.650

The variance estimate of the preceding subsection can be used in this instance, but it is based on only one degree of freedom. This estimate is:

$$(0.695 - 0.6725)^2 + (0.650 - 0.6725)^2 = 0.001012$$

An alternative procedure for the estimation of variance is the following. Let n_{1j} and p_{1j} be the number and proportion of respondents on the first survey in the jth region and similarly for n_{2j} and p_{2j}. Then the estimate of the population proportion from the first survey is

$$p_{1.} = \frac{n_{11}p_{11} + n_{12}p_{12} + \cdots + n_{18}p_{18}}{n_{1.}}$$

where $n_{1.}$ is the sum of $n_{11}, n_{12}, \ldots, n_{18}$. The variance of this estimate is given by

$$\frac{n_{11}^2 V(p_{11}) + n_{12}^2 V(p_{12}) + \cdots + n_{18}^2 V(p_{18})}{n_{1.}^2},$$

where $V(p_{11})$ is the variance of the estimate for the first geographic region, and similarly for the other seven regions. The variance of the estimate for the jth geographic region, $V(p_{1j})$, can now be estimated with one degree of freedom by

$$\tfrac{1}{2}(p_{1j} - p_{2j})^2 = [p_{1j} - \tfrac{1}{2}(p_{1j} + p_{2j})]^2 + [p_{2j} - \tfrac{1}{2}(p_{1j} + p_{2j})]^2$$

Consequently, we take as our estimate of the variance of the survey estimate

$$\frac{n_1{}^2(p_{11} - p_{21})^2 + \cdots + n_8{}^2(p_{18} - p_{28})^2}{\tfrac{1}{2}(n_1{}. + n_2{}.)^2},$$

where $n_j = \tfrac{1}{2}(n_{1j} + n_{2j})$. The average sample sizes were used in this final form because of the small discrepancies in the corresponding numbers between the two surveys. For the data of Table 10.5 this gives

$$p_1{}. = 0.695$$

$$p_2{}. = 0.650$$

$$\hat{V}(p_1{}.) = 0.000314$$

This same analysis can also be carried out by using comparisons within those interviewers who worked on both surveys instead of the comparisons within geographic regions.

This outlined procedure, using both the within-geographic-region and the within-interviewer comparisons, has been applied to the series of OPOR surveys. The identification number and the dates of these surveys are:

Survey Number	Date
21	8 January 1944
26	20 April 1944
27	20 June 1944 (delayed for D-Day)
28	7 June 1944
33	5 October 1944
34	26 October 1944
37	3 January 1945
38	1 February 1945
39	22 February 1945

The questions and surveys used in these analyses are set 'forth in Table 10.6, and the results of the analyses are given in Table 10.7.

Any attempt at interpretation of the results appearing in this table is extremely interesting because it forces the interpreter into formulating some fairly definite ideas concerning what is meant by the repetitive application of a quota sampling procedure. The simplest approach to this problem is through the within-interviewer comparison. In using this estimate of variance, we are assuming that everything remains unchanged from sample to sample with the single exception of the respondents selected by the interviewers. Therefore

TABLE 10.6

QUESTIONS, RESPONSE CATEGORIES, AND SURVEYS USED IN VARIANCE ANALYSES

Code Number	Question	Response Category	Survey Numbers
1	After this war is over, do you think every young man should be required to serve one year in the Army or Navy?	Yes	26, 27
2	Same as 1	No opinion	26, 27
3	Which country is the greatest military threat to the U.S., Germany or Japan?	Japan	26, 27
4	Do you believe in free speech to the extent of allowing radicals to hold meetings and express their views in this community?	Yes	26, 38
5	Would you like to see the United States join a league of nations after this war is over?	Yes	26, 27
6	Do you have a close relative in the armed forces at the present time?	No	26, 27
7	Same as 6	Yes	26, 27
8	In the war with Germany, do you feel that our chief enemy is the German people as a whole or the German government?	Government	26, 27
9	After the war, would you be willing to see Russia keep some Polish territory which the Russians now occupy, provided the Poles are given in exchange some German Territory?	No	37, 38
10	Do you think Russia can be trusted to cooperate with us when the war is over?	Yes	27, 28
11	Same as 10	No	33, 34
12	Same as 10	Yes	21, 39

the values given under the heading "interviewers" are measures of *within-interviewer variability* as regards selection of respondents. Actually, there may be a component due to population change included in these values since the two surveys used in any one computation were separated by a certain period of time. Also there were

TABLE 10.7

ESTIMATES OF VARIANCE FOR A NATIONAL QUOTA SAMPLE OBTAINED BY USING
COMPARISONS WITHIN GEOGRAPHIC REGIONS AND WITHIN INTERVIEWERS

(All variances are multiplied by 1,000,000)

			Estimated Variance from		
Code Number	Estimate [a] of Population Proportion	Binomial Model	Total Survey Results	Geographic Regions	Inter- viewers
1	0.6748	182	1012	314	337
2	0.0854	65	123	73	81
3	0.5402	205	25	365	368
4	0.4169	197	359	212	369
5	0.6889	177	90	424	332
6	0.2098	137	1	171	459
7	0.4512	204	167	223	241
8	0.7358	160	104	363	278
9	0.2354	142	105	89	292
10	0.5432	194	696	287	271
11	0.3453	191	15	131	213
12	0.5115	213	6	157	345
Average		172	224	234	300

[a] Average of two survey results.

some few instances in which interviewers differed and matched assignments had to be used.

Now consider the estimates given under the "geographic regions" heading. Provided there are no changes in the sampling points or in the interviewers for the two surveys used in computing an estimate of variance and there is no correlation between interviewers with respect to their variability from sample to sample, this procedure should give the same long-run results as does the within-interviewer procedure. However, if there are changes in sampling points or interviewers between the two surveys, then this estimate will include a component for each, as well as a component for population change. In the surveys used in preparing Table 10.7, the changes in sampling points and interviewers were of a very minor nature and consequently we should expect the two estimates to be comparable. Actually, an examination of this table shows the within-interviewer estimate to be larger than the within-geographic-region estimate in nine of the twelve instances. Although this result is not significantly different from what we would expect to find on a chance basis if each estimate tended to be larger in one-half of all comparisons, it does at least provoke some thought concerning whether or not there exists the possibility that the within-interviewer estimate is consistently larger than the within-geographic-region estimate. The appropriate amount of correlation between interviewers with respect to their variability from sample to sample could produce such a situation, but such a correlation seems extremely unlikely to arise since quota interviewers work in relative isolation. The only other source of such a result would seem to be differential changes in the population between the two surveys. No data were available in this instance to investigate this possibility.

Somewhat similar remarks can be made about the total survey comparisons, but since these estimates are based on only a single degree of freedom and since they are so obviously unstable, no further attention will be devoted to them.

As a final step in analyzing this table, we may compare the estimated variances with those predicted on the basis of the binomial model. This comparison was made in the following fashion. For each of the twelve items, the average of the within-interviewer and within-geographic-region estimates was divided by the binomial model estimate. This resulted in twelve ratios ranging in value from 0.90 to 2.30, the average value being 1.56. Thus the binomial model variance seems to underestimate the variance obtained from repetitions of the sampling procedure. The amount of underestimation is not great, however, especially if we are willing to assume that there was some change in the population between surveys.

Data similar to the foregoing have been reported by Moser and Stuart (2), except that their measures of sampling variability for quota samples are based entirely on between-interviewer comparisons. In their experimental study of quota sampling, they used twenty-four basic cells formed from the various combinations of four survey organizations, three towns, and two control levels. Within each cell there were two replicate samples of the same size and quota composition, selected and interviewed by two different interviewers. The variances for the quota samples were derived from the difference in observed proportions between pairs of interviewers working under similar experimental conditions. These variances were then compared with the variances of random samples drawn in the same towns. Table 36 of the Moser and Stuart article gives, for each of a number of questions, the ratio of the quota variance to the random variance. These ratios vary from 0.72 to 5.76, their average value being 2.22.

The average and variability of these ratios is considerably larger than for the ratios we have just obtained from the within-interviewer and within-geographic-region comparisons. The choice between these two approaches depends on the manner in which we wish to view the repetitive application of a quota sampling procedure. Further discussion of this point will be given in the next subsection.

C. Estimation of Variability from a Single Sample

Section 9.4 described a procedure whereby the repetitive application of a sampling method may be approximated by dividing up the total sample into a number of subsamples in such a manner that *each subsample is similar to the total sample*. From these subsamples estimates are then prepared just as would be done for the total sample. An estimate of variance is computed and adjusted for differences in size between the subsamples and the total sample. This same approach can be used for quota samples, though substantial difficulty may be encountered in defining exactly how the subsamples should be obtained and in applying this definition to a practical sample design.

Before we discuss the details of this situation, it will be advisable to make a few general remarks about the way in which the subsamples should be similar to the total sample. In the first place, the requirement of similarity does not mean that simple random samples are chosen from the total sample, as has so frequently been done in the past [see, for example, Link (3)]. All this type of subsampling accomplishes is to illustrate that the deductions drawn from the probability model for simple random sampling do hold in practice, the total sample being the population from which the random samples are drawn. It tells nothing about the particular sampling method under investigation. Instead, the total sample should be divided in such a manner that each subsample could have arisen as an independent sample from an application of the sampling procedure, all features remaining unchanged from the actual application except the matter of size. Thus the subsamples will include the same modes of stratification, the same procedures for selecting sample points, the same manner of selecting interviewers within sampling points, and the same procedure for assigning quotas to the interviewers.

Actually, the foregoing general remarks are not quite so simple to translate into practice as they may at first appear. The principal difficulty revolves around the question, "What sources of variation should be allowed to influence the subsamples?" A slightly hypothetical example will illustrate this difficulty to some extent. Since the interviewer is ordinarily regarded as the focal point in a quota sample design, he may also become the focal point for the division into subsamples. Assume, for example, that five subsamples are to be used in the estimation of sampling variability; that the quota categories are non-overlapping; and that the assignments to *each interviewer* for *each quota category* are in multiples of five. The term non-overlapping categories simply means that any one respondent falls into one and only one of these categories. Thus, if sex and age (under forty and over forty) quotas are used, the quotas will be given for the four categories male under forty, male over forty, female under forty, and female over forty rather than for the four categories male, female, under forty, and over forty. We can now go through the interviews obtained and, *within a quota category* and *within an interviewer*, assign one-fifth of the interviews at random to each of the five samples. From each of the five samples so defined, an estimate is computed. Under these circumstances, this estimate of variability will be a measure of *within-interviewer variability* only, and will, of course, be averaged over all interviewers used in the survey. Components due to variability between interviewers, between sampling points, and the like will not ordinarily affect this estimate. It is therefore similar to the preceding subsection estimate which was based on the within-interviewer comparisons.

Other methods of subdividing the total sample may also be considered, methods that will incorporate other sources of variability into the estimate. For example, let us suppose that within any given sampling point the number of interviewers used is a multiple of five and the subsamples are defined in the following manner. The interviewers within a sampling point are randomly and equally assigned to the five subsamples, and all interviews obtained by a single interviewer go into the same subsample. From each of the five samples so defined an estimate is prepared, and the variance between these five estimates is computed. This estimate contains not only a component for within-interviewer variability but also a component for between-interviewer variability. If desirable, we could even go one step farther and design the subsamples so that the estimate of variance would be influenced by the variability between sampling points, in addition to the within- and between-interviewer differences. As previously noted, the Moser and Stuart experiment (2) obtained estimates based only on between-interviewer variability.

A choice between these three methods of subdividing a quota sample must depend on the manner in which we wish to view the repetitive application of the sampling method, or alternatively, on the model being used. This choice in turn can be forced back to the point where we are deciding how to divide the total error of a quota sample estimate between bias and sampling variability. Many conflicting points of view could be stated and argued in this respect, but here we shall only state the preference of the Study of Sampling, this preference being as follows. A quota sampling procedure is defined by specifying

1. The set of fixed sampling points from which interviews are to be obtained. These sampling points are ordinarily obtained by making use of such stratification factors as geographic region of the country, city size, type of industry, type of farming activity, and the like.

2. A set of fixed interviewers located within the sampling points.

3. A set of fixed quotas assigned to the interviewers.

4. A standard set of instructions and training procedures for the interviewers that will tend to make somewhat uniform their working methods and objectives.

With these restrictions, variability from sample to sample arises only from the variability which an interviewer initiates in choosing his different sets of respondents. This variability can be measured from two or more successive surveys by the within-interviewer analysis used in the preceding subsection; or this variability can be measured from one survey by the first method, described in this subsection, of subdividing the total sample. In addition to this sampling variability, the total error of the survey will also contain a bias component due to the particular sampling points, quota assignments, and interviewers. The amount of this bias cannot be estimated from survey results alone but must be determined by using survey results in conjunction with known population parameters.

It should be noted that it will not be possible to take every quota sample and apply the subsample method of estimating variability. Many practical difficulties would be encountered. In particular, quotas would probably not be assigned to non-overlapping categories, and they would not ordinarily be multiples of the number of subsamples desired. These and other difficulties cannot be met very satisfactorily after the sample has been designed and selected. However, if these points are kept in mind when designing the sample, most of them can be avoided. In other words, if we wish to obtain a measure of sampling variance to attach to a quota sample estimate, it is necessary to consider this in design just as for probability model samples. As a matter of fact, too little attention has been given to this very point by the designers of probability model samples. It is of little value to have a theoretical expression for the variance of a sampling distribution if it is too costly to evaluate the expression after the sample has been drawn. Recently, very effective methods have been developed by Deming, Hartley, Keyfitz, Kish, and others.

10.4 COMPARISONS BETWEEN DIFFERENT QUOTA SAMPLING PROCEDURES

The preceding section has presented variance estimates for quota samples and has shown that, for the surveys examined, these variances can be reasonably well approximated by the familiar binomial model. With this result as a base, it is of interest to examine situations where different quota samples have been used to measure the same variable.

One set of data which can be used for this purpose has been provided by Cantril (4). The Office of Public Opinion Research at Princeton University collected a number of examples where two different polling organizations had asked the same opinion question within a relatively short time interval (the greatest interval being seventy days). The organizations contributing to these comparisons were OPOR, the American Institute of Public Opinion (AIPO), the Fortune Poll, and NORC. It should be noted that data from each of these organizations were used in the preceding section, the Fortune Poll having been conducted by Roper, and AIPO and Benson and Benson having used the same sample design.

Twenty different question comparisons were listed by Cantril, giving a total of 99 individual comparisons (each question having several response categories). Table 10.8 shows the distribution of differences. From these data Cantril concludes that the difference between the different polls is not unduly high, the average difference being 0.033. This conclusion is undoubtedly warranted if we consider how much effect a difference of this magnitude would have on the interpretation of the average opinion question. However, we may also ask whether or not such a distribution of differences is within the range that we would expect on the basis of sampling variability.

It is not difficult to see from even a crude examination of Table 10.8 that the observed differences are greater than we would expect on the basis of sampling variability as measured by the binomial model, where the reasonableness of this model has already been demonstrated for data taken from each of the organizations individually. For example, let the two estimates entering into a single comparison be $\hat{p}_1$ and $\hat{p}_2$, based on sample sizes of n_1 and n_2 respectively. Since $\hat{p}_1$ and $\hat{p}_2$ are

TABLE 10.8[a]

DISTRIBUTION OF RESULTS

Difference	Number of Occurrences
0.00	8
0.01	17
0.02	19
0.03	18
0.04	10
0.05	13
0.06	3
0.07	3
0.08	2
0.09	4
0.10	1
0.11	0
0.12	1

[a] Table 3 from Cantril (4).

assumed to follow the binomial model, we have the result that, for large n_1 and n_2, the difference $(\hat{p}_1 - \hat{p}_2)$ is approximately normally distributed with mean $(p_1 - p_2)$ and variance $(p_1 q_1/n_1) + (p_2 q_2/n_2)$ where p_1 is the expected value of $\hat{p}_1$ and p_2 is the expected value of $\hat{p}_2$. Consequently, if the expected value of $(\hat{p}_1 - \hat{p}_2)$ is equal to zero, we have the result that

$$\frac{(\hat{p}_1 - \hat{p}_2)^2}{(p_1 q_1/n_1) + (p_2 q_2/n_2)}$$

is distributed approximately as chi-square with one degree of freedom. In order to make a quick test with these data, we compute the average squared difference from Table 10.8 and use it in the numerator of this expression. The largest possible value of the denominator will occur when $p_1 = p_2 = 0.50$ and when n_1 and n_2 are taken to be the smallest pair of sample sizes entering into the comparisons. The average squared difference from Table 10.8 is equal to 0.001708 and the largest possible value of the denominator is equal to

$$(0.50)^2 \left(\frac{1}{1200} + \frac{1}{2500} \right) = 0.000308$$

The ratio of the two, $(0.001708)/(0.000308) = 5.55$, should approximate chi-square with one degree of freedom. The probability of ob-

ining a chi-square value this large or larger by chance is 0.019, as-
ming that $p_1 - p_2 = 0$.

Since this test is extremely conservative, it seems likely that Table
1.8 clearly indicates differences that are larger than we would ex-
ct on the basis of sampling variability alone. This observation
as borne out by a more detailed examination of the comparisons
ading to this table. Each individual comparison was examined in
is fashion, the appropriate values of p_1, p_2, n_1, and n_2 being used
each instance. Of 67 chi-square values thus computed, 26 were
yond the 0.001 value for chi-square. There can be little doubt in
is instance that the observed differences far exceed those that would
e expected from sampling variability according to the binomial model.
his result could be due to one or more of the following four factors.

1. The true sampling variability for each sample is greater than
at predicted by the binomial model.

2. The different sampling procedures involved in the comparisons
e subject to differing amounts of systematic error.

3. The measurement procedures produced different results. This
ems unlikely in view of the fact that identical questions were asked
nd that there was no reason for presuming the interviewing staffs to
e substantially different.

4. The population distribution was changing with time, thus adding
systematic component to the observed differences.

is impossible to state the relative amounts which each of these con-
ibutes to the observed differences, there being no way to separate out
eir effects. However, in view of the previous discussions, the best
dgment would seem to be that factor 2 is probably the primary con-
ibutor, with smaller amounts coming from factors 1, 3, and 4.

Assuming that the major contribution to the observed discrepancy
omes from factor 2, we may make a rough estimate of this systematic
ifference. Taking the rough computations for the data of Table 10.8,
e have

$$\text{Average squared difference} = \text{variance of difference}$$
$$+ (\text{systematic difference})^2$$
$$0.001708 = 0.000308 + (\text{systematic difference})^2$$

herefore, an estimate of the systematic difference is 0.037 or almost
percentage points on the average.

The state-by-state predictions of the popular vote in the 1948 presi-
ential elections were a second situation to which this type of analysis

has been applied. In Appendices *C* and *D* of *The Pre-election Po|| of 1948* (1) will be found summaries of the state survey data c which were based the final election reports of the Crossley Poll an AIPO. These data do not lend themselves to the present analys| quite as well as the preceding data because the reported figures refe in many instances, to a combination of surveys taken over an interv| of time rather than to a single survey and because of differences i 'questions, methods of determining non-voters, and other similar fa| tors. The results of the analysis are quite interesting, however, i comparison with those already given for the opinion questions.

In adapting the data, the following steps were taken. Only thos| states were used in which there was not a substantial Thurmond vot| No mail ballot results were used, and the raw survey results wer employed in the analysis (i.e., the results before adjustment). Th| effective sample size was regarded as the total sample less those ir dividuals who were classified as non-voters. The computations wer| carried out in terms of the two categories, Truman and Other. Wit| these rules of procedure, each of 37 states provided a two-by-two tab| for which a value of chi-square was computed. The quartered dis| tribution of these values is given in Table 10.9.

TABLE 10.9

QUARTERED DISTRIBUTION [a] OF χ^2 VALUES OBTAINED FROM STATE-BY-STAT| COMPARISONS OF CROSSLEY AND GALLUP, 1948 ELECTION DATA

Quartered	Number of Occurrences
First	17
Second	8
Third	5
Fourth	7
Total	37

[a] See Table 10.3 for description of "quartered distribution."

Table 10.9 shows much better agreement between the Crossley an| Gallup results than we would expect on the basis of chance, almos| half the chi-square values falling in the first quartile of the theoreti| cal distribution. There was only one of these 37 values that was sig| nificant at the 5 per cent level of significance. This high level o| agreement between the two sets of data is even more striking whe| we consider that procedural differences would tend to increase th| observed variability over that expected on the basis of samplin|

one. This result may be due to any number of causes, or it may
due to some peculiarity of the analysis, but it stands in sharp con-
ast to that obtained from the previous set of data, in which there was
strong indication that factors other than sampling were contributing
the observed differences.

10.5 SUMMARY AND CONCLUSIONS

Under the assumption that the repeated application of a specific
uota sampling method will produce a sampling distribution, in this
apter various methods for estimating the variance of these distri-
itions have been presented and applied; we have also discussed
hat is meant by the repetitive application of a quota procedure.
he methods used were:

1. A straightforward estimate of variance based on two or more
plications of the method.
2. An estimate of variance from two or more applications of the
ethod, based on the stratification features of the designs.
3. An estimation procedure based on designing a single sample as
number of independent subsamples.

lthough the third procedure may appear the most appealing for
actical application, it actually requires great care in the formulation
the design and, consequently, the second procedure will probably
found the most useful.

In general, the applications of these methods have produced results
at can be approximated in a rough fashion by the variance of the
inomial model for random sampling when it is multiplied by a suit-
ble factor of the order of 1.5. This result is, of course, closely linked
the particular sets of data used in these analyses and to the manner
which we have viewed the repetitive nature of a quota sampling
rocedure. Further extensions must be made with great care and on
e basis of analyses of the particular survey operations to which
ey are to be applied.

The contents of this chapter, by themselves, tell us nothing about
e systematic error associated with a quota sampling procedure.
for do they imply that all quota procedures have the same amount
f systematic error. As a matter of fact, some evidence is presented
show that the quota samples used by different organizations have
iffering amounts of systematic error.

REFERENCES

1. Frederick Mosteller, Herbert Hyman, Philip J. McCarthy, Eli S. Marks, a David B. Truman, *The Pre-election Polls of 1948,* Bulletin 60, Social Scie Research Council, New York, 1949.
2. C. A. Moser and A. Stuart, "An Experimental Study of Quota Samplin *J. Roy. Statist. Soc.,* 116 (1953), 349–405.
3. Henry C. Link, "How Many Interviews are Necessary for Results of a C tain Accuracy," *J. Appl. Psychol.,* 21 (1937), 1–17.
4. Hadley Cantril, "Do Different Polls Get the Same Results?" *Publ. Op. Qua* 9 (1945), 61–69.

CHAPTER 11

Problems of Accessibility
and Cooperation

11.1 INTRODUCTION

A. Problems of Accessibility and Cooperation in the Execution of a Survey Design

There are many practical difficulties that arise between the time at which the "paper plans" of a sample survey are laid down and the time at which the completed questionnaires are ready to be sent into the analysis programs. Though a general discussion of these problems will be given in Part III of this book, there are two primary sources of difficulty which deserve special attention; these will be treated in this chapter.

In conducting attitude and opinion surveys, we have to cope with many vagaries of human behavior and with the nature tendency of some people to be preoccupied with their own affairs. Two kinds of difficulties arise:

1. *Accessibility.* With any given form of approach (i.e., personal interview, telephone or mail) and with any given form of sampling, there are always some people who are relatively more inaccessible (i.e., more difficult to contact) than others.

2. *Cooperation.* With any given form of approach, there are always some people who will not cooperate by answering questions, at least not at the time they are approached. Though accessibility is usually measured in terms of the number of calls or attempts required to make contact for the first time with an individual or household in a fixed-address method of sampling, the term can be given a much broader meaning, and this will be done in the following section.

These two characteristics of potential respondents will influence the conduct of any sample survey. The manner in which they are dealt with will affect the cost of the survey, as well as the systematic error and sampling variability of the survey results. This fact must be recognized in survey design. The succeeding portion of this chapter will discuss the ways in which this can be done.

B. Accessibility and Cooperation Are Relative to the Sampling Method and Method of Approach

The accessibility of an individual and the degree of cooperation which we may expect from him are conditioned somewhat by the method of sampling and by the method of approach that are being used in a particular survey situation. Thus, in a *fixed-address, personal-interview* type of survey, there is little difficulty encountered in identifying the relatively inaccessible, hard-to-reach type of person or household. The interviewer usually calls at a home address and attempts either to interview any responsible person in the household or to interview a specified individual. The difficulty encountered in obtaining or the failure to obtain a required interview shows quite clearly in appropriately kept survey records. Moreover, some of the more important reasons why contact is obtained with difficulty, or not at all, are well known. These are:

1. There are many people, particularly in urban communities, who spend a substantial portion of the day away from home. This is especially true for young childless couples where both the husband and wife are working.

2. Vacation periods draw many people away from their customary homes.

3. Illness and urgent household duties make it impossible to interview some people, even when they are at home.

4. In the rural regions, weather conditions (leading to impassable roads) and great distances may make contact relatively impossible.

5. In urban regions, the doormen or locked entries of high-class apartment houses may bar entrance to an interviewer.

Although it has not been generally recognized, *quota sampling procedures* are also subject to the effects of inaccessibility. For example, if interviewers are instructed to make their interviews in the home, all the points set forth in the preceding paragraph apply. The only difference lies in the fact that quota interviewers do not ordinarily keep a record of unsuccessful calls, nor do they call back at a dwelling place where no person was found at home. One quota survey in which records were kept will be reported in Chapter 12. In addition to interviewing people in their homes, quota interviewers may try to obtain their respondents on the streets, in stores, in public buildings, and in similar places. In this type of approach, there is still a relatively inaccessible group, including those individuals who spend most of their time in the home.

As a final illustration of the relativity of the inaccessible group, we may note that if a *mail* or *telephone* approach is being used, those individuals who do not have a permanent mail address or who do not have a telephone will be completely inaccessible to the survey.

Just as accessibility is relative to various features of the survey design, so also will the size and composition of the group whose members refuse to cooperate vary from situation to situation. Some subjects of investigation will tend to produce more people who refuse than will other subjects. Even with a fixed subject of investigation, there are relative degrees of cooperation. Some people refuse a general approach but succumb to a more skillful, personalized approach. Others cooperate only if paid, and some individuals almost always refuse to grant an interview. Some persons only refuse to answer certain questions. A mail questionnaire may reach a busy executive who would not be available to an interviewer, and so on.

All these points make it difficult to formulate very precise generalizations from the results of specific studies of accessibility. Nevertheless, survey appraisal and design must take these factors into account.

C. Accessibility and Cooperation Are Frequently Relative to Time

Some of the considerations set forth in the preceding discussion suggest that the accessibility of an individual, or the degree of cooperation that may be expected from him, will vary with respect to time. Thus, a man hurrying to catch a plane will not be willing to grant an interview, but the same man may be quite cooperative when he has arrived at the airport and is waiting for his plane. Urban populations are more difficult to find at home during the height of the summer vacation period than they are during the winter months.

The transitory nature of accessibility and cooperation is well illustrated by data derived from the Elmira Study, described in Section 9.2A. This Study was carried out as a panel operation, the first interview being made in June of 1948. As noted in Section 9.2A, this first interview attempt resulted in 1029 completed interviews and 238 non-interviews. In this instance, the interviewers were instructed to make at least two intelligent callbacks (i.e., on the basis of appointment or other information) in order to obtain an interview with the designated respondent. In some instances as many as eight calls

were made. Also, uncooperative individuals were approached by three
different interviewers. The respondents lost from the original sample
were not included in the next two regular interviews with the panel
(July and October 1948). Vigorous efforts were made to interview
them in November, however, to determine the effect of their omission
on the forecast of the presidential vote in Elmira. The results of this
special attempt are given in Table 11.1.

TABLE 11.1

RESULTS OF NOVEMBER ATTEMPT TO INTERVIEW PERSONS MISSED IN THE
JUNE SURVEY

Reasons for Omission from the June Survey

Result of November Attempt	Re-fused	Not at Home	Out of Town	Ill	Does Not Speak English	Mis-take	Total
Interviewed	68	58	3	32	12	8	181
Mortality:							
Refused	16	11	0	0	0	0	27
On a trip out of town	1	0	0	0	0	1	2
Moved out of town	3	4	0	0	0	0	7
Cannot locate	0	5	0	0	0	0	5
Ill	1	4	0	1	0	0	6
Mistake	3	5	0	0	0	2	10
Total	92	87	3	33	12	11	238

The November success in obtaining interviews with the original
June losses was probably due in large part to the changing accessibility and cooperative characteristics of the mortality individuals,
though there were a number of other contributing factors. In the first
place, though a vigorous attempt had been made in June, still greater
effort was expended in the November survey. Thus interpreters were
used for those persons who did not speak English. In the second
place, the questionnaire was very short, and interest in the election
and failure of the election forecasts was high. Finally, conditions
changed somewhat for those who had been ill, out of town, or persistently not at home. It may be of interest to note that the greater
effort expended in November did not have to extend to the number
of calls required to obtain an interview, at least as far as those actually interviewed were concerned. The figures in Table 11.2 illustrate this very well. Though they must be interpreted with some

TABLE 11.2

MPARISON OF AVERAGE NUMBER OF CALLS MADE IN JUNE (BEFORE LOSS
ASSIFICATION) WITH AVERAGE NUMBER OF CALLS REQUIRED TO OBTAIN
NOVEMBER INTERVIEW

Reason for June Loss	Number of Cases	Average Number of Calls Made in June before Classified as Loss	Average Number of Calls Required to Obtain November Interview
ople who refused in June and were interviewed in November	68	2.8	2.1
ople who were not at home in June and were interviewed in November	58	4.3	2.9

re because of possible differences in technique between June and
ovember, the figures would seem to indicate that increasing the num-
·r of calls is not enough, by itself, to do away with mortality.

One further bit of evidence relating to the transitory nature of ac-
·ssibility is provided by the panel nature of this study. The follow-
·g results were obtained for two additional attempts at interview
ith the original 1029 respondents:

First interview   1029 respondents
Second interview   5.6 per cent of the 1029 never contacted
Third interview   4.7 per cent of the 1029 never contacted
Second and third   0.5 per cent of the 1029 never contacted
   interviews   for both interviews

onsequently, we see that there is very little overlap between the
roup not contacted on the second interview and the group not con-
·cted on the third interview.

Similar circumstances, in which an individual's accessibility and
·gree of cooperation change with respect to time, will be met in al-
·ost any survey situation and must be provided for in the survey
·esign.

). Plan for This Chapter

There are few people who are completely inaccessible if we are
·illing to expend enough time and money to contact them. Perhaps
·e only exceptions are those individuals who are continually on the
·ove or who are located in institutions (e.g., prisons) into which in-

terviewers cannot ordinarily gain entrance. Likewise, there may
few permanently uncooperative if sufficient care is taken with th
who are reluctant to respond. The extent to which we can go
order to interview a chosen sample is well illustrated by an accou
given by Deming (1, Chap. 12) of a sample survey taken in Gree
We quote:

> Some of the towns and villages selected for the sample were accessi
> only with great difficulty, but no substitutions of either village or househ
> were permitted. Ofttimes a village could be reached only after a da
> journey by jeep, supplemented by transportation by burro or on foot. Bo
> furnished transportation to the islands and to coastal towns inaccessi
> otherwise because of swamp or mountains. Aeroplanes were used occasi
> ally for dispatching observers and for supervision of the work.

The usual survey conducted in the United States does not go to the
extremes. For example, the Bureau of Agricultural Econom:
enumerative survey of January 1947, described by Brooks (2), miss
a portion of the designated segments because weather conditic
made many roads impassable. It is quite clear that these conditic
would have improved with time, or that if extraordinary effort h
been expended the interviews could have been obtained.

For most attitude and opinion surveys conducted on the basis
probability model sampling, there is a question of how many times
is worth while to call back at a designated address to obtain an inte
view with a desired respondent. The decision concerning the cut
point must be based on the following four items:

1. The number of respondents who are not contacted and, in pa
ticular, the proportion they constitute of the total sample.
2. The differences with respéct to the subject of investigation b
tween those who are contacted and those who are not contacted.
3. The accuracy required in the survey results.
4. The cost of obtaining interviews with those who are relative
difficult to contact.

The first two points jointly determine the effect the non-contacte
individuals will have on the survey results, the third determines t
magnitude of the effect that can be tolerated, and the fourth i
fluences the cost of the survey. Similar considerations apply to tho
who refuse to grant an interview.

The succeeding sections of this chapter will be devoted to a di
cussion of these points. Most of the data and discussion will pertai
to fixed-address methods of sampling, using the personal intervie
approach, but other forms of sampling will be referred to where it

ppropriate. Although the problem of non-response to mail question-
aires is closely related to this discussion, no attempt will be made
discuss it in this book. The interested reader may find an ex-
nsive discussion and bibliography in a paper by Clausen and Ford
3). A general discussion of non-availability and refusals will be
und in Parten (4, pp. 409–417).

11.2 INACCESSIBILITY

The Magnitude and Characteristics of the Inaccessible Group

In fixed-address samples, an interviewer may be instructed to call
ack at a designated address until the proper person is found at home.
f careful records are kept of this operation, these records can be
nalyzed to determine the proportion of homes at which the desired
erson is found on the first call, on the second call, on subsequent
alls, or not at all. Such data, by themselves, do not particularly
elp reduce the difficulty in subsequent samples. They only serve to
efine the magnitude of the problem for purposes of planning and
upervision. However, if they can be related to certain features of
he field operation of the survey, or to the choice of an estimation
rocedure, future surveys may profit considerably.

It is important to distinguish two approaches to a household: (a)
n some surveys the interviewer may ask his questions of any adult
n the designated dwelling unit, especially if he is seeking information
bout the entire household or about all members in the household.
(b) In other surveys the interviewer is required to talk with a speci-
ied individual (i.e., an individual chosen at random, the housewife,
he farm operator, etc.), especially if he is asking about personal opin-
ons or about matters not equally well known by all the adults in the
ousehold. The specific individual is usually determined after a
reliminary inquiry about the household in a conversation with some-
ne who "comes to the door." Enough experience has now been developed to give us a very good idea of what to expect in our own particular situation. Some representative sets of data to illustrate this
experience have been assembled and are presented in Tables 11.3 and
11.4 on pages 242 and 243.

Although the data in Table 11.4 come from many widely diverse
survey organizations whose practices may be expected to differ, they
seem to be remarkably consistent in demonstrating the general magnitude of the accessibility problem. The non-contacted group runs
between 3 and 8 per cent of the total attempts, with the exception of

TABLE 11.3

IDENTIFICATION OF SURVEYS PROVIDING DATA ON ACCESSIBILITY

Survey Number	Individual to Be Interviewed	Urban or Rural	Locale of Survey	Time	Source of Data
1	Randomly selected adult from d.u.[a]	Urban	Elmira, New York	Summer 1948	Elmira Study, see Secti 9.2A
2	Randomly selected adult from d.u.	Urban and rural	California, Illinois, and New York states	Spring 1948	NORC
3	Randomly selected adult from d.u.	Urban and rural	Nationwide	Fall 1948	SRC, University of Mic igan (5, p. 373; 6)
4	Any responsible adult in d.u.	Urban and rural	Nationwide	Spring 1951	Current Population Surv of the Bureau of t Census (CPS) (7)
5	Any responsible adult in d.u.	Urban	Nationwide	Fall 1943	Hilgard and Payne (8)
6	Any responsible adult in d.u.	Urban	Nationwide	Winter 1947	Market Research Compar of America
7	Any responsible adult in d.u.	Rural	Nationwide	Winter 1947	Market Research Compar of America
8	Head of spending unit	Urban and rural	Nationwide	Winter 1950	Survey of Consumer F nances (9), conducted b SRC
9	Housewife	Urban	Milwaukee County	Spring 1946	McCall Corporation, co ducted by Alfred Poli Research (10), see Se tion 8.2D
10	Farm operator	Rural	New York State	Summer 1951	New York State Farm Su vey, see Section 9.3A
11	Systematically selected individ- ual from Na- tional Register (16 years of age or over)	Urban	London	Spring 1950	Durbin and Stuart (11), a experimental study

[a] Dwelling unit.

number 2. It would appear that this figure is considerably larger tha the others because only three calls were made in this survey and mor than three were made in the others. It should be noted that the non contacted figure does not represent all survey losses but only tha portion of potential respondents who were never reached. The tota mortality of a survey will be discussed in a later portion of this chap ter. The right-hand portion of this table documents the effort (ir terms of calls) necessary to contact those individuals who were ac tually interviewed. These values illustrate very well the rather ob vious facts that urban populations are somewhat more difficult to contact than rural populations and that it is considerably more dif ficult to contact a randomly selected individual in a dwelling unit than it is to contact any responsible person in the dwelling unit.

TABLE 11.4

NON-CONTACTED RESPONDENTS AND NUMBER OF CALLS REQUIRED FOR
COMPLETED INTERVIEWS

Survey Number	Total Attempts [a]	Per Cent Not Contacted	Total Interviews	Per Cent of Those Interviewed Contacted on		
				First Call	Second Call	Third or Later Call
1	1,256	7.2	1,029	38.0	32.9	29.1
2	4,207	13.7	3,065	43.8	35.4	20.8 [b]
3	736	...	610	36.9	30.8	32.3
4	25,000	6.0	...	...	...	...
5	...	...	3,265	63.5	22.2	14.3
6	3,006	7.0	2,796 [c]	72.0	19.4	8.6
7	2,076	3.0	2,014 [c]	83.5	10.3	6.2
8	3,512	4.5	...	...	...	...
9	2,074	6.7	1,734	77.6	16.2	6.2
10	1,709	...	1,530	72.2	20.8	7.0
11	1,376	6.4	1,183	42.3	40.8	16.8

[a] The total-attempt figures represent, insofar as it is possible to ascertain, all instances where an interview should have been obtained. They do not include attempts at vacant dwelling units, attempts where there was no person satisfying the survey definition of a respondent, and other similar types of attempts. They do include attempts classified as refusal, ill, out of town and the like.

[b] This survey, in general, required its interviewers to make only three calls at each assigned address.

[c] These values represent all households reached in the survey, and the call percentages are based on them rather than on the number of completed interviews.

Over and above determining the magnitude of the hard-to-reach group and the effort necessary to make contact in any specific study, we can compare the characteristics of households in which it is difficult to find a responsible person at home with the characteristics of those households for which this is easy; or we can compare hard-to-reach individuals with readily accessible individuals. The results of such comparisons tend to confirm and quantify facts that can usually be stated on *a priori* basis. Thus it is quite clear that we are more likely to find a responsible person on the first call at a home where there are preschool children than at a home where there are no children and both husband and wife work; that young men and women are less likely to be at home during daytime interviewing hours than

are older men and women; and that women, as a class, are easier t
find at home than men.

One set of data illustrating these facts is given in a study by Hi
gard and Payne (8). Their data are reproduced in Table 11.5.

TABLE 11.5

CHARACTERISTICS OF URBAN HOUSEHOLDS INTERVIEWED ON FIRST, SECON
AND LATER CALLS

Household Characteristic	Households Interviewed on First Call	Households Interviewed on Second Call	Households Interviewed on Third or Later Call	All Househol Interview
Number of urban interviews	2072	726	467	3265
Per cent	100.0	100.0	100.0	100.0
Respondent reporting on household purchases				
1. Responsible person, not employed outside home	78.2	57.8	46.4	69.1
2. Responsible person, employed outside home	21.8	42.2	53.6	30.9
Household composition				
1. Having children under two years of age	17.2	9.5	6.2	13.9
2. Having older children only	37.6	34.8	32.3	36.3
3. Having no children	45.2	55.7	61.5	49.8
Size of household				
1. One person	6.3	13.1	15.2	9.1
2. Two persons	24.6	29.5	34.6	27.1
3. Three persons	25.6	23.2	21.0	24.4
4. Four persons	19.9	16.5	16.5	18.7
5. Five or more persons	23.6	17.7	12.7	20.7
Average size	3.56	3.11	2.84	3.35

this particular survey had ended with the first interview, there woul
have been too few families with housewives otherwise employed, to
few families without young children, and too few small households
These distinctions are, of course, intensified if we are attempting t
obtain an interview with a specified individual rather than with an
responsible person. Additional data on these points have been mad
available by Kiser (12), by Williams (13), and by Gaudet an
Wilson (14). Extensive analyses in this respect of the Elmira Stud
have been made by the Study of Sampling. In addition, there is
wealth of information on this subject (most of it unanalyzed) in th
files of survey organizations.

The contents of Table 11.4 also have some obvious implications fo
quota sampling procedures. Just as individuals have relative degree
of accessibility for the fixed-address interviewer, so also do individ
uals have varying degrees of accessibility for the quota interviewer
However, the problem of accessibility in relation to quota sampling i

t ordinarily as apparent as in fixed-address sampling because of the
llowing reasons:

1. Quota interviewers will not ordinarily attempt to obtain inter-
ews with those who will obviously be difficult to contact.
2. If an attempt at contact is unsuccessful, it will be ignored and,
effect, a substitution is immediately made.
3. No accounting is ordinarily expected from quota interviewers to
ow how many unsuccessful attempts at contact they actually do
ake.

At the request of the Study of Sampling, the National Opinion Re-
arch Center attempted to supply some data on these and related
ints. A complete account of this study will be found in Chapter 12.
nce these interviewers were instructed to attempt to make all their
lls in the home, it would seem to be of interest at this point to in-
ire concerning the proportion of home attempts they reported as
ding in finding no one at home. If this proportion were substan-
ally lower than for fixed-address samples, it would indicate that
ota interviewers tend to select the more accessible homes (and,
erefore, individuals), even on their first attempts. In this particu-
r study, a national quota sample, 88 interviewers reported a total
2319 attempts at interview, of which 1223 (or 53 per cent) ulti-
ately ended in completed interviews. Twenty-two per cent of 1992
me attempts were classified as "no one at home."

The figures given in Table 11.4 for the first calls of a fixed-address
mple show that no responsible person is found at home for between
0 and 40 per cent of the addresses. If we relax the condition "no
esponsible person at home" to "no one at home," the appropriate
gure would be somewhat lower than this. Although there are no
ata to show how much lower this value would become, the 22 per
ent quoted above for a quota sample seems to be somewhat lower
han we might expect in a fixed-address sample. Consequently, for
his particular survey, there would seem to be evidence that in their
riginal selection of households the quota interviewers tended in some
egree to choose the more accessible ones. This, of course, tells us
othing about the substitutions made for these "no one at homes,"
ince a quota interviewer does not call back at such a home.

In rebuttal to statements that quota interviewers obtain only the
ore readily accessible individuals in the population, it is sometimes
laimed that the selected individuals must conform to assigned quo-
as, and that these quotas insure that the proper proportion of hard-
o-reach individuals is obtained. For example, first calls in a fixed-

address sampling procedure are heavily weighted toward women, b the quota interviewer must secure the proper proportion of men a women. Similar observations may be made about other factual cha acteristics of the respondents such as age and socio-economic stat (or controls such as rent and small geographic area which are repla ing socio-economic status). In reply, it has been argued that with any quota control category only the more readily accessible indivi uals will be obtained. Ultimately, this chain of argument leads the following proposition. An individual is relatively hard to rea not because of any special attribute which he does or does not posse but simply because of the net effect of many characteristics and a tivities related to his age, sex, economic status, and the like. Althou it is true that quota controls such as age, sex, and economic status w remove some of the effects of inaccessibility, it is clear that in sor situations they can remove only a small portion of its effects. With any group of persons of the same age, sex, and economic status, the are different degrees of availability. For example, individuals va with respect to their "going out" patterns (e.g., to shop, to visit frienc to the movies, etc.). It is difficult to predict just how these differenc affect a particular subject of investigation. Consequently, no speci statement of the effects of accessibility on quota sampling can yet formulated for general application.

B. Accessibility in Relation to Attitudes, Opinions and Other Subjects of Investigation

The primary conclusion to be drawn from the discussion and da of the preceding section is the obvious one—that in a fixed-addre system of sampling there will always be, at the time of the surve certain households and individuals who are more difficult to conta than others, and that there are certain factual characteristics f which these households and individuals differ rather markedly fro those easy to contact. Moreover, it is not difficult to predict on a *a priori* basis those factual or demographic characteristics for whic differences will be obtained.

Because people who have different degrees of accessibility to th fixed-address interviewer (as measured by the number of calls r quired to obtain contact) have different factual characteristics, it ca be expected that differences will also frequently arise with respe to the subject matter of the investigation. However, it is not eas to predict which subjects of investigation will show differences, an to make quantitative evaluations of the differences in advance is al most impossible.

There are a large number of examples that show that such differences do exist, but they would only serve to illustrate the already mentioned point for specific variables. Each time a new survey is taken, a new situation arises. Consequently, we shall content ourselves by presenting a summary of results obtained in one particular survey. This survey was conducted by NORC in the spring of 1948 in the three states, New York, Illinois, and California. A probability model sample was used and 1339 interviews were completed on the first call, 1083 on the second, and 636 on the third. Approximately 28 per cent of the originally assigned respondents were not interviewed. This is somewhat greater mortality than is experienced in many similar surveys, and it weakens to a degree the quantitative conclusions that may be drawn from the analysis. The survey results were analyzed according to the call number on which the interview was obtained. The responses given to a large number of questions by those interviewed on the first call, on the second call, and on the third call were compared by means of the chi-square distribution. The end results of this analysis are given in Table 11.6, and two illustrative examples are presented in Table 11.7 on page 248.

TABLE 11.6

DISTRIBUTION OF χ^2 VALUES OBTAINED FROM TESTING THE SIGNIFICANCE OF RESPONSES GIVEN BY PERSONS INTERVIEWED ON FIRST, SECOND, AND THIRD CALLS

	Frequencies of Occurrence for			
Probability of Obtaining a χ^2 Larger Than That Observed	Characteristics of Respondents [a]	Information Questions [b]	Factual Questions [c]	Attitude and Opinion Questions [d]
0.50–1.00	0	0	2	4
0.20–0.50	0	0	1	3
0.10–0.20	0	0	1	1
0.05–0.10	0	1	1	1
0.01–0.05	0	4	0	3
0.001–0.01	2	1	0	0
Less than 0.001	4	0	0	0
Total	6	6	5	12

[a] Age, sex, education, economic status, city size, and occupation of chief family worker.

[b] Knowledge of current affairs.

[c] Past voting behavior, etc.

[d] Vote intention, attitude on domestic and foreign relations, etc.

TABLE 11.7

χ^2 COMPARISON OF INDIVIDUALS INTERVIEWED ON FIRST, SECOND AND THIRD CALLS

(Figures in parentheses are the number of cases)

Characteristics of Respondent	Call on Which Interview Was Obtained		
	First	Second	Third
Education	(1320)	(1077)	(629)
At least some college	18.8	21.4	20.7
Completed or some high school	44.3	47.3	52.0
Grammar school graduate or less	36.9	31.3	27.3

$\chi^2_{4\text{d.f.}} = 20.6$, Pr $(\chi^2 > 20.6) < 0.001$

U.S. contribution to the U.N.	(1339)	(1081)	(634)
All it should be	53.1	52.2	52.4
Should have done more	30.5	31.5	34.4
Do not know	16.4	16.3	13.2

$\chi^2_{4\text{d.f.}} = 5.3$, Pr $(\chi^2 > 5.3) = 0.20$

It should be noted that a small probability of obtaining a chi-square value larger than the observed one means that it is unlikely that the three groups being compared can have the same composition with respect to the variable under consideration. The two examples given in Table 11.7 provide some indication of the magnitude of differences producing various probabilities, education giving an extremely small probability and the opinion comparison giving a more moderate value.

These tables illustrate the points that have already been made. Extremely small probabilities are associated with chi-square values computed for the listed factual characteristics of respondents. Information questions also tend to show this behavior, which is to be expected since responses to this type of question are usually highly correlated with such variables as age, sex, and education. The factual questions and the attitude and opinion questions show more of a spread on the probability scale, however. This more or less amounts to saying that sometimes there is a substantial difference between the callback groups on these types of questions and sometimes there is not much of a difference.

Data such as the preceding, although interesting, are not actually of much value to the user of sample surveys unless they can aid in evaluating the effects of the non-contacted on a specific set of survey results, or can aid in the design of future sample surveys. Many approaches have been suggested for accomplishing these ends, and a discussion of them will be given in the next subsection. Before proceeding into the details of this discussion, we shall present one set of data that does illustrate the effects of the non-contacted. As can be noted from Table 11.1, the first wave of the Elmira Study failed to contact 87 individuals, of whom 58 were found and interviewed after the 1948 presidential election. A comparison of the June interviewed and June non-contacted will be found in Table 11.8. It should be noted that, for the two questions on voting, not all the 1029 June respondents were available for questioning in the post-election survey.

The difference column in this table shows the effect of the non-contacted on some of the survey results. It should be noted that this represents *only* the effect of the non-contacted and not the effect of the entire sample mortality. The net effect of all mortality will be discussed in Section 11.4. The differences range from 0 to 1.8 percentage points, the average absolute difference over all questions and characteristics being 0.7 of a percentage point (only the maximum difference having been taken from each question). Whether or not such differences are important depends on the uses to which the survey data are to be put, and there is little to be said in a specific way about this problem; general comments relating to this problem will be found in Parts I and III.

C. Methods of Dealing with Inaccessibility in Sample Surveys

The material presented in the preceding two sections has indicated that hard-to-reach households and individuals exist for any particular study, and that they differ in reasonably obvious ways from those who are relatively easy to contact. Moreover, the differences may be of such a magnitude that this group must be taken into account in planning the survey, in the preparation of estimates from the sample, and in the determination of the accuracy of such estimates. The present section will be devoted to a discussion of the ways and means of accomplishing this. Admittedly the problem is intimately connected with the time, cost, and accuracy requirements of a survey. Yet there is, at the present time, very little readily available data to show, for example, the effects of making callbacks on the total survey cost. Lacking adequate cost information, we shall simply assume that whatever cuts down on the required number of callbacks or on the cost of making the callbacks will be worth while.

1. *Reducing the Number of Non-contacted Respondents.* One of the first points at which we may begin to take account of the hard-

TABLE 11.8

COMPARISON OF JUNE "NOT HOMES" WITH JUNE RESPONDENTS

(Figures given with a decimal point are percentages of the figures in parentheses)

Characteristic of Respondents	Interviewed in June	June Not-at-Homes	Interviewed Plus Not-at-Homes [a]	Differ-ence [b]
Total number of cases	1029	87	1116	
"Did you vote on November 2?"	(944)	(58)	(1002)	
Yes	72.4	75.9	72.7	−0.3
No	27.6	24.1	27.3	+0.3
(Voters) "Whom did you vote for?"	(683)	(44)	(727)	
Dewey	61.2	38.6	59.4	+1.8
Truman	32.6	47.7	33.8	−1.2
Wallace	0.4	0.0	0.4	0.0
Other	0.3	0.0	0.3	0.0
Not stated	5.4	13.6	6.1	−0.7
Sex	(1029)	(58)	(1087)	
Male	43.9	55.2	44.8	−0.9
Female	56.1	44.8	55.2	+0.9
Age	(1029)	(56)	(1085)	
Less than 35	36.1	30.4	35.7	+0.4
35–64	52.2	64.3	53.1	−0.9
65 and over	11.7	5.4	11.2	+0.5
Economic level	(993)	(54)	(1047)	
A	0.3	1.8	0.4	−0.1
B	9.4	11.1	9.5	−0.1
C	71.9	75.9	72.2	−0.3
D	18.4	11.1	17.8	+0.6
Education	(1029)	(56)	(1085)	
Less than grammar school graduate	10.3	16.1	10.8	−0.5
Grammar school graduate and some high school	43.4	30.4	42.4	+1.0
High school graduate and college	46.3	53.6	46.9	−0.6
Marital status	(1029)	(57)	(1086)	
Single	11.9	14.0	12.1	−0.2
Married	76.8	80.7	77.1	−0.3
Other	11.3	5.3	10.8	+0.5
Religion	(1024)	(56)	(1080)	
Catholic	29.0	39.3	29.8	−0.8
Protestant	67.0	53.6	66.0	+1.0
Jewish	1.5	3.6	1.7	−0.2
Other answer	2.5	3.6	2.6	−0.1
Telephone ownership	(1020)	(56)	(1076)	
Yes	71.2	80.4	71.9	−0.7
No	28.8	19.6	28.1	+0.7
Automobile ownership	(1026)	(56)	(1082)	
Yes	61.9	60.7	61.8	+0.1
No	38.1	39.3	38.2	−0.1
Race	(1019)	(58)	(1077)	
White	98.1	98.3	98.1	0.0
Negro	1.7	1.7	1.7	0.0
Other	0.2	0.0	0.2	0.0

[a] Except for "candidate voted for," these values were obtained by weighting the first two columns in the proportion 1029:87. This assumes that the 58 are a random sample of the 87. In the case of "candidate voted for," the proportion voting in each group was also used in adjustment.

[b] (Interviewed in June) less (interviewed plus not-at-homes).

to-reach group is in the definition and selection of sampling units. If a number of small, widely spaced units are to be covered by the same interviewer, then the cost of callbacks may become acute. Traveling back and forth between the groups to pick up the unsuccessful attempts will be both costly and time consuming. Sometimes money and time can be saved by having the units appropriately located with respect to one another. For example, suppose an interviewer has two units so located that he must pass by one in order to reach the other. An efficient form of operation might then be for him to canvass first the unit nearest his place of residence; drive on to the second unit and canvass it, at the same time attempting to make callbacks; and, finally, make callbacks in the first unit on the return trip home. The extent to which such devices are feasible naturally depends not only on the location of the units but also on the number of interviews to be obtained in each. Units of a certain size may prove to be more efficient than units of any other size. Further research is needed to clarify the situation.

Appropriate timing of first calls and the intelligent use of information on a household's or individual's "at home" pattern may also pay dividends by increasing the proportion of successful first calls, and also by decreasing the number of subsequent calls needed to obtain an interview. For example, it is quite obvious that most working men are not going to be home during either the midmorning or the midafternoon periods. The only exceptions would be men who work night shifts and men, such as farmers, who work near their homes. Therefore first calls should be made at a time when the largest proportion of individuals will be at home and will not be occupied with duties that preclude their granting an interview. Thus the most productive time of call for any population that includes a relatively high proportion of working men and women will be in the early evening hours or at appropriately chosen portions of the week end. Data on the problem of "best" interviewing time can easily be obtained by keeping appropriate survey records. One such set was assembled during the course of the New York State Survey of Farmers (see Section 9.3A) and is given in Table 11.9.

It may be noted from this table that first calls made after 6 P.M. were somewhat more productive than first calls made before 6 P.M. (the difference being significant at the 0.01 level). However, the increase in proportion of successful first calls may not be great enough in this instance to be of any practical utility. These data refer, of course, to a very specialized population, namely farm operators in New York State. Very similar data, derived from a survey conducted

TABLE 11.9

Time of First Call	Number of First Calls	Percentage Successful First Calls
Before noon	528	72.0
Noon–4 P.M.	609	68.3
4–6 P.M.	167	71.8
6–8 P.M.	134	84.3
After 8 P.M.	36	86.1
Not reported	56	80.4

in London, may be found in a paper by Durbin and Stuart (11).
They, of course, reported a much lower proportion of successful first
calls, and there was more of a difference between various time periods.

In addition to determining such "best" times for making first calls
there are also many practical problems that must be faced. For ex
ample, an interviewing staff of housewives reacts very unfavorabl
to evening interviewing. They have their own households to care fo
and want to be home when their husbands and children are at home

Closely related to this problem of the "best" periods for making
first call is that of minimizing the number of calls required subse
quent to an unsuccessful first one. Various devices must be used t
discover the most productive time for making a callback. If no on
is at home, a neighbor may be able to help in the timing of a secon
call. Appointments made by telephone may be useful in certain in
stances. The principal point is that once an address has been visited
every effort should be made to secure supplementary information t
aid in the making of future calls.

2. *Incorporating the Bias of Those Not Contacted into the Sam
pling Errors of the Survey Estimates.* After a particular survey ha
been completed and the proportion or number of non-contacted indi
viduals has been determined, it is always possible to estimate th
maximum bias that the non-contacted may introduce into the surve
results. If so desired, this maximum bias may then be included i
the measure of sampling error ·assigned to an estimate. A simpl
example will serve to illustrate the line of reasoning and the procedur

From Table 11.8, we observe that in the Elmira Study (first wave
1029 respondents were interviewed and 87 were not contacted. W
shall here ignore the effects of other sample losses. Suppose that w
were trying to estimate the proportion of individuals who had com

leted grammar school or had received some high school education.
he proportion of such individuals among those interviewed is ob-
rved to be 0.434, and this is used as the estimate. The maximum
ias in this estimate that could be introduced by the non-contacted
ould occur in one direction if *none* of them fell into this category,
nd in the other direction if *all* of them fell into this category. There-
re, we know that no matter what the characteristics of the non-
ontacted group may be, the estimate from the combination of re-
pondents and non-contacted must lie between

$$(0.922)(0.434) + (0.078)(0) = 0.400$$

nd

$$(0.922)(0.434) + (0.078)(1) = 0.478$$

here 0.922 and 0.078 are the relative sizes of the groups of respond-
nts and those non-contacted, respectively. In this instance, the
aximum bias in either direction is $(0.478 - 0.434) = 0.044$, or 4.4
ercentage points. Actually, this value of 0.044 depends on the par-
icular sample estimate obtained and would be somewhat different
r different values of the estimate. This effect will in general be
ather small and is ignored in this discussion.

If so desired, the results of the preceding analysis can be incor-
orated into an over-all measure of the error of the survey estimate.
ssume that repeated random samples of size 1116 (1029 + 87) are
eing drawn from a population in which 92.2 per cent of potential
espondents will be contacted and 7.8 per cent will not be contacted.
hen the mean square error of a survey estimate made only from
he contacted respondents will be of the form

$$\text{Mean square error} = \frac{p'q'}{1029} + (\text{bias})^2$$

here p' is the true proportion in the group that can be contacted
nd the bias is the difference between p' and the corresponding pro-
ortion in the entire population. We have an estimate of p' from the
ample and an *extreme* estimate of the bias. Therefore, we state that
he largest possible estimate of the mean square error subject to the
estriction noted at the end of the preceding paragraph, is

$$\frac{(0.434)(0.566)}{1029} + (0.044)^2$$

$$= 0.000239 + 0.001936$$

$$= 0.002175$$

In this example the bias term dominates the estimate of the mean square error, being eight times as large as the sampling variability. As will be seen from Table 11.8, the observed bias for the present data was only 0.01, giving a contribution to the estimated mean square error of 0.0001.

It may be noted that other forms of estimate can be used, forms that will reduce the maximum estimate of the mean square error. For example, let us arbitrarily impute a value of 0.50 to the non-contacted group. Then the estimate is of the form

$$(0.922)(\hat{p}') + (0.078)(0.50)$$

where p' is the ordinary estimate from the contacted group. The mean square error of this estimate is

$$(0.922)^2 \frac{p'q'}{1029} + (\text{bias})^2.$$

The maximum value of the bias, which now does not depend on p', is equal to 0.039. Therefore, the largest possible estimate of the mean square error is

$$(0.922)^2 \frac{(0.434)(0.566)}{1029} + (0.039)^2$$

$$= 0.000203 + 0.001521$$

$$= 0.001724$$

as compared with the preceding value of 0.002175. The arbitrary value of 0.50 works in this analysis since its use reduces the maximum amount that the true value for the non-contacted group may differ from the value going into the estimate from 1.0 to 0.50.

The foregoing type of analysis, with the additional complication that the number of non-contacted is allowed to vary randomly from sample to sample, has been used by Birnbaum and Sirken (15, 16) in a procedure for determining the number of callbacks that should be made in a particular sample survey. Their summary of the work is as follows:

A technique is presented for the treatment of errors introduced into sampling surveys due to the non-availability of respondents. The expected cost and variance of the sample survey are expressed as functions of sample size and of the number of callbacks made on the non-availables. A method is then presented which optimizes precision for a given cost by playing sampling error against the bias resulting from non-availables.

This is a well-written and straightforward approach to the problem of callbacks, but several difficulties will be met in application. These are as follows.

The procedure uses an even more extreme approach to maximum bias than that set forth in the preceding paragraphs, since the computations are carried out before the sample is drawn and no knowledge is assumed concerning the true or estimated proportion in the group of contacted respondents. In other words, instead of a maximum bias figure of 0.044, their approach calls for a maximum bias figure of 0.078, the proportion of non-contacted respondents. This is about ten times the average bias observed in Table 11.8, and it is felt that discrepancies of this magnitude will frequently be found in practice. This means that the mean square error will be completely dominated by the bias term if the samples are of any appreciable size, even more than in our numerical example. It would appear that appreciable gains could be made by some such device as that used in the preceding discussion, where an arbitrary value of 0.50 was imputed to the non-contacted group.

The second difficulty arises from the fact that the application of the theory requires information relating to the costs of making callbacks. Very few published data on this aspect of sample surveys were found by the Study of Sampling. Sample surveys should be designed in the future so that they will provide this type of information, and the resulting data should be made available to all users of survey techniques.

3. *Use of Data Obtained from Contacted Respondents to Predict the Characteristics of the Non-contacted or to Adjust for the Effects of the Non-contacted.* In the literature, several procedures have been set forth whose primary aim is to predict from that portion of the sample actually reached and interviewed the characteristics of the non-contacted individuals. Although these procedures have been suggested primarily in the treatment of non-response to mail questionnaires, they are equally applicable to fixed-address, personal-interview surveys. Clausen and Ford (3) have used simple extrapolation procedures based on the successive returns to a mail questionnaire, and Hendricks (17) has given a more complex, computational procedure for accomplishing the same objectives.

These approaches assume that there exists some sort of functional relation between the call number at which an interview is obtained (or the time period in which a mail questionnaire is returned) and the variable to be measured in the survey. The functional relationship is estimated by graphical or computational procedures and is then used for prediction. A simple example will show the graphical approach. The data, derived from the Elmira Study, are given in Table 11.10. These data are next plotted as in Figure 11.1.

TABLE 11.10

PROPORTION OF INDIVIDUALS HAVING A FAVORABLE ATTITUDE TOWARD LABOR UNIONS, BY CALL NUMBER

Call on Which Interview Obtained	Number of Cases	Proportion with Favorable Attitude	Cumulated Per Cent Interviewed	Cumulated Proportion Favorable
First	391	0.545	35.0	0.545
Second	338	0.548	65.3	0.546
Third	300	0.592	92.2	0.559
Not contacted	87	...	...	...

The line or curve fitted to the points is extrapolated to 100 per cent interviewed, and the corresponding proportion favorable is the value used for the entire sample. In this instance, the value is 0.558. There are a number of more or less obvious difficulties connected with an analysis such as this, the principal ones being:

1. Even if the data indicate a definite trend, there will be some error associated with the extrapolation. There is no satisfactory way of determining the magnitude of the error and its subsequent effects on the sample estimate. Moreover, in many instances there will not be any well-defined trend, and it will not be at all clear what procedure should then be used. The data of Table 11.10 are probably too well

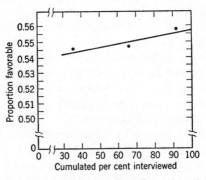

FIG. 11.1. Graphical extrapolation to adjust for non-contacted individuals.

balanced in this respect to be generally indicative of most situations.

2. Successive calls must be made in accordance with a known model so as to bring out the functional relation, if it exists. A simple illustration of this would be furnished by a situation in which first calls were made at randomly distributed times, second calls on the nonresponses were also made at randomly distributed times, etc. Any trend relationship between first and second calls might be destroyed by using appointments and other information in timing the second calls.

3. Although this procedure might work reasonably well for making estimates from the entire sample, complications would undoubtedly arise if the main purpose in drawing the sample were to compare certain population subgroups.

In view of these comments, great care would have to be exercised in carrying out this extrapolation, and its use is not recommended except under exceptional circumstances.

Somewhat different approaches to inaccessibility have been suggested with respect to quota sampling. One such suggestion was made by Dr. H. O. Hartley, relative to the sampling of households rather than individuals, and will be found in the discussion of a paper by Yates (18). The appropriate quotation is as follows:

The method is simply to eliminate the bias by introducing another set of strata. All households may be classified into groups according to the length of daytime period (time of interviewing) during which an adult may be found at home. There are households where there is always somebody in, which we may call the 100 per cent homes; and there will be, on the other hand, households where only during 10 per cent of the daytime somebody is in, the 10 per cent at homes; there will be 90 per cent at homes; 80 per cent at homes, and so on. Now, a rather larger random or quasi-random sample of households is selected as before, but now only one call is made at each house of the sample, all not in at the call drop out of the sample. There are definite rules of chance for this game of hide and seek, just as accurate as Dr. Yates' law of randomness!! Of the 100 per cent at homes, none of the random sample is being lost, of the (say) 40 per cent at homes we lose, in the average, 60 per cent of the original frequency in the random sample but—and this is the point—the remaining 40 per cent can now be regarded as a quasi-random sample within the stratum of the 40 per cent at homes. All the households in the 40 per cent at home group had approximately an equal chance of being in this sample.

Now, in order to utilize this theoretical device, important additional information is required—we must be able to classify all actually interviewed households as to whether they are 100 per cent at homes, 90 per cent at homes and so on. In practice, therefore, the interviewers would have to add to their questionnaire an appropriate question as to how often it occurs that there is nobody at home. This is certainly an awkward question to answer, but a question to which the answer should be of similar reliability to the answers which the sampling inquiry attempts to analyze. If this is done, all interviewed households can be classified in their appropriate at home stratum.

Once the data have been obtained in the form Dr. Hartley suggests, computations can be applied to give the sample estimates.

The same effect might also be obtained by appropriate control of the time periods in which the interviewing is to take place (assuming that all interviews are to be made in the home). Of course, the simplest solution to this problem would be to conduct all interviewing in the evening when most people are at home and to supplement the sample with some individuals who are at home during the daytime but away during the evening. This program may not always be possible, however, because of interviewers' preferences for working hours. A possible compromise might be as follows. Suppose it were determined that 15 per cent of the women are away from home during the day. Then the interviewers might be given two samples, one to get 85 per cent of the females with home interviews during the day, the other to obtain the remaining 15 per cent of the females among women who are not home during the day but are seen in the evening while the interviewers are also interviewing male respondents.

Finally, in quota sampling, we may try to accomplish these same aims by conducting some of the interviews in the home and by conducting the remainder on the street, in stores and offices, etc. In other words, individuals who are hard to reach for home interviews will be easy to reach relative to places away from home. The Study of Sampling has made a number of studies aimed at comparing the characteristics of those people who were interviewed in the home with the characteristics of those who were interviewed in other places. One such study was carried out on data made available to the Study by NORC, these data relating to a survey conducted for the American Library Association (19). The 2114 completed interviews were obtained in 17 cities, 73 per cent of them in the home and the remaining 27 per cent elsewhere. As might well have been predicted in advance of the analysis, those interviewed outside the home tended to be male and the main earners of their family. Moreover, they were somewhat more highly educated, were more predominantly white-collar workers, and came from a slightly higher economic status than those interviewed in the home. These characteristics are in agreement with the results cited for comparisons between individuals interviewed on first calls with those interviewed on later calls, suggesting that quotas might be determined for the two categories, home interviews and interviews elsewhere. At the present time, however, there is no obvious way available for the actual determination of these quotas.

In effect, these considerations of time and place of interview in connection with a quota sample design may lead to the establishment of an additional control on the sample. This additional control may reduce but not remove objections that have been raised against a quota method of sampling. It will still be necessary to have prior information for the determination of quotas, and the ultimate selection of respondents will still be dependent on the judgment of the interviewers, even though their choice may be restricted by one further control.

Procedures very similar to the foregoing have been described by Politz and Simmons (20) and by the American Institute of Public Opinion * in connection with fixed-address methods of sampling. In the Politz-Simmons approach, we start from a fixed-address sample of individuals, and each person contacted on the first call is asked concerning his presence or absence from home on each of the preceding five nights (at the same time as that at which contact is made). The information collected by this process is used to adjust the sample results. The difficulties arising from the application of such a procedure are primarily of an administrative nature. It is necessary to ask additional questions concerning presence or absence from the home, and this places an extra burden on the interviewers and respondents. Moreover, work is added in the analysis because the sample results must be weighted.

Still another approach, not unrelated to those already described, was tried on an experimental basis by Durbin and Stuart (21). They selected random and clustered samples in each of two boroughs of London, two large English towns and two small English towns. Every effort was made to interview the informants originally selected from the Electoral Registers, and careful records were kept of the results of each attempt to obtain an interview. The results of 1260 attempts were distributed as given in Table 11.11. Substitutes were taken when the individual whose name was drawn from the Electoral Register no longer lived at the stated address and an adult of the same sex did live at the address.

The experimental phase of the study, as far as callbacks were concerned, consisted of the following two steps.

* Report to Editors, June 1949.

TABLE 11.11

RESULTS OF ATTEMPTED INTERVIEWS

Result	Per Cent
Interview obtained	70.6
Refused or partially refused	7.1
Informant removed (substitute taken)	4.7
Informant removed (no substitute taken)	5.6
Informant absent from home for long period	5.7
Other causes of failure (death, illness, no such address, unsuitable for interview)	6.3
Total	100.0

1. On the first call, information was obtained that would permit the application of the Politz-Simmons scheme.

2. Following the completion of the field work for the random samples, a single interviewer in each area selected and interviewed a quota sample which matched, on sex, age, and social-income grade, the characteristics of individuals not interviewed on the first call of the random samples. These quota sample interviewers had not worked on the random samples.

A number of comparisons were made on the basis of these data, the usual procedure being to take all interviews obtained in the random samples as a standard and to see whether various other samples could be regarded as random subsamples from this standard. The Politz-Simmons scheme did not work at all well, but the authors did not feel they had given it a fair test since all first calls were made during the day rather than being scattered throughout the day and evening. The principal conclusions regarding the other phases were:

The results of the composite sample thus obtained (first calls plus quota replacements) did not justify the additional cost incurred; and even when the quota sample replacement was made after the second or third call, the results were still inferior to those of the random sample based on three calls . . . The three-call random sample alone gave results which were satisfactorily close to those of the all-calls sample.

These conclusions must be qualified by the fact that the tests of the three-call and quota substitutes methods were not comparably exacting and did not reflect the (unknown) characteristics of 20 per cent of the sample still missing after all the calls were made.

4. *Subsampling the Non-contacted Respondents.* As a final method of dealing with the hard-to-reach individuals, we may refer to the

work of Hansen and Hurwitz (22). Their approach is somewhat similar to that of Birnbaum and Sirken, but instead of incorporating the possible bias of non-response into the sampling error, it seeks to accomplish the same end by subsampling the non-contacted individuals (after a stated number of callbacks). It consists of allocating interviews between first calls and subsequent calls in order to obtain a sample estimate with a stated standard error at minimum cost. This method depends not only on knowing the relative costs of first and subsequent calls, but also on knowing the population variance of the quantity being estimated and the population variance for the elements that will not be reached on the first call. This assumes that we choose a sample, call on all elements of the sample, and then subsample those elements not reached on the first call. This subsample must be enumerated in its entirety, no matter how many callbacks are necessary. Another possibility would seem to be to extend this procedure to more than two calls. Thus, we should study the sample estimate obtained on first calls (together with the proportion of total sample covered), the estimate obtained on the second call (together with the proportion of the total sample covered in this call), etc., and determine the subsampling rates that should be used in each successive call for satisfying some particular criterion. We might, for example, sample 50 per cent of those not covered on the first call, 25 per cent of those for whom a second call was attempted without success, and so on. Procedures incorporating some of these ideas have been discussed by Deming (23), Simmons (24), and El-Badry (25). From a practical point of view, all these involve knowing relative costs of each successive call. It is to be hoped that data on this aspect of the problem will become available in the near future.

This method suffers from the same administrative difficulties associated with the Politz-Simmons method, these difficulties having been described in the preceding section.

11.3 REFUSALS

A. Magnitude and Characteristics of the Refusal Group

As noted in the introduction to this chapter, any survey will find individuals who refuse either to be interviewed or to answer certain types of questions. Whether or not an individual refuses to grant an interview depends on a large number of variables such as the following: (a) the form of approach (personal interview, mail, or telephone), (b) the type of information being requested and whether or not the respondent is informed of this content in advance of the questioning, (c) the group affiliations of the respondent (e.g., rural or urban and native or foreign-born), (d) the respondent's attitude toward the organization sponsoring the investigation, (e) the efforts made to overcome resistance to furnishing information, and (f) the circumstances under which the interview is attempted. Before discussing these various factors, it will be well to present some representative data on the magnitude and characteristics of the refusal group for several specific studies. The data on magnitude are given in Table 11.12.

The figures given in Table 11.12 do little more than indicate that we may ordinarily expect to obtain refusal rates ranging up to and somewhat over 10 per cent. Though they do afford an opportunity for comparing rates between methods of sampling (quota and fixed address) and subject of investigation (opinion, unemployment, financial, magazine readership, and personal marital affairs), there are not enough data to make any one comparison conclusive. Nevertheless, the magnitude of the refusal group, if considered in conjunction with the characteristics of refusals, may make the problem an important one for certain forms of investigation. This has been particularly noted in relation to studies of income, wealth, consumer expenditures, and like subjects of investigation. Comments on this particular topic may be found in the work of the Bureau of Labor Statistics (28) and in publications of the National Bureau of Economic Research (29). Information on refusal rates in a number of European countries will be found in a paper by Wilson (30).

The Elmira Study afforded a unique opportunity to examine the characteristics and effects of the refusal group, since some of the original refusals were later interviewed. The data that were available from this study are given in Table 11.13 (see Table 11.8 and accompanying discussion).

The first two columns in Table 11.13 present a comparison of the characteristics of the interviewed and refusal groups, whereas the first and last show the effect of the refusal group on some of the survey results. With respect to characteristics, it is of interest to compare the contents of this table with those of Table 11.8. It will be seen that the group of individuals who refused to grant an interview tended to contain a higher proportion of females than did the not-at-home group, a higher proportion of older people, and a smaller proportion of married persons. The net effects of this group on survey results are shown in the last column. The differences range in absolute value from zero to 1.5 percentage points, the average of the largest absolute

TABLE 11.12

REFUSAL RATES FOR A NUMBER OF SAMPLE SURVEYS

Per Cent Refusals	Number of Respondents Contacted [a]	Source of Data
0.4	2097	Margaret Hogg (26). This study was carried out on a fixed-address basis, in New Haven in 1931, for the purpose of collecting unemployment statistics. Not only was every attempt made to recapture the refusals, but all persons contacted had a personal interest in the subject of the survey, unemployment being so widespread at that time.
8.9	1648	Clyde V. Kiser and P. K. Whelpton (27). This refusal rate occurred in securing interviews from a population of couples carefully selected as to age, age at marriage, religion, education, and color (white), with special sampling ratios for couples with various numbers of children in the family. The study was carried out in Indianapolis and was sponsored by a group of prominent local citizens. Though the subject was one of a very personal nature and the questionnaire was quite long, there was a compensating effect in that the respondents were paid for their cooperation. The interviewers were women with graduate training in psychology, sociology, and social case work and with successful experience as interviewers.
8.2	1933	McCall Corporation (10). See Table 11.3 and Section 8.2D. This study was carried out in Milwaukee by Alfred Politz Research on a fixed-address basis. Interviews were conducted in the home and with women only. The report stated that many women felt that the interview was an attempt to sell magazine subscriptions and that interviews were refused for that reason.
4.9	1036	McCall Corporation (10). See Tables 11.3 and 11.4 and Section 8.2D. This study was conducted parallel to the one quoted above and was in all respects similar to that, except the sampling was carried out on a quota control basis rather than on a fixed-address basis.
13.0	1879	A national quota sample conducted by NORC in December 1947. See Chapter 12.
6.8	3354	1950 Survey of Consumer Finances (9), conducted by SRC, University of Michigan. Nationwide sample. See Tables 11.3 and 11.4.
10.5	3630	A fixed-address sample conducted by NORC in three states, California, Illinois, and New York. See Tables 11.3 and 11.4.
7.7	1177	A city sample (Elmira) conducted on a fixed-address basis. See Section 9.2A and Tables 11.3 and 11.4.
8.0	1286	Durbin and Stuart (11). See Table 11.3. An experimental study conducted in London, comparing experienced and inexperienced interviewers. The rate given here is over-all, but it is important to note that the inexperienced interviewers had a significantly higher refusal rate than did the experienced interviewers.

[a] These figures represent the total number of contacted individuals. In general, they are the sum of interviews and refusals.

TABLE 11.13

Comparison of June Refusals with June Respondents

(Figures given with a decimal point are percentages of the figures in parenthese•

Characteristic of Respondents	Interviewed in June	June Refusals	Interviewed Plus Refusals [a]	Diffe enc
Total number of cases	1029	92		
"Did you vote on November 2?"	(944)	(68)	(1014)	
Yes	72.4	64.7	71.8	+0
No	27.6	35.3	28.2	−0.
(Voters) "Whom did you vote for?"	(683)	(44)	(727)	
Dewey	61.2	45.4	60.0	+1.
Truman	32.6	43.2	33.4	−0.
Wallace	0.4	0.0	0.4	0
Other	0.3	0.0	0.3	0.
Not stated	5.4	11.4	5.9	−0.
Sex	(1029)	(68)	(1097)	
Male	43.9	35.3	43.2	+0.
Female	56.1	64.7	56.8	−0.
Age	(1029)	(67)	(1096)	
Less than 35	36.1	17.9	34.6	+1.
35–64	52.2	59.7	52.8	−0.
65 and over	11.7	22.4	12.6	−0.
Economic level	(993)	(66)	(1059)	
A	0.3	0.0	0.3	0.
B	9.4	10.6	9.5	−0.
C	71.9	74.2	72.1	−0.
D	18.4	15.2	18.1	+0.
Education	(1029)	(66)	(1095)	
Less than grammar school graduate	10.3	16.7	10.8	−0.
Grammar school graduate and some high school	43.4	37.9	42.9	+0.
High school graduate and college	46.3	45.4	46.2	+0.
Marital status	(1029)	(68)	(1097)	
Single	11.9	26.5	13.1	−1.
Married	76.8	61.8	75.6	+1.
Other	11.3	11.8	11.3	0.
Religion	(1024)	(66)	(1090)	
Catholic	29.0	42.4	30.1	−1.
Protestant	67.0	54.5	66.0	+1.
Jewish	1.5	0.0	1.4	+0.
Other answer	2.5	3.0	2.5	0.
Telephone ownership	(1020)	(67)	(1087)	
Yes	71.2	82.1	72.1	−0.
No	28.8	17.9	27.9	+0.
Automobile ownership	(1026)	(67)	(1093)	
Yes	61.9	55.2	61.4	+0.
No	38.1	44.8	38.6	−0.
Race	(1019)	(68)	(1087)	
White	98.1	98.5	98.1	0.
Negro	1.7	1.5	1.7	0.
Other	0.2	0.0	0.2	0.

[a] Except for "candidate voted for," these values were obtained by weighting t▮ first two columns in the proportion 1029:92. This assumes that the 68 are a rando• sample of the 92. For "candidate voted for," the proportion voting in each grou• was also used in adjustment.

[b] (Interviewed in June) less (interviewed plus refusals).

differences (one from each question) being 0.8 of a percentage point. This is about the same as for the not-at-homes. Here again, as in the case of the not-at-homes, the problem of whether or not such differences are important depends on the uses to which the survey data are to be put.

B. Possible Approaches to the Problem of Refusals

The only satisfactory solutions to the problem of refusals are to reduce their number to such an extent that they can have little or no effect on sample analyses, or to incorporate the bias due to refusals into the sampling error of the survey estimates. If this latter course of action is adopted, the method described in the preceding section in relation to not-at-homes is applicable. Although the Study of Sampling has been unable to obtain data that would show the relative effectiveness of various devices in reducing the refusal rate, a list of possible devices can be readily set forth. Included would be such items as the selection and training of interviewers so that they will be able to secure and maintain the proper rapport, the assignment of respondents to interviewers (e.g., Negro interviewers to Negro respondents), the presentation of the survey to the potential respondents (identification of the sponsoring organization and advance notification of the content of the questionnaire), the use of payments and premiums to secure cooperation, the making of future appointments if the respondent refuses because of current pressures, and the follow-up of refusals by another interviewer or by a supervisor. In this connection it is interesting to note that several individuals have commented that the most effective means of reducing the number of refusals is by building up the confidence of the interviewers so that they never even think of the possibility of obtaining a refusal.

Even after a positive refusal has been obtained, an interviewer can do much to aid in the analysis of the sample results by keeping a record of all possible characteristics of the individual who refused. At the very least, the interviewer will be able to note down sex and an estimate of age and economic status. It may even happen that an individual will be willing to give the factual information required by the questionnaire when he is unwilling to reply to the attitude and opinion items. In some instances this information may be used to make estimates for those who refuse. One method for doing this was reported and applied in a Bureau of Labor Statistics publication (28).

In connection with the selection and training of interviewers, the paper by Durbin and Stuart (11) reports on an experimental comparison between the response rates obtained by experienced and inexperienced interviewers. The inexperienced interviewers were students from the London School of Economics and the experienced interviewers were from the Government Social Survey and from the British Institute of Public Opinion. Their concluding paragraph expresses the problem very well.

Though the inquiry has demonstrated the inferiority of the students in obtaining interviews when compared with professional interviewers, it tells us nothing of the causes of the differences and whether they can easily be remedied. Is it simply a matter of inexperience, or are the differences due in part at least to deeper causes, such as the students' youthfulness or the personality characteristics of people who go to universities? To investigate these matters further inquiries are needed, in which, for instance, some of the students would be given a short course of training before starting on the field work proper. It is hoped to include projects along these lines in the future research programme of the Division.

11.4 TOTAL SAMPLE LOSSES

The preceding sections have discussed separately the problems of not-at-homes and refusals. Though these are not the only individuals who are lost in the conduct of sample surveys, they do account for the major portion of the total sample losses. In this section we shall briefly discuss this total-loss group.

Table 11.14 gives some representative figures showing the extent of

TABLE 11.14

TOTAL LOSSES FOR A NUMBER OF SAMPLE SURVEYS

Per Cent Losses	Total Number of Attempts	Identification of the Survey
21.7	1380	SRC study on public use of the library (31, p. 63). Randomly selected adult from chosen dwelling unit in 33 cities and towns.
17.1	736	SRC, University of Michigan (5, p. 373; 6). Se Tables 11.3 and 11.4.
18.8	1267	A city sample (Elmira) conducted on a fixed address basis. See Section 9.2A and Tables 11.3 11.4, and 11.12.
27.6	4240	A fixed-address sample conducted by NORC in three states, California, Illinois, and New York See Tables 11.3, 11.4, and 11.12.
10.5	1709	New York State Survey of farm operators. Se Section 9.3A and Table 11.3.
14.0	1376	Durbin and Stuart (11). See Tables 11.3, 11.4 and 11.12.

all losses of potential respondents. It will be observed that the values in this table in general range between 10 and 20 per cent of the total number of attempts in a sample survey, and this seems to be reasonably indicative of what we may expect under present practices.

The comments that we can make about the total-loss group are not much different from those that have been made concerning individuals who are not found at home or who refuse to give an interview. Any reduction in the magnitude of the total group must come about through attacks on the separate components, and the methods of such attacks have been discussed in some detail in the separate sections of this chapter. The maximum possible effect of the mortality group may be incorporated into the sampling error of the survey estimates as was demonstrated in Section 11.2C, and the method of Birnbaum and Sirken (15) may be used in determining sample size under these circumstances. Heneman and Patterson (32) have expressed the point of view that total losses can be substantially reduced through interviewer training and have reported experiences to support this contention. There is little doubt that much can be accomplished in this direction, though the reader should note that their low rates were obtained in surveys that used a short interview and did not require the interviewer to see a specified individual in each household. Moreover, the interviewers were working in a small area under close supervision.

In order to show the effects of the total survey loss in comparison with the effects of the not-at-homes and the refusals, we have prepared Table 11.15 to summarize the experience of the Elmira Study (see Tables 11.8 and 11.13). In this study a large proportion of the total-loss group were interviewed at a later date (58 of 87 not-at-homes, 68 of 92 refusals, and 55 of 59 other losses). Under the assumption that the interviewed losses are representative of all losses, an assumption that cannot be checked readily, these various groups have been combined in proportion to actual losses in making up Table 11.15. In general, it will be seen that the effects of the three groups tend to cumulate, though the not-at-homes and the refusals balance one another to some extent in sex and about the question "Did you vote on November 2?" If balancing of this kind were very frequent, we might argue that attempts to interview members of the inaccessible group should not be pushed beyond a certain point. The remainder will then serve to balance against the group of refusals which, for a specific survey, will ordinarily be irreducible. Certainly nothing simple can be done to insure that this balancing will occur. However, it is unfortunately true that there may be situations in which, for a time

TABLE 11.15

COMPARISON OF THE EFFECTS OF THE VARIOUS LOSS GROUPS ON SOME
RESULTS OF THE ELMIRA STUDY

	Effect of Weighing Together Results for the Interviewed Group and Results for			
Question	Those Not at Home	Those Who Refused	The Combination of Not-at-Homes and Refusals	The Total-Loss Group
"Did you vote on November 2?"				
Yes	−0.3	+0.6	+0.3	+1.6
No	+0.3	−0.6	−0.3	−1.6
(Voters) "Whom did you vote for?"				
Dewey	+1.8	+1.2	+2.8	+2.9
Truman	−1.2	−0.8	−1.9	−1.8
Wallace	0.0	0.0	+0.1	+0.1
Other	0.0	0.0	0.0	+0.1
Not stated	−0.7	−0.5	−1.0	−1.3
Sex				
Male	−0.9	+0.7	−0.2	−0.2
Female	+0.9	−0.7	+0.2	+0.2
Age				
Less than 35	+0.4	+1.5	+1.8	+3.5
35–64	−0.9	−0.6	−1.4	−1.8
65 and over	+0.5	−0.9	−0.4	−1.7
Economic level				
A	−0.1	0.0	−0.1	−0.1
B	−0.1	−0.1	−0.2	−0.1
C	−0.3	−0.2	−0.5	0.0
D	+0.6	+0.3	+0.8	+0.2
Education				
Less than grammar school graduate	−0.5	−0.5	−0.9	−2.1
Grammar school graduate and some high school	+1.0	+0.5	+1.4	+1.3
High school graduate and college	−0.6	+0.1	−0.5	+0.8
Marital status				
Single	−0.2	−1.2	−1.3	−1.3
Married	−0.3	+1.2	+0.9	+0.7
Other	+0.5	0.0	+0.4	+0.5
Religion				
Catholic	−0.8	−1.1	−1.8	−2.4
Protestant	+1.0	+1.0	+1.9	+3.2
Jewish	−0.2	+0.1	0.0	−0.2
Other answer	−0.1	0.0	−0.1	−0.6
Telephone ownership				
Yes	−0.7	−0.9	−1.5	−0.4
No	+0.7	+0.9	+1.5	+0.4
Automobile ownership				
Yes	+0.1	+0.5	+0.6	+1.5
No	−0.1	−0.5	−0.6	−1.5
Race				
White	0.0	0.0	0.0	+0.1
Negro	0.0	0.0	0.0	0.0
Other	0.0	0.0	0.0	−0.1
Average of the largest absolute differences (one difference from each question)	0.7	0.8	1.1	1.8

at least, further efforts to contact people who have not been interviewed will increase rather than decrease the bias.

11.5 SUMMARY AND SUGGESTIONS FOR FURTHER RESEARCH

In the preceding sections some of the available data relating to the problems of callbacks and refusals have been summarized. These data are mainly of a descriptive nature and serve more to define the problems than to solve them. Also, several models have been presented for treating callbacks and refusals in special instances, models that require more information than is presently available for their effective application. In this section a summary will be presented of the suggestions that have been made for the collection of additional data and for the development of new methods of attack. Logically, the method defines the need for data and so should be presented first. However, the discussion seems to fit in better with what has gone before if the data are discussed first.

In connection with callbacks under a fixed-address method of sampling, one of the fundamental problems is the cost of making a callback compared with the cost of an original interview or a substitute interview. In many instances these relative costs should determine the exact sampling procedure that is to be followed. Unfortunately little data has been published on this aspect of the problem, and so attempts should be made to collect pertinent information. Such information could be obtained by asking interviewers to keep appropriate time logs on a number of surveys. The results of one such study, for a quota design, are described in Chapter 12. These data, combined with the analysis of Hansen and Hurwitz (22) or with the Birnbaum-Sirken approach (15) would allow us to make the most effective allocation of interviews among the various calls.

Another item of information that is valuable for fixed-address sampling, as well as for quota control sampling, is the proportion of people to be found at home during different periods of the day. For example, if it were known that only 60 per cent of the first calls made during the hours of 9 A.M. through 5 P.M. found an eligible person at home whereas the corresponding percentage for first calls after 5 P.M. was 80, then it might be possible to arrange for more evening calls and thereby increase the proportion of first-call successes. Similarly, for a quota control sample, the data might be used to determine the times at which interviewing should be done. One direct way of obtaining such information would be to ask a sample of individuals what times they were at home on the preceding day. This information could apply either to the entire time period during which interviewing might be attempted or to certain fixed time periods, each respondent being asked only about one such period. This latter procedure would simplify the questioning of respondents, and if the individual time periods were distributed randomly among enough respondents, adequate information could be obtained. These data on the distribution of times at which people can be found at home could easily be supplemented by a study of where people can be interviewed during the periods when they are not at home.

As for refusals, there is a great need for data to help reduce the refusal rate, for example data that pertain to the characteristics and proportions of refusals for various subjects of investigation and methods of sampling. It would also be valuable to know the effectiveness of various devices used for the reduction of refusal rates, such as payment for an interview, selection and training of interviewers, preparatory approach (e.g., appointments by telephone), and the like.

Once information on these points has been collected, we must consider its efficient use. We must apply methods and models already in existence to practical situations and develop new methods and models. At this point we may encounter one of the most difficult problems facing the survey designer, that of getting users of survey estimates to specify the error they can tolerate. As stressed repeatedly throughout this book, a rational approach to the determination of allowable error in surveys designed to measure attitudes and opinions is most urgently needed.

REFERENCES

1. W. Edwards Deming, *Some Theory of Sampling,* John Wiley & Sons, New York, 1950.
2. Emerson M. Brooks, "The General Enumerative Surveys," *Agric. Econ. Res.,* 1 (1949), 37–48.
3. John A. Clausen and Robert N. Ford, "Controlling Bias in Mail Questionnaires," *J. Am. Statist. Assoc.,* 42 (1947), 497–511.
4. Mildred Parten, *Surveys, Polls, and Samples,* Harper & Brothers, New York, 1950.
5. Frederick Mosteller, Herbert Hyman, Philip J. McCarthy, Eli S. Marks, and David B. Truman, *The Pre-election Polls of 1948,* Bulletin 60, Social Science Research Council, New York, 1949.
6. Angus Campbell and Robert L. Kahn, *The People Elect a President,* The Survey Research Center, University of Michigan, Ann Arbor, 1952.

U.S. Bureau of the Census, *Current Population Reports,* Series P-60, No. 9, March 1952.

Ernest H. Hilgard and Stanley F. Payne, "Those Not at Home: Riddle for Pollsters," *Publ. Op. Quart.,* 8 (1944), 254–261.

U.S. Federal Reserve Board, "Methods of the Survey of Consumer Finances," *Fed. Reserve Bull.,* July 1950, 795–809.

McCall Corporation, *A Qualitative Study of Magazines: Who Reads Them and Why: The Third in a Series of Continuing Studies,* New York, 1946.

J. Durbin and A. Stuart, "Differences in Response Rates of Experienced and Inexperienced Interviewers," *J. Roy. Statist. Soc.,* 114 (1951), 164–206.

Clyde V. Kiser, "Pitfalls in Sampling for Population Study," *J. Am. Statist. Assoc.,* 29 (1934), 250–256.

Robert Williams, "Probability Sampling in the Field: Case Study," *Publ. Op. Quart.,* 14 (1950), 316–330.

Hazel Gaudet and E. C. Wilson, "Who Escapes the Personal Investigators?" *J. Appl. Psychol.,* 24 (1940), 773–777.

Z. W. Birnbaum and Monroe G. Sirken, "Bias Due to Non-availability in Sampling Surveys," *J. Am. Statist. Assoc.,* 45 (1950), 98–110.

Z. W. Birnbaum and Monroe G. Sirken, "On the Total Error Due to Non-interview and to Random Sampling," *Int. J. Op. Att. Res.,* 4 (1950), 179–191.

Walter A. Hendricks, "Adjustment for Bias by Non-Response in Mailed Surveys," *Agric. Econ. Res.,* 1 (1949), 52–56.

F. Yates, "A Review of Recent Statistical Developments in Sampling and Sampling Surveys," *J. Roy. Statist. Soc.,* 109 (1946), 12–43.

National Opinion Research Center, *What . . . Where . . . Why . . . Do People Read?,* Report No. 28, Denver, 1946.

Alfred Politz and Willard Simmons, "An Attempt to Get the 'Not at Homes' into the Sample without Callbacks," *J. Am. Statist. Assoc.,* 44 (1949), 9–31.

J. Durbin and A. Stuart, "Callbacks and Clustering in Sample Surveys: An Experimental Study," *J. Roy. Statist. Soc.,* 117 (1954), 388–428.

Morris H. Hansen and William N. Hurwitz, "The Problem of Non-Response in Sample Surveys," *J. Am. Statist. Assoc.,* 41 (1946), 517–528.

W. Edwards Deming, "On a Probability Mechanism to Attain an Economic Balance Between the Resultant Error of Response and the Bias of Non-response," *J. Am. Statist. Assoc.,* 48 (1953), 743–772.

Willard R. Simmons, "A Plan to Account for 'Not-at-Homes' by Combining Weighting and Callbacks," *J. Marketing,* 11 (1954), 42–54.

M. A. El-Badry, "A Sampling Procedure for Mailed Questionnaires," *J. Am. Statist. Assoc.,* 51 (1956), 209–227.

Margaret H. Hogg, "Sources of Incomparability and Error in Employment-Unemployment Surveys," *J. Am. Statist. Assoc.,* 25 (1930), 285–294.

Clyde V. Kiser and P. K. Whelpton, "Social and Psychological Factors Affecting Fertility. V. The Sampling Plan, Selection and the Representativeness of Couples in the Inflated Sample," *Milbank Mem. Fd. Quart. Bull.,* 24 (1946), 49–93.

U.S. Bureau of Labor Statistics, *Family Spending and Saving in Wartime,* Bulletin 822, U.S. Government Printing Office, Washington, D. C., 1945.

National Bureau of Economic Research, *Studies in Income and Wealth,* Vol. 5, New York, 1943.

30. Elmo C. Wilson, "Adapting Probability Sampling in Western Europe," *Pu Op. Quart.*, 14 (1950), 215–223.

31. Angus Campbell and Charles A. Metzner, *Public Use of the Library c Other Sources of Information,* The Survey Research Center, University Michigan, Ann Arbor, 1950.

32. Herbert G. Heneman and Donald G. Patterson, "Refusal Rates and Int viewer Quality," *Int. J. Op. Att. Res.,* 3 (1949), 392–398.

CHAPTER 12

An Analysis of the Field
Operations of a National
Quota Sample Survey

12.1 PRINCIPAL RESULTS OF THE ANALYSIS

In order to obtain data for a detailed analysis of quota sampling methods, a simple operating record or "log" was obtained from 88 interviewers who took part in a national survey in December 1947. The survey was conducted by the National Opinion Research Center and was typical of the Center's quota sampling methods and other procedures. The interviewers recorded the time they spent in each attempt to obtain a complete interview, the result of the attempt, and other data about it. From the sequence of entries in the interviewers' log, we can trace their progress toward filling the quotas of interviews with people of various characteristics whom they were assigned to find for the survey. These logs are a first step toward direct observation of interviewers at work and experimentation in interviewing, both of which are needed for an adequate understanding of the operating characteristics of surveys conducted by interviewing in the field. The principal results of the analysis serve as a detailed example of such operations and provide data for rough estimates of what may be expected in similar situations.

1. The system of quota assignments appeared to work without adding any major operating difficulties to the technical and personnel problems that are common to a large class of censuses and surveys. Quota assignments did not seem to handicap the interviewers very much, even when the filling of some categories narrowed their work to a search for a few remaining classes of people. For example, the proportion of attempts resulting in completed interviews did not fall away substantially toward the end of the work, nor did the time spent in

seeking respondents increase markedly in relation to the number of completed interviews obtained.

2. The field operations were completed quickly. Most of the interviewers began their work within two days after they received their instructions and a supply of interview forms. Most of them completed their interviews in three days or less from the time they started interviewing.

3. The 88 interviewers made 2319 attempts in completing 1223 interviews and worked 1340 hours, including the time they spent studying the instructions for the survey and filling out reports. The log itself took about 180 hours of this total. The balance of 1160 hours averages 30 minutes per attempt and just under 1 hour per interview completed.

4. The log recorded 191 hours spent by interviewers in seeking the next respondent, i.e., between the end of one attempt to obtain an interview and contact with a person who might be interviewed. This count did not include time spent in going from one district or locality to another or in traveling between the interviewer's home and the district in which the interviewing was done. Also excluded is time spent between interviews to write reports, stop for meals or refreshments, and do other things not directly aimed at starting another interview. The time spent in seeking respondents averaged five minutes per attempt, about one-sixth of all the time that interviewers worked apart from the time they spent on the log itself.

5. The time spent in actual interviewing, and in related contacts with persons for the purpose of finding respondents and gaining their cooperation, totaled 520 hours. This is less than half the total time devoted to the survey, about 45 per cent of it, and averages almost 14 minutes per attempt or 25 minutes per completed interview. The interview was shorter than most NORC survey interviews. (The survey was chosen for this reason as one in which the work of keeping the log would not unduly increase the load on the interviewers.)

6. To obtain 1223 completed interviews it was necessary to make 2319 attempts. Forty-seven per cent of the attempts were not successful because no one was home when the interviewer called at a dwelling; or the person approached refused to be interviewed or was too young (under 21) or belonged in a quota classification for which the assigned number of interviews had been completed. In addition some interviews were terminated after they had been started but before all the questions had been answered. Some of the unproductive attempts were directed to persons who were to be excluded from the sample, even if the interviewers had found it possible to obtain a com

te interview. Hence, their loss represented only a loss of time.
her attempts, however, if they had been successful, would have pro-
ced interviews that were needed for the sample. Their loss had
ne biasing effect on the results of the survey. The nature and mag-
ude of the bias have not been determined but could be studied by
king repeated calls and trying to gain cooperation from people
o refuse at the first call. The problem is similar to that for fixed-
dress sampling discussed in Chapter 11, pages 241–266.

The time spent in seeking and interviewing persons who gave a
mplete interview was 28½ minutes per interview. If the time spent
unsuccessful attempts is added, the average was 35 minutes per com-
eted interview. Thus the loss of time was about 20 per cent of the
ne devoted directly to finding and interviewing respondents. This
mparison takes no account of time devoted to travel and other ac-
vities, since there is no clear basis for determining how it was in-
eased by the unsuccessful attempts; but, including all the time the
terviewers worked (apart from keeping the log), the time lost in
successful attempts was more than 10 per cent.

7. About seven-eighths of all attempts were made by calling at pri-
te homes and apartments, and one-eighth were made in offices,
ores, and public places. A little more than one-fifth of the home
lls were unsuccessful because no one was at home. This propor-
on is lower than that found on the first call in surveys for which
e sample is designated by a list of addresses or dwelling units.
onetheless, it is large enough to show that those quota surveys in
hich most of the interviewing is done in the home do not escape
uch of the cost that is added by people who are hard to reach (see
hapter 11), even though no further effort is made to find them after
e first unsuccessful attempt.

8. There was no great difference between home calls and attempts
sewhere in the average time per attempt or the proportion of at-
mpts that resulted in completed interviews.

9. There were differences between regions of the United States and
tween communities of various sizes in average times and propor-
ons of attempts that were successful. These differences may be ex-
cted to change somewhat with the season of the year, and in part
ey are due to factors other than the characteristics of the region or
pe of community.

10. There are greater differences between interviewers and, of
urse, the greatest variation of all between interviews and attempts
y the same interviewer. These differences suggest that there may be
pportunities for savings and improvements by further analysis of

interviewer performance, though most of the variation may be inherent in the human behavior that is involved—first, in the formation and expression of opinion by individuals and, second, in the process of interviewing.

11. The average times and proportions of successful attempts also vary with the time of day, day of the week, and type of respondent. There was no overwhelmingly best time to interview or most difficult type of person to find, at least so far as the quota categories of age, sex, race, economic level, and farm or suburban residence were concerned.

12. The survey analyzed in this study did not utilize any system of distributing quotas by blocks or other local areas within the community, a practice that has since become fairly common in quota sampling. However, it did instruct interviewers to scatter their interviews. Roughly, one-fifth of the completed interviews were within the length of a city block from the preceding completed interview. The proportion of attempts was higher since unsuccessful attempts were usually followed by another attempt at a neighboring home or location. Roughly half of all attempts in cities were within a block of the preceding attempt. These proportions understate the proportion of interviews or attempts that are within the length of a city block, since some of those that did not have an interview or attempt preceding them within the distance did have one following within the distance. The total mileage reported by interviewers who drove their own cars was 3200 miles. No account was kept of bus rides and walking distances, but it is clear that this variety of quota sampling involves a great deal of local travel, more than 3 miles per completed interview.

A considered comparison of this quota sampling procedure and other sample designs would include many other questions of cost, accuracy, operational feasibility, and particular conditions. The general findings and detailed results of this analysis, however, do represent the kind of information that is needed for all similar sampling operations. Much of it can be obtained as part of the regular system of reports and even as part of the interview record. The more general theory of sampling operations cannot be applied adequately without such an analysis of operations.

12.2 BACKGROUND AND PREPARATIONS

Early in the development of the Study of Sampling, it was recognized that relatively little information was available about the actual ork of quota interviewers or the particular methods by which they fill their assignments in opinion surveys. Since rather specific data ere needed about quota operations by the Study of Sampling, we ggested to a number of organizations that they undertake a special udy of the work and experiences of their interviewers in the field. e talked with interviewers and supervisors, asking them about the ethods that they use in finding the sample of respondents and obining interviews to complete their quotas. We explored with them e possible ways in which their methods of working might affect the sults of the survey, either favorably or unfavorably.

In these conferences, we found general agreement that a study of ota interviewing could not be carried out by sending someone with ch interviewer to observe what he does and record various facts out time and conditions to use in analyzing his methods. An interewer would feel uncomfortable if he were watched. His performce would be affected, making the study less valuable for obtaining realistic understanding of the work and experience of the quota terviewers.

Another possibility was to have each interviewer keep a record as e went about his interviewing. He would keep a running account, a "log," listing each attempt to find a respondent, where it was, e time it took, whether he got an interview or refusal or some other sult, and whatever additional facts would help make up a fair and ccurate picture of his work. The head of one research organization ld us that his interviewers would resent being requested to keep ack of their time. He said he would not dare to ask them to do . Others indicated a favorable interest in the proposal, especially Ir. Paul Sheatsley of NORC. He said he had planned for some time make just such a study. Consequently, arrangements were made conduct the study with NORC interviewers as a trial or experiment at would serve as an example for similar studies to be conducted in ther organizations.

A pretest was made by NORC in Denver in July 1947. Nineteen terviewers participated in the study. Each had a quota of approxiately thirty interviews to obtain. The pretest demonstrated that it as feasible to undertake a study of this kind and provided quite a it of information that was helpful in preparing for the study taken a national basis. It showed that there are substantial differences etween interviewers, making it desirable to analyze the result of the ational study in terms of a number of interviewer characteristics, as ell as in terms of type of community in which the interviewer oper- tes, section of the country, and a variety of other factors.

Following the Denver pretest, we waited for a good opportunity
incorporate the study in a regular national opinion survey. The stu
was launched in December 1947 as part of a survey that had a sche
ule of questions short enough to permit the addition of the interview
log without overloading the interviewers.

A record form for the log was sent to the interviewers with the
regular instructions and materials. Appropriate instructions for kee
ing the log were included. When the completed interviewer logs we
returned, they were analyzed by the staff of the Study of Sampliı
with general advice and assistance from the staff of NORC.*

12.3 THE SURVEY, QUESTIONS, AND INTERVIEWERS

The survey for which the log was kept was a study of opinion c
domestic and international questions. The field operations were co
ducted in the period December 5 to 15 and were affected by the sea
sonal conditions of winter weather and Christmas shopping. No othe
circumstances of an unusual nature were observed during the perio
that the survey was in the field. The interviewer assignments weı
similar to those usually made by NORC in conducting its surveys, eı
cept that the interview schedule was relatively short.

The schedule consisted of eighteen questions plus the usual factuı
items. One of the questions was a multiple-choice question, and th
others involved answering "yes," "no," "don't know," or a similar se
of alternatives. Four of the questions provided that, when certaı
answers were obtained, the interviewer would ask additional ques
tions to obtain a better picture of the respondent's opinion. A typ
ical attitude question is the following: "Do you think there is any
thing the United States can do to make the United Nations more suc
cessful?" In addition to the questions about attitudes and opinion
there were four factual questions concerning the respondent's occupa
tion, education, age, and rent. The interviewer himself observed an
recorded five other facts, about sex, race, economic level, size of com
munity, and address. The name of the place, the date, and the signa
ture of the interviewer were also entered.

This description of the questions and schedule has been given t

* Except for this advice and information, the analysis has been made inde
pendently of NORC. The interest of NORC in the study, its cooperation, and iı
willingness to have the results analyzed by another organization reflect a ver
commendable attitude toward research on opinion survey methods and plac
the opinion research profession in debt to NORC for this contribution to iı
knowledge of survey methodology.

provide a reasonably accurate statement of the amount of work to be done in the interview. It is important that studies of interviewer performance provide a description of the questions that are asked and other interviewing that is performed, so that the amount of time recorded for the actual interviews can be related to the amount of work done.

Interviewers were allowed a week in which to complete their interviews and mail the schedules to the Central Office. In a few instances, interviewers received their assignments late and were given additional time to complete them.

The interviewers were drawn from the regular staff of the NORC. No effort was made to select or reject any particular interviewer for reasons related to the study. In other words, the choice of interviewers and assignment of work to them were made precisely as they would have been made if the survey had been conducted without the interviewer log.

The interviewers included some who had been on the staff of NORC for more than 100 surveys and others who were making their first survey for NORC. They varied in age, experience, occupation, education, and other characteristics. Most of them were housewives who did not have other jobs and welcomed an opportunity to work part time, or occasionally full time, for such an organization as NORC. Undoubtedly, they differ in many ways from other groups of people, and their performance can be expected also to differ from that of workers in other occupations. Their reasons for doing this kind of work, their interest in it, and their reactions to it vary greatly from one individual to another. In a subsequent section of this report, we shall give quotations of their reactions to the study as they expressed them in their reports.

12.4 INFORMATION TO BE OBTAINED

The principal purpose of this study was to find out how quota interviewers do their work, what problems arise, what conditions they encounter, how long it takes them, how many times they have to try in order to obtain their assigned number of interviews, what difficulties make some of the efforts unsuccessful, and how often these difficulties arise. We wanted to see in what order interviewers usually fill their quotas, that is, whether they obtain the men before the women and younger people before older people. We wanted to know whether they found it harder to reach certain categories such as the A and D economic levels or young men or some other group. We wished to find out how much the quota restrictions increased the amount of work they had to do by narrowing down the possibilities. We might expect that as they neared the end of the assignment they would have to turn away from many possible interviews because they already had enough individuals to fill that part of their quota assignments.

We also wanted to discover for a typical quota sample how large a proportion of all attempts were unsuccessful because no one was found at home. Then we could compare quota sampling with other types of sampling in this respect. Similarly, we were interested in knowing the proportion of attempts that were unsuccessful for other reasons, such as refusals.

In addition to these questions about the results of the interviewers' efforts, we wished to obtain representative data about the length of time that is taken for seeking out contacts, interviewing, and other work. These data are helpful in studying the relative efficiency and cost of different types of sampling methods and survey procedures. Of course, surveys are not all alike and differ from one to another in the time required to complete them, depending on the nature of the interview, the nature of the sample, the rules that are followed about scattering the interviews, and a number of other important factors. However, an accumulation of information about the time required for various parts of survey operations, with appropriate descriptions of the operations and the circumstances under which they are performed, will provide a basis for estimating the time requirements of different sampling schemes much more accurately than has been possible in the past.

12.5 THE LOG, A RECORD OF FIELD OPERATIONS

These general purposes led to the development of the log form on which the interviewer recorded each attempt he made to obtain the respondents for interviewing. It included a record of the time spent in traveling from the interviewer's home or other starting place to the district in which he worked. Then, successively for each attempt, were recorded the time at which he started the attempt, the type of person he was seeking in terms of sex, age (under 40 years of age or 40 years and over), economic level (A, B, C, or D, expressed in terms of rental classes which differ from one place to another to take account of the local rent level), and whether he was looking for farmers or Negroes in the case of those interviewers who had been assigned quotas of farmers and Negroes. Following this the interviewer recorded the time at which he began his contact. Ordinarily this was the time at which he arrived at the home or other place where he would attempt an interview. He recorded the place, that is, whether the interview was at the home of the respondent, or some other place, the address, the type of person with whom he made contact, the time the contact ended, and the result of the attempt. The result of the attempt was classified in one of six groups:

NH   No one found at home.

Ref   The person approached refused to be interviewed.

DFQ   The person approached belonged in a quota category that had already been filled and hence was rejected by the interviewer.

Rej   The person approached was rejected after the interview had begun because he was unsuitable for reasons other than the quota, such as being too young, in the Armed Forces, a non-resident, or unable to speak English.

Term   The interview was begun but was terminated before it was completed.

I   A completed interview.

After recording the information for an attempt, the interviewer noted on the sheet any time that elapsed before he started the next attempt. This time might be devoted to any one of several purposes. It might be devoted to reviewing the instructions or to time out for a cup of coffee or personal shopping. In other instances the time might be used for travel to another district or for some other activity necessary for the survey but not directly involved in establishing contact with a potential respondent.

In this way, the interviewer wrote in the interviewer's log, in succession, each attempt he made and the results of the attempt. Unusual conditions or incidents were recorded as notes on the back of the form. Explanations were given when the amount of time taken was different from the usual. A number of other observations were made in notes entered on the interviewer's log.

The interviewers were instructed to report separately on their regular time sheets the time required to make up the log and were paid for the time involved in the preparation of the log.

12.6 GENERAL REACTIONS AND COMMENTS
OF THE INTERVIEWERS

In any study of this kind it must be recognized that the application of the instrument, in this case the log, will affect the behavior of individuals whose work is being studied. This is true even though considerable part of the purpose of the log was to find out not about the interviewers but about the people who were being interviewed and about the sheer physical requirements of the job in terms of time and travel. It was inevitable that some interviewers would suspect that the purpose of the log was to reduce their rate of pay or to check up on their performance and that it might be used to their disadvantage. Others apparently found it interesting and saw its possibilities for scientific purposes. Most of them, however, regarded it as incidental to their regular work and, therefore, did not react toward it very noticeably one way or another.

An analysis of the report forms showed that approximately fourteen of the ninety interviewers who returned the report form expressed favorable reactions to the log, sixteen expressed strongly unfavorable reactions, and the remaining sixty were either indifferent or mildly unfavorable. Favorable remarks of the following type were made:

"It's a good idea and I enjoyed it after I got started."
"It was a very simple procedure. I didn't mind it a bit."

Examples of the mildly unfavorable comments are:

"An interesting experiment, but it did slow up the job."
"It didn't seem too difficult, though it was a nuisance."
"I didn't mind it this time but surely would hate to do it on all surveys."

The highly critical group made remarks such as these:

"I did not enjoy this but if it helps I guess I can endure it. I'm not at all methodical and recording something every time I breathe is painful to me."
"We have enough forms to keep now. Too much paper work."
"Never has there been such a difficult assignment given interviewers. I'm sizzling!"
"Had a tendency to take my thoughts from the interview. It took a great deal of time for small results."

These comments reveal a great variety of reactions and, undoubtedly, reflect fundamental differences among interviewers in their general mental attitudes, habits of work, and attitudes toward their

ob, as well as in their ability to record carefully the detailed data
that were called for on the log sheet.

There was some evidence that the keeping of the log had a stimulating effect on the interviewers and actually speeded up their work.
It is doubtful that this would continue to be true if the log were a
regular feature of their work. Unquestionably, the initial effect was
to make them more aware of the time they were taking in doing the
work, which may have affected their methods and procedures. However, it is believed that the amount of such stimulation was not very
great and not enough to disturb unduly the results of the study.
But this stimulation is a factor that should be kept in mind if we
attempt to use the results of the study in a very precise manner.

The interviewers, for the most part, were not accustomed to keeping detailed records or fond of that kind of work. Many of them
had chosen interviewing in preference to other kinds of work because
they had more than average ability to talk with people and because
of their interest in human contacts. Keeping time records tends to
be somewhat alien to these aptitudes and interests. Staff members
of NORC report that they have noticed in the past that interviewers
who get the best rapport and the fullest response to questions are
often less careful in filling out the details of a questionnaire, and
those who excel at errorless records are often weaker in their relationship with respondents.

Naturally a survey organization seeks people who can establish
good relations with individuals chosen for interview. Such interviewers can be taught accuracy, whereas it is difficult to teach people who are accurate but not successful in their social relations to do
effective interviewing work. The inevitable result is that people selected because of their ability to handle people tend to react unfavorably to additional paper work. These tendencies present something of a dilemma to a survey organization. It is important to have
interviewers who are successful in getting cooperation from the respondents and in eliciting full and frank statements of attitudes and
opinions, so that the results of the survey may truthfully reflect
what people think. On the other hand, if these successful interviewers are not adept at catching and recording fully the responses they
obtain, the value of obtaining them may be very largely lost. It is
even possible that the answers that are recorded on the survey forms
will be less accurate than those the somewhat less skillful but more
methodical interviewer would obtain.

Studies of the bias of question asking and response recording that

were undertaken concurrently with this Study throw light on thi
problem of interviewer selection (1). In the present study this prob
lem has some important consequences. If interviewers had been se
lected with greater emphasis on their ability to make accurate rec
ords, they might have had a larger proportion of their attempts end
ing unsuccessfully in refusals or unsatisfactory schedules. An;
change in the policies of selecting interviewers would, of course, b
reflected in changes in the rates and average times shown in thi
report. The results of this study are most representative, therefore
of survey organizations similar to NORC, with similar personnel op
erating under similar conditions. Allowance must be made for an;
differences between the situation in which this information may b
applied in the future and the situation as it was at the time of th
survey in which the log was recorded.

A number of interviewers reported that they found it particularly
burdensome to make the entries on the log when they encountered a
long string of not-at-homes in a short space of time. In apartmen
houses, for instance, they might ring three or four different doorbell
in the space of a minute, and some of them confessed that they di
not make the appropriate entries in every case. One commented, "I
takes two minutes to write down what you spend thirty seconds do
ing." This is not a serious omission, since the amount of time in
volved is very small, but it does tend to diminish somewhat th
proportion of all attempts that should be recorded as not at hom
and increase the average time per attempt. It is likely these omis
sions occurred only occasionally. Omissions and inaccurate work ar
to be expected in a study of this kind, since it was of secondary in
terest to the interviewers. Omissions are especially likely when th
item that is omitted appears to be insignificant, as a few seconds
spent in ringing a doorbell may well seem to be. In future work
it is important to explain adequately to the interviewers the impor
tance of some of these seemingly minor items, so that they will not
feel they are wasting their time in making entries for them.

12.7 HOW THE SURVEY PROGRESSED

A few general facts about the progress of the survey will be help
ful as background for the more detailed analyses that follow. The
assignments were mailed on Wednesday, December 3, 1947. About
three-fourths of the interviewers received their assignments on the
fifth or sixth. This gave them an opportunity to work during the

eek end, when they could reach some of the respondents who
ight not be at home on weekdays. The assignments varied from
to 25 interviews, averaging 14. Table 12.1 shows for each date the
ate at which assignments were received, work started, work com-
leted, and interviews obtained.

TABLE 12.1

DAILY PROGRESS OF THE SURVEY

| Day and Date, December 1947 | Date on Which Interviewers | | | Number of Interviewers Who Worked This Day [a] | Number of Interviews Completed | Interviews per Interviewer Working |
	Received Work	Started Inter- viewing	Completed Interviews			
ri. 5	19	5	0	5	16	3.2
t. 6	42	27	3	32	178	5.6
n. 7	2	11	6	30	158	5.9
on. 8	11	23	20	54	316	5.9
es. 9	4	13	25	48	292	6.1
ed. 10	1	4	21	29	160	5.5
hur. 11	1	2	6	10	45	4.5
ri. 12	1	1	3	5	27	5.4
t. 13	0	1	2	3	9	3.0
n. 14	0	0	0	1	7	7.0
on. 15	0	0	1	1	8	8.0
ate not recorded	7	1	1	1	7	7.0
otal	88	88	88	219	1223	5.2

[a] All the interviewers worked every day from the date on which they started until they finished their
ssignments, except ten who did not work on Sunday the 7th, three who did not work on the 8th, and
vo who did not work on the 9th.

Ninety-one interviewers participated in the work of the survey.
)f these, two did not fill out the log and one log was excluded from
he study because the interviewer was quite atypical in methods of
vork and in the record that was compiled.

A total of 2319 attempts were reported on the logs of the 88 interviewers whose logs were tabulated. Of these, 1223 resulted in completed interviews.

Most of the interviewers completed their assignments within four
lays after they received them. The last interviewer to complete his
work finished on December 15, twelve days after the assignments
were mailed. Eighty-four of the 88 interviewers completed their
work in four days.

Each interviewer was left free to determine how many hours he
would work each day. Most of the work was done at a rate of one
to six hours a day, i.e., less than a full-time job would require.

Normally, interviewers did not start work before nine o'clock in the morning because respondents are likely to be preoccupied wit household tasks or other business before this hour. Interviewe: started work before noon about as often as they started after noo Ordinarily, they did not work beyond nine o'clock at night. Man of them did not work in the evenings at all if they could obtain th respondents they sought during the daytime. Almost a third of th working days included some work after 6 P.M.

12.8 THE QUOTAS OF INTERVIEWS ASSIGNED

The 88 interviewers who submitted logs lived in 70 localitie throughout the United States ranging in size from the New Yor Metropolitan District to rural areas. The distribution of the sampl by size of place and by quota categories was as shown in Table 12. on page 288. The distribution of the assignments by regions is show in Table 12.3 on page 289. The two accompanying examples ar fairly typical of the quotas that were assigned.

EXAMPLE 1

QUOTA ASSIGNED TO INTERVIEWER FOR THE ST. LOUIS METROPOLITAN DISTRICT

	Men	Women	Total
Quotas by age			
Under 40	3	4	7
40 and older	4	4	8
Total	7	8	15
Quotas by race and economic level [a]			
A ($96.51 or more)	0	0	0
B ($44.01–96.50)	1	1	2
C ($18.01–44.00)	3	4	7
D ($18.00 or less)	2	2	4
Negro	1	1	2
Total assignment of interviews	7	8	15

[a] The economic levels were defined in terms of rentals (or equivalent rent fo an owner-occupied dwelling). The particular rental classes varied from on locality to another. The Negro quota was not subdivided by economic leve It was also specified that six of the interviews should be from the suburbs.

EXAMPLE 2

QUOTA ASSIGNED TO INTERVIEWER FOR RURAL NON-FARM RESPONDENTS NEAR
CLARION, IOWA

Rural Non-farm Residents	Men	Women	Total
Quotas by age			
Under 40	1	1	2
40 and older	2	2	4
Quotas by race and economic level *a*			
A	0	0	0
B	0	1	1
C	2	1	3
D	1	1	2
Negro	0	0	0

Farm Residents	Men	Women	Total
Quotas by age			
Under 40	2	2	4
40 and older	2	3	5
Quotas by race *a*			
White	4	5	9
Negro	0	0	0
Total assignment of interviews	7	8	15

a For small communities economic levels were not defined in terms of rent but by relative standards.

TABLE 12.2

SUMMARY OF QUOTA ASSIGNMENTS BY SIZE OF PLACE

		Size of Place, in Thousands								
Quota Category		Rural	2.5–10	10–25	25–50	50–250	250–500	500–1000	1,000–10,000	Total
Farm residents	M	83 [a]	6	2	3	4	2	2	0	102
	F	85 [a]	6	2	2	3	4	2	0	104
Non-farm White by economic level										
A	M	2	0	0	0	1	1	2	3	9
	F	3	1	0	0	0	1	2	3	10
B	M	8	5	6	4	8	10	5	21	67
	F	12	5	7	4	11	6	5	24	74
C	M	41	18	19	13	36	29	20	80	256
	F	35	18	23	15	36	32	23	84	266
D	M	22	9	14	7	22	17	15	49	155
	F	28	12	12	8	21	21	11	49	162
Non-farm Negro	M	5	1	2	3	4	1	1	11	28
	F	8	0	3	0	5	1	2	11	30
Total quota	M	161	39	43	30	75	60	45	164	617
	F	171	42	47	29	76	65	45	171	646
Total		332	81	90	59	151	125	90	335	1,263
Age										
Under 40	M	69	15	18	13	37	26	19	75	272
	F	74	20	21	14	36	32	22	83	302
40 and over	M	92	24	25	17	38	34	26	89	345
	F	97	22	26	15	40	33	23	88	344
Number of interviewers to whom these quotas were assigned		22	6	6	5	13	9	7	23	91

[a] The rural farm quota was divided further, as follows: White male 69, White female 72, Negro male 10, Negro female 9.

TABLE 12.3

SUMMARY OF ECONOMIC, RACE, AND FARM QUOTAS BY REGION

Region	Economic Level				Negro	White Farm	Negro Farm	Total
	A	B	C	D				
New England	2	13	48	28	2	11	0	104
Middle Atlantic	4	31	118	71	13	6	0	243
East North Central	4	33	122	74	11	38	0	282
West North Central	3	13	50	31	5	36	0	138
South Atlantic	2	16	56	35	7	29	11	156
East South Central	1	6	21	14	11	26	0	79
West South Central	0	9	34	19	6	24	8	100
Mountain	1	4	16	10	0	8	0	39
Pacific	2	16	57	35	3	9	0	122
Total	19	141	522	317	58	187	19	1,263

Region	Males [a]		Females [a]		Total	Number of Interviewers
	Under 40	40 and older	Under 40	40 and older		
New England	20	25	23	25	93	7
Middle Atlantic	52	65	57	63	237	16
East North Central	51	68	58	67	244	20
West North Central	21	29	25	27	102	10
South Atlantic	26	31	27	32	116	12
East South Central	11	14	14	14	53	7
West South Central	13	19	16	20	68	8
Mountain	7	8	8	8	31	3
Pacific	25	30	28	30	113	8
Total	226	289	256	286	1,057	91

[a] Exclusive of farm quotas.

The number of interviews assigned to one interviewer varied from seven to twenty-five but most of the assignments were from twelve to fifteen.

Interviews in Quota	Interviewers
7	4
8	1
9	2
10	5
11	8
12	12
13	11
14	16
15	11
16	3
17	4
18	2
19	6
20	4
21	0
22	0
23	0
24	1
25	1
Total	91

From these facts it can be seen that a large and complicated quota of interviews widely distributed over the United States was assigned. The work was completed in eleven days. The number of interviews assigned to each interviewer was small enough to permit him to complete his work in three or four days and even in two. It is doubtful that allowance of a longer period of time for the field work would have contributed to the accuracy or completeness of the survey. If larger quotas had been assigned to each interviewer, however, it might have been less difficult to complete the survey. With a large number of interviews to fill, each interviewer would have had somewhat more latitude in accepting potential respondents. The effects of chance in encountering eligible respondents would have been reduced. Obviously, if a quota of only one interview had been assigned, the chances would have been only about 0.5 that the first cooperative adult person would fit the quota with respect to sex, possibly less than 0.05 that he would fit with respect to the combination of sex, age, and economic status. In a larger quota these restrictions become less important. Only the last of the interviews must be conducted under such highly restrictive circumstances. No special analysis of this point was undertaken here, since variation in the size of assigned quotas was quite small. It was thought that special circumstances might be present with very large or very small quotas.

12.9 THE HURDLES—KINDS OF UNSUCCESSFUL ATTEMPTS

Each interviewer set out to fill the assigned quota of interviews by making calls at homes in various parts of the district and in some instances by approaching persons in stores, offices, public buildings, or on the street. About five out of nine of these calls and contacts were successful in producing completed interviews. The other attempts were unproductive for a variety of reasons and were classified in the log as "not at home," refusals, "didn't fit quota," rejected, and terminated. Any one of these classifications, of course, could have been divided to exhibit more specifically the circumstances that made' the attempt unproductive, but this would have introduced troublesome problems of making the classification comparable from one interviewer to the next and it would have made the number of instances of each type too small for reliable analysis. This section will show the relative importance of each type of unsuccessful attempt and the rate of shrinkage or *attrition* separately by region, size of place, and type of interviewer. A study of the attrition rates may be helpful in searching for possible sources of bias and in improving the field operations of surveys. At each attempt to start and complete an interview, an interviewer may encounter any one of a number of possible difficulties. First of all, when interviews are sought at a residence, the interviewer may find no one at home when he rings the doorbell. Upon getting in contact with a possible respondent, the interviewer may find that this person refuses to be interviewed because he or she is, at the moment, preoccupied with personal business, family duties, or housework, or going out of the house for business or pleasure. The prospective respondent may also refuse to be interviewed because he is not in sympathy with the survey or is unwilling to state his views.

After clearing the hurdle of refusals, the interviewer has to determine whether the potential respondent is qualified for interviewing in relation to the quota categories that remain to be filled. In the first few attempts, of course, none of the quota categories is filled and a person of either sex, of any age above twenty, and any economic status will be acceptable, providing that person is not a visitor from outside the interviewer's district. Toward the end of the interviewing, however, the interviewer may have completed the number of assigned interviews with females, with persons under forty years of age, and for all economic levels except the lowest class, D. He will then have to look for a male, forty years of age or older, in the economic class D. Then the quota requirements will definitely restrict his opportunities to find a suitable respondent.

TABLE 12.4

THE HURDLES FROM INITIAL ATTEMPT TO COMPLETED INTERVIEW

Hurdles to Completed Interview	Number	Per Cent
Attempts to obtain interviews	2319	100
Less those in which no one was at home	443	19
	—	—
Remainder with a contact made with some individual	1876	81
Less those who refused to be interviewed	244	11
	—	—
Remainder willing to be interviewed	1632	70
Less those in which the person was too young or did not fit an unfilled quota category for some other reason	280	12
	—	—
Remainder, interviews begun	1352	58
Less those who terminated interview before it was completed or who were rejected by the interviewer after the interview was started	129	5
	—	—
Remainder, completed interviews	1223	53

Finally, after clearing the hurdles of obtaining a willing respondent who qualifies under the quota restrictions, the interviewer may find it necessary to reject the respondent for other reasons, such as inability to answer the questionnaire or conviction that the respondent is not replying truthfully. In some instances, interviews that are started are terminated before they are completed for a variety of reasons. Consequently, the last two hurdles are those of rejected and terminated interviews.

After all these hurdles had been cleared in the NORC survey, there remained a group of respondents who were qualified under the quota restrictions and who answered the questions that were put to them, yielding completed interviews. The numbers of attempts that failed to clear each hurdle are shown in Table 12.4.

12.10 ATTRITION RATES

The purpose of this section is to examine the successive losses or attrition at these various hurdles and to examine the relation of the regions, the size of the city, the time of day, the day of the week, and various interviewer characteristics to these types of attrition. We shall also seek to determine the cost in time of these types of attrition, so that it can be used in the designing of sampling surveys and can be compared with other types of interviewing and survey procedures. Obviously, these loss rates depend in part on the nature of the survey, particularly in the case of refusals. (See Chapter 11, pp. 265–266.)

There appear to be definite differences in attrition rates between various parts of the country. Table 12.5 shows for each region and major geographic division the percentage of the attempts reaching

TABLE 12.5

ATTRITION RATES AT EACH HURDLE, BY REGIONS

Major Geographic Division and Region	Number of Attempts	Percentage Attrition [a]				Per Cent of Attempts Successful
		No One at Home	Refused	Did Not Fit Quota	Rejected or Terminated	
East						
New England	243	18	13	31	12	44
Middle Atlantic	502	19	21	13	20	45
North Central						
East North Central	493	21	12	13	5	57
West North Central	228	21	7	18	1	60
South						
South Atlantic	272	16	9	15	9	59
East South Central	109	16	9	17	9	59
West South Central	173	16	9	20	5	58
West						
Mountain	67	7	19	18	2	60
Pacific	232	24	13	18	14	47
All regions	2319	19	13	17	10	53
East	745	19	18	19	17	45
North Central	721	21	10	15	4	58
South	554	16	9	17	7	59
West	299	20	15	18	11	49

[a] Each percentage is computed from the number of losses of the kind indicated at the head of its column and the number of attempts remaining after one excludes those that were unsuccessful for reasons represented by columns to the left of it. Hence the percentages do not add to 100. The regions are those used by the Census of Population.

the hurdle that did not succeed in passing it. Some of the differences that are shown may be due to chance or irrelevant factors, or to the particular characteristics of the interviewers who were recruited for the field staff in each region. There are certain differences, however, that appear to be related, at least in part, to characteristics of the population of each region and to the regional weather in December. The East shows a high rate of refusals, rejections, and terminations and, in comparison, the North Central and the South show low rates. The similarity of the rates of not-at-home and did-not-fit-quota suggests that there may be no substantial regional differences in these types of attrition, though there may be important local differences within the regions.

Attrition rates are given in Table 12.6 for rural interviews and a classification of cities by size. Attempts in rural districts or urban places of less than 10,000 population appear to have the lowest attrition rate. Apart from this there is little evidence of a relation between attrition rates and the size of cities over 10,000. This kind of analysis of attrition can be carried down to areas within a city, but to get enough experience for comparison we must group areas of similar type, such as apartment house areas, high and low rent areas, and areas with high percentages of home ownership.

Attrition may be expected to vary by the day of the week and time of day. Table 12.7 shows the rates by the date and day of the week. The variation shown on the last four days is probably due to chance and special circumstances. Sunday seems to differ remarkably little from other days of the week, though it clearly has lower rates of not-at-homes and refusals. Sunday in the summer time might show higher rates.

Table 12.8 presents attrition rates by the time of day at which the interviewer started to seek a respondent. Great differences in the rate of not-at-homes might be expected, and actually there are definite differences, of perhaps a lesser degree than we would guess in advance. Attempts after 4:00 P.M. appear to have about half the attrition of this kind that earlier attempts have. The dinner hour is, of course, the time that shows the lowest rate of all. Refusal rates seem to be lower after 2:00 P.M. than before. The rates for did-not-fit-quota rise during the afternoon, whereas rejection and termination rates decline. The latter may reflect the pressure of housework and other demands on the respondent's time in the early morning, at noon, and in the evening. The final rate of successful attempts varies very little and seems to be highest before 10:00 A.M. and after 5:00 P.M. All these variations must be analyzed further to determine

TABLE 12.6

ATTRITION RATES AT EACH HURDLE, BY RURAL LOCATION AND SIZE OF CITY

| Size of Place [a] | Number of Attempts | Percentage Attrition [b] | | | | Per Cent of Attempts Successful |
		No One at Home	Re-fused	Did Not Fit Quota	Rejected or Terminated	
Rural	528	16	9	17	5	60
2,500–	131	15	5	18	6	62
10,000–	192	15	21	16	15	48
25,000–	130	25	5	33	5	45
50,000–	272	23	11	17	14	49
250,000–	272	19	15	20	18	46
500,000– 000,000	166	17	16	17	5	54
nd over	628	21	17	13	10	52
ll localities	2319	19	13	17	10	53

[a] The classification is that given to the entire quotas of the interviewer. Hence me rural interviews are included in the city-size classes. Rural includes farms d non-farm population. Areas of 50,000 or more include all their Metropolitan stricts as defined by the Census.

[b] See footnote a in Table 12.5.

TABLE 12.7

ATTRITION RATES AT EACH HURDLE, BY DATE OF ATTEMPT

| ay and Date, ecember 1947 | Number of Attempts | Percentage Attrition [a] | | | | Per Cent of Attempts Successful |
		No One at Home	Re-fused	Did Not Fit Quota	Rejected or Terminated	
i. 5	34	32 [b]	9 [b]	10 [b]	16 [b]	47 [b]
t. 6	349	19	16	16	10	51
n. 7	282	13	11	16	13	56
on. 8	618	23	15	17	6	51
es. 9	549	17	12	17	12	53
ed. 10	304	18	13	22	6	53
ur. 11	85	21	9	18	10	53
i. 12	37	5 [b]	3 [b]	12 [b]	10 [b]	73 [b]
t. 13	16	25 [b]	8 [b]	18 [b]	0 [b]	56 [b]
n. 14	13	23 [b]	33 [b]	0 [b]	0 [b]	54 [b]
on. 15	22	41 [b]	0 [b]	8 [b]	33 [b]	36 [b]
ate not recorded	12	33 [b]	12 [b]	0 [b]	0 [b]	58 [b]
tire survey	2319	19	13	17	10	53

[a] See footnote a in Table 12.5.

[b] These rates and percentages are calculated from fewer than fifty attempts and e subject to large sampling errors.

TABLE 12.8

ATTRITION RATES AT EACH HURDLE BY TIME OF DAY

| Time, A.M. | Number of Attempts | Percentage Attrition [a] | | | | Per Cent of Attempts Successful |
		No One at Home	Re-fused	Did Not Fit Quota	Rejected or Terminated	
8:00– 8:59	5	0 [b]	0 [b]	20 [b]	0 [b]	80 [b]
9:00– 9:59	94	19	11	3	10	63
10:00–10:59	254	26	13	10	11	51
11:00–11:59	240	22	18	13	8	51
Time, P.M.						
12:00–12:59	217	17	14	16	13	52
1:00– 1:59	282	21	18	18	7	49
2:00– 2:59	315	21	12	21	5	53
3:00– 3:59	349	21	11	22	9	50
4:00– 4:59	266	14	8	22	14	53
5:00– 5:59	101	10	10	16	7	63
6:00– 6:59	55	5	15	16	14	58
7:00– 7:59	96	10	10	17	12	58
8:00– 8:59	32	22 [b]	12 [b]	5 [b]	14 [b]	56 [b]
9:00– 9:59	13	0 [b]	23 [b]	40 [b]	0 [b]	46 [b]
Entire survey	2319	19	13	17	10	53

[a] See footnote a in Table 12.5.

[b] These rates and percentages are calculated from fewer than fifty attempts and are subject to large sampling errors.

what factors other than time of day may be involved. Some interviewers start early, others late. Some work evenings and others do not. Week end interviewing involves different conditions. After factors associated with these complications are accounted for, the remaining effect of the time of day may be different than that shown in Table 12.8. Until enough data can be accumulated for such an analysis, the general conclusion may be that time of day has a small effect on attrition rates.

All observed attrition rates vary quite a bit by chance. They also vary somewhat with the interviewer's skill in attaining cooperation and holding down the refusal rate. They vary with his ability to locate persons who will meet the quota requirements. It is possible that some interviewers do not apply the quota requirements strictly and precisely but are willing to accept persons close to an unfilled quota category in age or economic status, for example, as satisfactory for the category. There is no way of determining to what extent this relaxing of the quota requirements may be practiced. A

xamination of the ages reported on the schedules might throw light
pon it. (It was not thought worth while to undertake such an ex-
mination for this study.)

The interviewers' experience in survey work, the ratings on the
uality of their work in recent surveys, and other interviewer charac-
eristics were examined to see to what extent they appeared to be
elated to the various attrition rates. The results of some of these
nalyses are given in Tables 12.9 to 12.13.

Experience in previous surveys appears to have little effect on at-
rition rates, except that the refusal rate is higher for interviewers
rho have had no previous experience with NORC and the rate for
id-not-fit-quota rises with experience.

TABLE 12.9

ATTRITION RATES AT EACH HURDLE, BY INTERVIEWER'S EXPERIENCE IN
NORC SURVEYS

| Number of Previous Surveys or NORC | Number of Interviewers | Number of Attempts | Percentage Attrition [a] | | | | Per Cent of Attempts Successful |
			No One at Home	Re-fused	Did Not Fit Quota	Rejected or Terminated	
None	5	164	15	27	12	11	49
1–3	14	311	18	15	15	10	53
4–9	15	388	18	13	18	7	54
10–19	17	459	20	10	18	13	52
20–29	13	325	21	9	20	7	54
30–49	7	216	21	18	24	6	46
50–99	14	402	20	11	15	12	53
00 and over	3	54	15	4	5	2	76
ll attempts	88	2319	19	13	17	10	53

[a] See footnote a in Table 12.5.

TABLE 12.10

ATTRITION RATES AT EACH HURDLE, BY RATING OF INTERVIEWER ON QUALITY
OF WORK IN LAST THREE SURVEYS

(Rating 1 is lowest, 5 highest)

| Rating | Number of Interviewers | Number of Attempts | Percentage Attrition [a] | | | | Per Cent of Attempts Successful |
			No One at Home	Re-fused	Did Not Fit Quota	Rejected or Terminated	
1	2	64	20	2	34	15	44
2	17	459	18	16	19	11	50
3	26	677	20	11	15	11	54
4	28	709	20	11	15	8	55
5	6	178	18	14	26	9	47
Not rated	9	232	16	21	12	8	54
ll attempts	88	2319	19	13	17	10	53

[a] See footnote a in Table 12.5.

TABLE 12.11

Attrition Rates at Each Hurdle, by Judgment Ratings of Interviewer by Expensiveness in Previous Surveys

| Rating | Number of Interviewers | Number of Attempts | Percentage Attrition [a] | | | | Per Cent of Attempts Successful |
			No One at Home	Re-fused	Did Not Fit Quota	Rejected or Terminated	
Expensive	7	213	18	13	25	15	46
Average	51	1324	19	13	15	9	54
Inexpensive	10	289	25	13	15	6	52
Not rated	20	493	15	12	21	10	53
All attempts	88	2319	19	13	17	10	53

[a] See footnote a in Table 12.5.

TABLE 12.12

Attrition Rates at Each Hurdle, by Education of Interviewer

| Education | Number of Interviewers | Number of Attempts | Percentage Attrition [a] | | | | Per Cent of Attempts Successful |
			No One at Home	Re-fused	Did Not Fit Quota	Rejected or Terminated	
Completed high school	12	265	16	9	8	9	64
Some college	37	963	20	14	16	6	54
Completed college	24	699	19	13	23	15	47
Graduate training	10	249	18	14	17	4	55
Not reported	5	143	17	17	14	15	50
All attempts	88	2319	19	13	17	10	53

[a] See footnote a in Table 12.5.

TABLE 12.13

Attrition Rates at Each Hurdle, by Age of Interviewer

| Age | Number of Interviewers | Number of Attempts | Percentage Attrition [a] | | | | Per Cent of Attempts Successful |
			No One at Home	Re-fused	Did Not Fit Quota	Rejected or Terminated	
20–24	8	281	22	23	17	11	44
25–29	8	152	19	13	14	3	59
30–34	8	196	23	13	14	24	46
35–39	15	371	17	11	22	7	56
40–44	17	462	22	11	18	8	55
45–49	12	301	23	10	20	7	54
50–59	13	370	26	12	15	9	52
60 and over	2	43	21 [b]	0 [b]	0 [b]	0 [b]	72 [b]
Not reported	5	143	22	17	14	15	50
All attempts	88	2319	19	13	17	10	52

[a] See footnote a in Table 12.5.
[b] These rates and percentages are calculated from fewer than 50 attempts and are subject to large sampling errors.

The rating of the interviewer's work on the three preceding surveys also shows little relation to attrition rates, except in the case of the two interviewers who were rated in the lowest grade. It is possible that the data from the logs of these interviewers are subject to a greater degree of error than data from the other logs. Similarly, the informal judgment rating on expensiveness reveals only a modest difference in attrition rates, and the number of interviewers in the extreme groups is too small to exclude the possibility that most of the difference is due to other individual characteristics of the interviewers who happen to be rated in each class.

Attrition rates vary with the education of the interviewer, but in a direction opposite to what might be expected. Interviewers whose education did not extend beyond high school had the lowest attrition rates and college graduates the highest.

The age of the interviewer appears to have little relation to attrition, except that interviewers who were under 25 years of age had higher refusal rates.

All the comparisons of attrition rates by various characteristics of interviewers are limited to interviewers who had been selected and trained and whose records were satisfactory. If there is a strong relation between a personal characteristic and attrition, it is likely to lead to a higher proportion of rejections of applicants for the interviewing staff or even a lower rate of applications because experience in work of a similar nature discourages an otherwise likely applicant. Those interviewers who are selected and who continue in interviewing may make up for unfavorable factors by exceeding the average on other factors favorable to low attrition rates and acceptable quality of work. If natural processes of selection have had effects such as these, a vigorous effort to recruit interviewers may reveal relationships between interviewer performance and characteristics quite different from the modest relationships exhibited by previous experience and analyses such as the foregoing.

12.11 EFFECTS OF ATTRITION

How serious were these losses through attrition? A general measure of their cost in time can be computed. If every attempt had been successful, the total time required for the completion of the survey would have been reduced by the elimination of time lost for attempts in which no one was found at home or there were refusals or other losses. This is a very unrealistic assumption, but it appears to set an upper limit on the amount of time that might be

saved by more skillful performance of the interviewer's work and
better luck in finding qualified respondents. On this basis, 47 per
cent of the attempts could have been avoided. Assuming that the
"other" time was reduced in proportion to the reduction in time spent
in seeking respondents and in contact with them, the total time
spent on the survey would have been reduced 18 per cent. This
percentage seems remarkably low in view of the extreme assumption
that has been made. It is small because the unsuccessful attempts
did not involve very much, if any, interviewing time. The contact
time that was spent on completed interviews amounted to 65 per
cent of the total time spent on all attempts. (This is exclusive of
time out.) The contact time for unsuccessful attempts was only 2
per cent of all time spent. It may be assumed, therefore, that the
saving through the complete elimination of attrition will be less than
18 per cent, since the time that must go into the study of instructions and filling out of reports will not be reduced very much by the
elimination of attrition. It may even be increased by the special
steps necessary to avoid the losses.

On the basis of time, therefore, it would appear that attrition is
not a very serious problem. However, the strain on the interviewer
and also the distortion of the sample through the omission of people
who are away from home, or who refuse to cooperate, may be quite
important in many surveys. No data on these effects of attrition
were available in the current study. It should be noted that some
of the individuals who were missed in unsuccessful attempts were
not eligible to be interviewed or were not needed to fill the quotas.
Apart from these people who were not qualified to be in the survey
there were others who were qualified and whose omission may have
biased the survey results. Unfortunately, we know very little if anything about them. Further studies of the characteristics of such individuals would be quite valuable.

There is some danger of overestimating the differences that exist
between the opinions and attitudes of people who are temporarily
away from home or who are inclined to refuse to be interviewed and
those of people who are available and cooperate. Refusal as well as
absence from home are ordinarily rather temporary factors in the
life of an individual who may be approached as a respondent. At
another time, he will be found at home or in a mood to cooperate.
Only a small fraction of the population is away from home a large
part of the time or persistently non-cooperative. Even if their attitudes are substantially different from those of other persons, these
persons form so small a proportion of the sample that the effect of

ncluding or excluding them will usually be negligible. On this
)oint, see Chapter 11, page 268.

12.12 OPERATION AND EFFECTS OF THE QUOTA ASSIGNMENTS

One of the principal purposes of this study was to find out something about the ways in which the quota requirements introduced
difficulties into the interviewers' work and how serious these difficulties were. We were also interested in finding out how rapidly the
unfilled balances became restrictive as the interviewers' work neared
completion. It was possible to study this in several ways.

First of all, an attempt was made to determine whether certain
types of persons required to fill quota categories were especially difficult to locate. We thought, for example, that young men or older
men might be the more difficult to locate and that most of the interviewers would end up searching for respondents of the few most
difficult types. The results of the analysis dispelled any notion that
there are strong differences in difficulty of finding various kinds of
people. At least such differences as do exist apparently can be overcome by looking more actively for the difficult types of individuals
from the beginning of the survey.

It is significant, for example, that of the 86 interviews completed
on the last attempt, thereby filling the interviewers' quotas, 42 were
men and 44 were women. They were about equally divided by age
as well as by sex; 40 were under 40 years of age and 46 were 40 years
of age and over. It would appear that if there are difficult types,
they vary from one place to another. It is also interesting that of
these 86 attempts, 70 were home contacts. Thus the percentage of
home contacts in the final attempt was 81, almost identical with that
for the entire survey, indicating that there was no pressure to roam
the streets looking for individuals who could not be found in the
calls at homes. Many interviewers, of course, completed their entire assignment without making any interviews on the street, in offices, or in other places outside the home.

The extent to which the quota restrictions increased the work of
the interviewers can be estimated by taking the number of attempts,
apart from those that were unsuccessful because no one was at home
or the person refused, and calculating the percentage that were unsuccessful because the individual did not fit the quota. On this basis
280 out of 2319 attempts were unsuccessful because quota restrictions would not permit the interviewing of one more person of the

type contacted. This ratio is probably an underestimate. Apparently in some instances the losses were classified as rejections, though the instructions stated clearly that this category was for individuals rejected for reasons other than quota reasons. Also a few interviewers exceeded their quotas. It would appear, therefore, that 12 per cent of attempts and 5 to 6 per cent of the total time were lost as a result of the quota restrictions. There may have been some additional loss of time in travel, looking for particular types of individuals who would fit the quota categories that remained to be filled, but our data do not permit an extraction of this portion of the time It is probably relatively small compared with the total time spent for the survey.

For the whole survey, the loss for quota reasons was about equal to the loss through not finding anyone at home and about the same as the combined loss from refusals and rejections.

The effect of quota restrictions was negligible at the beginning of the survey but increased as the interviewer neared the completion of his assignment. The last attempt necessarily was a completed interview since it finished the job. Therefore, all last attempts should be excluded from the calculation of the rate at which the quota restrictions became operative. Twenty-three out of 88 attempts that were next to the last were unsuccessful because of quota reasons. This is 26 per cent of such attempts. The percentages for the third, fourth, fifth, sixth, and seventh from the last ran 18, 26, 13, 19, 9 per cent, respectively. Taking the five attempts that preceded the successful interview, we find that 20 per cent of the attempts were unsuccessful because of quota reasons. For all attempts, 12 per cent were unsuccessful for this reason. It would appear, therefore, that the percentage increased as the interviewing neared its conclusion, but not as much as might be expected.

Oddly enough, the increase in the losses due to quota reasons was offset in part by decreases in losses due to other reasons as the interviewing progressed. The percentage of attempts that resulted in completed interviews fell from 53 per cent for all attempts to 48 per cent for next-to-last attempts. Excluding the last four attempts, 52 per cent of all attempts were successful. Thus it would appear that the tightening up of the effects of the quota restrictions toward the end of the work did not cut down very much the yield of completed interviews per 100 attempts, the number remaining at 47 to 50 up to the last attempt which was, of necessity, 100.

Table 12.14 exhibits the relation of attrition to the quota situation just previous to the attempt. The classification is by the age and

TABLE 12.14

SUCCESS OF ATTEMPTS IN RELATION TO AGE-SEX QUOTA CATEGORIES
COMPLETED PRIOR TO THE ATTEMPT

Age-Sex Quota Categories Completed	Number of Attempts	Completed Interviews (I)	Per Cent of Attempts Successful	Did Not Fit Quota (DFQ)	DFQ per I
ne	1317	723	55	111	0.15
ales under 40	171	84	49	20	0.24
ales over 40	127	62	49	24	0.39
males under 40	149	79	53	20	0.25
males 40 and older	42	26	62 [a]	2	0.08 [a]
tal, one category	489	251	51	66	0.26
males	87	43	49	11	0.13 [a]
females	54	29	54	13	0.24 [a]
under 40	101	51	51	21	0.21
40 or older	17	8	47 [a]	5	0.29 [a]
ales under 40 and females 40 and older	31	14	45 [a]	11	0.03 [a]
ales 40 and older and females under 40	33	23	70 [a]	0	0.00 [a]
tal, two categories	323	168	52	51	0.31
except males under 40	47	21	45 [a]	16	0.76 [a]
except males 40 and older	49	23	47 [a]	11	0.48 [a]
except females under 40	15	10	67 [a]	3	0.30 [a]
except females 40 and older	76	27	36	21	0.78
tal, three categories	187	81	43	51	0.63
rand total [b]	2316	1223	53	279	0.23

[a] These percentages and proportions are calculated from fewer than fifty
tempts or interviews and are subject to large sampling errors.
[b] Exclusive of three attempts after all categories were filled.

sex requirements that have been satisfied and does not take accou
of the farm, race, and economic categories.

More than half the attempts were made before any one of t
four age-sex categories (males under 40, females under 40, males
or older, females 40 or older) had been filled. Of these 1317 a
tempts, 723 resulted in completed interviews. This is 55 per ce
However, 111 of these attempts were marked as not fulfilling t
quota restrictions, presumably because one or more of the farm, rac
or economic categories had been filled or did not have any inte
views assigned to it in the quota. Some interviewers had no assig
ment for economic class A, and many had none for Negroes a
farmers. If these 111 had all been completed interviews, the percen
age of successful attempts would rise to 63 per cent. However,
is unrealistic to suppose that in the absence of any quota restri
tions all 111 contacts would have been successful. Some of the
would have been lost by refusal or rejection. This is confirmed b
the fact that if all DFQ contacts are counted as successes, the pe
centage of success increases as the quota situation becomes more r
strictive. If only three-fifths of the DFQ contacts are assumed
lead to successful interviews, the success rate remains almost cor
stant at 60 per cent. This assumption, doubtless, understates th
percentage of completed interviews that would have been obtaine
if quota restrictions had not interfered.

It was thought that perhaps one effect of quota restrictions woul
be to increase the time per attempt, as the quota was progressivel
filled, especially the time spent seeking a suitable respondent. How
ever, as will be seen from Table 12.15, the average time spent seek

TABLE 12.15

COMPARISON OF THE RESULT OF ATTEMPT AND AVERAGE SEEKING TIME FC
FIRST THREE AND LAST THREE ATTEMPTS

| | Attempts | | Average Seeking Time Minutes | |
Result of Attempt	First Three	Last Three	First Three Attempts	Last Thre Attempt
Completed interviews	162	178	6.1	5.9
No one at home	59	26	3.3	3.7
Did not fit quota	5	40	5.4	5.2
Rejected, refused, or terminated	39	20	5.1	4.6
All attempts	264	264	5.4	5.5

ng a contact for the last three attempts prior to the final attempt was only one-eighth of a minute greater than the average time spent or the first three attempts.

An extensive analysis was made of the relation of the type of person contacted to the date and hour of day when the contacts were made, since it was felt this would throw light on the effect of quota restrictions. On the whole, this analysis showed very little in the way of demonstrable differences. As we might expect, a somewhat larger proportion of women were interviewed in the middle of the day, and a large proportion of the men were interviewed on Sunday.

12.13 SCATTERING THE INTERVIEWS

Since people who live in different neighborhoods tend to differ in their attitudes and opinions, it is desirable to obtain a more representative sample of these variations by spreading the interviews among a number of neighborhoods instead of taking them all in one place. The quota requirements tend to spread the interviews geographically because they require fixed numbers of interviews from people in different economic groups. In addition, the interviewers are instructed specifically to avoid bunching their interviews in any locality. They are requested not to make more than three calls in any block or more than six interviews in any one neighborhood. These efforts to obtain a selection from several neighborhoods necessarily increase the distance traveled per interview, but such efforts are justified by the contribution they make to the representativeness of the sample.

No attempt was made to record the distance interviewers traveled in covering their assignments, but 1300 miles of travel by automobile were reported on their expense sheets. There was no very practical method by which interviewers could measure the distances they traveled while walking. However, the addresses were recorded in the log, and it was possible to make a rough classification of the attempts and interviews in relation to the address of the preceding attempt and the preceding completed interview.

Table 12.16 shows a classification of attempts by address in relation to the address of the preceding attempt. Thirty-five per cent of the attempts were made "within a block," i.e., on the same street as the preceding attempt and with house numbers differing by less than 100. This is a very rough measure since some house numbering systems run less than 100 to the block. The average time between contacts reported for such attempts was only 1.9 minutes, about a third

TABLE 12.16

Average Seeking Time by Address in Relation to Address of Preceding Attempt

Address in Relation to Address of Preceding Attempt	Number of Attempts to Obtain an Interview			Average Time Spent Seeking Contacts, Minutes		
	At Home	In Other Places	All Places [a]	At Home	In Other Places	All Places
Within a block	732	73	813	1.9	2.1	1.9
On same street	45	9	56	5.5	3.7	5.3
Different street	548	106	670	6.2	7.1	6.2
First attempt this day	181	31	212	7.9	9.0	8.1
One or both on rural route	319	27	355	7.2	3.4	6.9
In a town without street addresses	151	31	188	5.9	7.5	6.1
Not recorded	14	6	25	5.6	2.7	4.4
Total	1990	283	2319	4.9	5.6	4.9

[a] Includes 46 for which place of interview was not recorded.

of the average time for attempts on the same street but with house numbers differing by 100 or more. Some of the contacts on a different street may have been in the same block but just around the corner.

In analyzing the figures in Table 12.16, we must bear in mind the fact that many calls will be made in the same block because the first attempt in the block was unsuccessful. Consequently we have also compared the address of each attempt with the address of the last previous successful interview. On this basis, 378 completed interviews were obtained within a block of the last preceding attempt, 204 of them being within a block of the last preceding interview. Thirty more interviews were obtained on the same street as the last preceding interview but more than a block apart. In 482 successful interviews, the address was on a different street than that of the preceding interview. There were 187 interviews that were counted separately because they were the first for the day and were almost always on a different street from the last interview on the previous day. In 204 instances the interview or the preceding interview or both were on a rural route, and in 113 instances neither was on a rural route but one or both was given with the name of the community instead of a street address. Almost all the latter were in small towns or villages.

Out of approximately 900 interviews for which street addresses were given, 22 per cent were close enough to the preceding interview to be classified as less than a block apart, accepting house numbers as a crude measure of distance; addresses around the corner are not

cluded in this estimate. This is a smaller percentage than would
obtained by a block segment or block subsampling procedure with
a average of say 1.3 or more respondents per block. In many in-
ances the addresses were on different blocks, although they were
ss than a block length apart.

Similarly the rural and small-town interviews were spaced out at
arious distances. The average travel time between them is decep-
ve since it merges a few very long trips with many short ones.
lso, some of the rural interviews were made in town instead of at
e farm or the home of rural non-farm respondents.

The attrition rates for rural and small-town attempts, given in
able 12.17, were a little lower but not very much lower than those

TABLE 12.17

ATTRITION RATES FOR RURAL AND SMALL-TOWN ATTEMPTS

Attempts, Located in Relation to Preceding Completed Interview	Percentage of Attempts That Were Unsuccessful
Within a block	54
On same street	32 [a]
Different street	46
First attempt this day	41
One or both on a rural route	43
In a town without street addresses	40
Total	47

[a] Based on only 56 attempts.

or the rest of the survey. Refusal rates were somewhat higher, but
sses to ineligibility under the remaining quota requirements were
uch smaller than those attempts that were made within a block
f the preceding completed interview.

As we might expect, the time required for seeking the contact in-
reased with an increase in the apparent distance, as Table 12.16
hows.

Interviewing times did not vary appreciably with distance, except
hat the time required for interviews made within a block of the pre-
eding interview or on the same street was a little shorter. This
nay be due to a combination of causes. For example, there may
e a tendency on the part of interviewers who do not scatter their

interviews to be a little less thorough in their interviewing. Tab▌
12.18 shows the average contact times.

TABLE 12.18

AVERAGE CONTACT TIME BY ADDRESS IN RELATION TO PRECEDING ATTEM▌

Address in Relation to Preceding Attempt	Home Attempts		Attempts Elsewhere▌	
	Contact Time	Seeking and Contact Time	Contact Time	Seeking an▌ Contact Tir▌
Within a block	10.9	12.8	13.1	15.2
On same street ᵃ	15.3	20.8	16.0	19.7
Different street	13.9	20.1	19.5	26.6
First attempt this day	14.5	22.4	22.9	28.9
One or both on rural route	14.4	21.6	15.3	18.7
In a town without street addresses, either or both	15.1	21.0	17.3	24.8
Not recorded	9.1	14.7	16.7	19.3
Total survey	12.9	17.7	17.4	22.9

ᵃ Based on only 56 attempts.

Of all the attempts, 1990 were reported as having been made a▌
the home of the respondent, and 283 at other places, whereas 4▌
were not recorded. The interviews made elsewhere than in the re▌
spondent's home show a little lower refusal rate and lower losse▌
from did-not-fit-quota than do home attempts, even after attempt▌
that found no one home are excluded from the computation. Whe▌
interviewers seek respondents in public places, they are able to avoi▌
certain losses for quota reasons and non-cooperation merely b▌
"sizing up" the persons they see before selecting them.

12.14 SEEKING RESPONDENTS

We now turn to a further analysis of time rates. The interviewe▌
log enables us to separate time spent in seeking an interview from▌
time spent in contact with potential respondents or waiting for then▌
to come to the door. The time spent in seeking was partly spent i▌
traveling from one place to another, but also partly in making in▌
quiries about where certain types of persons might be found. It i▌
clear from several of the interviewers' reports that a portion of th▌
time spent in seeking was included in the entries for "time out."▌
From the explanations of time out recorded in the log, we ca▌

ake a rough estimate of this additional seeking time, though most
f it is not distinguished from travel between districts and other ac-
vities. All together, between 210 and 320 hours were spent in
eeking contacts compared with 520 hours of contact time. The
otal amount of time recorded in the logs for survey business was
etween 940 and 1000 hours. The relation of seeking time to the
utcome of the attempt, size of the community, and characteristics
f the interview will be analyzed on the basis of the 191 hours of
ime specifically reported on the log as spent in seeking respondents.

The amount of time spent in seeking respondents varies somewhat
rom region to region, between interviewers, and from one type of
esult to another. For all attempts the average seeking time was 4.9
inutes. For those attempts that resulted in completed interviews,
; was slightly higher, 5.8 minutes, whereas the seeking times for at-
empts that resulted in not-at-homes and refusals were 3.4 and 3.7
inutes respectively. The differences in average time are not great
nd probably reflect differences between the situations in which we
re likely to obtain completed interviews and those in which we are
nore likely to be unsuccessful for various reasons. Thus, in an
partment house the time spent in seeking the next attempt may be
elatively small but the probability of finding no one at home rather
arge, whereas in the suburbs the time required to go from one house
o another is greater than in the more densely settled areas and, at
he same time, the probability of finding a cooperative respondent
ome is larger. Because a number of factors of this kind may ac-
ount for substantially all the differences in seeking time, we may
easonably conclude that the average amount of time in seeking con-
acts reflects various factors and situations but does not show any
eculiar factors of great importance other than interviewer differ-
nces and the scattering of interviews shown in Table 12.16.

12.15 CONTACT TIME

The time involved in actual contact with a respondent or a po-
ential respondent varies greatly according to the outcome of the effort, which is exactly as we might expect; but the regularity of the
increase is impressive, as shown in Table 12.19.

In analyzing these contact time figures, a distinction can be made
between the completed and terminated interviews on one hand and
other types of outcome on the other. The length of the questionnaire
affects only the former. In making any estimates for a different type
of survey, therefore, the contact time for interviews and terminated

TABLE 12.19

RESULT OF ATTEMPT AND AVERAGE CONTACT TIME

Result of Attempt to Obtain Interview	Average Contact Time, Minutes
No one at home	2.0
Refused	2.8
Did not fit quota	3.5
Rejected	4.3
Terminated	14.8
Interview completed	22.8

interviews should be adjusted to the length of the prospective que tionnaire, but contact time for the other outcomes of an attem should not be adjusted for the change in length.

12.16 AGGREGATE TIME FOR THE WHOLE JOB

In addition to the time spent on seeking interviews and in th contacts themselves, time was taken out for reading instruction writing up reports, traveling between districts, and a variety of othe purposes connected with the survey, as well as for personal busines and relaxation. Table 12.20 presents an analysis of the time ou as reported on the log. It should be noted that this table does no include time for travel from home to the district in which the inter viewer started work for the day or for returning home from th district, nor is time spent in studying instructions and making re ports included. Only partial information is available from the lo

TABLE 12.20

TIME REPORTED ON THE LOG APART FROM SEEKING AND CONTACT TIME

Type of Activity	Minutes per Attempt
Work on survey (checking schedules, reviewing specifications, etc.)	0.5
Travel between attempts (but not to and from home)	1.1
Luncheon or dinner, coffee, rest, etc.	1.2
Combinations of two or more of the types above	2.9
Personal business, shopping, etc.	0.8
Combinations of personal business with other types	0.9
Type of activity not recorded	1.1
All time between attempts	8.4

on this last activity. Some information is given by regular time reports. The average time spent by interviewers for these purposes was about two hours per interviewer. There may have been additional time that was not charged on the time sheets. The total time reported for payment is divided approximately as follows: seeking time 14 per cent, contact time 39 per cent, and other interviewer time, 47 per cent.

The total time was analyzed in relation to the result of the attempt and various other factors. Tables 12.21 to 12.23 show some

TABLE 12.21

TOTAL SEEKING AND CONTACT TIME BY RESULT OF ATTEMPT, NOT INCLUDING TIME BETWEEN ATTEMPTS

Result of Attempt to Obtain Interview	Average, Minutes	Standard Deviation, Minutes	Coefficient of Variation
No one at home	5.4	5.7	107
Refused	6.5	6.7	101
Did not fit quota	8.4	8.8	105
Rejected	9.5	9.2	97
Terminated	17.8	16.1	90
Interview completed	28.5	14.0	49
All attempts	18.4	15.7	85

of the results of this analysis. The average time spent in seeking and contact increases with the apparent amount of interviewing, from 5.4 minutes for attempts that found no one at home to 28.5 minutes for completed interviews. These averages represent distributions of actual times that have standard deviations about equal to the average itself, except for completed interviews and all attempts. The time required for interviewing is less variable than that for seeking, as we should expect from the standard set of questions used for interviewing and the diverse situations in which the interviewer seeks to find a respondent.

The average times show no very regular relation to size of place. Other factors appear to introduce quite a bit of irregular variation. Regional differences in average time per completed interview are of moderate size and show the New England, West South Central, and Mountain regions high and the Pacific region low. New England is high because it has high attrition rates, and the other regions are high because of travel difficulties and other factors.

TABLE 12.22

AVERAGE TIME PER COMPLETED INTERVIEW, AND PER ATTEMPT, BY SIZE OF PLACE

Minutes per Completed Interview [a]

Size of Place	Number of Completed Interviews	Seeking	Contact	Time Out between Attempts	Total
Rural	319	10.8	24.8	12.6	48.2
2,500–	81	10.2	26.6	19.7	56.5
10,000–	92	8.1	27.4	18.3	53.8
25,000–	59	22.7	34.0	20.5	77.2
50,000–	134	9.5	28.2	17.2	54.8
250,000–	124	8.1	24.1	17.0	49.2
500,000–	90	9.6	28.2	26.0	63.9
1,000,000–	324	6.1	22.4	13.1	41.5
Total	1223	9.4	25.5	16.0	50.8

Size of Place	Number of Attempts	Minutes per Attempt			Total
Rural	528	6.5	15.0	7.6	29.1
2,500–	131	6.3	16.4	12.2	34.9
10,000–	192	3.9	13.1	8.8	25.8
25,000–	130	10.3	15.5	9.3	35.0
50,000–	272	4.7	13.9	8.5	27.0
250,000–	272	3.7	11.0	7.8	22.4
500,000–	166	5.2	15.3	14.1	34.6
1,000,000–	628	3.1	11.5	6.7	21.4
Total	2319	4.9	13.4	8.4	26.8

[a] Including time spent on unsuccessful attempts.

12.17 DIFFERENCES BETWEEN INTERVIEWERS

In any study of this kind, there is always a temptation to analyze the performance of interviewers very impersonally as if they were more or less identical human beings. Actually, as every field supervisor knows, each interviewer exhibits individual variations in the general procedure of work and reveals characteristic strengths and weaknesses. The interviewers differ in ability with respect to the various operations necessary in the performance of their work. Consequently, we should not expect identical results from different inter-

TABLE 12.23

AVERAGE TIME PER COMPLETED INTERVIEW, BY REGION

Minutes per Completed Interview [a]

Region	Number of Completed Interviews	Seeking	Contact	Time Out between Attempts	Total
New England	106	13.6	25.8	19.6	59.0
Middle Atlantic	225	7.3	24.3	15.6	47.2
East North Central	282	7.1	23.9	16.5	47.4
West North Central	137	7.8	26.2	11.8	45.8
South Atlantic	160	9.3	25.5	18.5	53.2
East South Central	64	10.8	23.1	17.3	51.2
West South Central	101	17.0	32.4	20.4	69.8
Mountain	40	10.4	32.1	12.9	55.3
Pacific	108	9.1	23.5	9.0	41.7

[a] Including time spent on unsuccessful attempts.

viewers, though as a result of the selection of interviewers for their competence in the work, a certain amount of uniformity is to be expected and is actually observed.

It is very important in comparing the interviewers to recognize the differences that exist between the circumstances in which one of them works and the circumstances in which another works. However, it is not correct to assign all differences to variations in circumstances. Actually, as we have seen, such factors as the size of the community show less variation than interviewer differences. Also, the regional differences that are observed are partly a matter of the conditions in the region in which the interviewer is working and partly a matter of regional differences between interviewers, for example, in the rate at which they are accustomed to work.

In a number of the preceding sections, an indication has been given of the variation between interviewers in the attrition rates and time rates reported for this survey. Here some additional analyses will be given.

First of all, some explanation can be found in comments that were made by the interviewers in their reports. Some interviewers said they had had more refusals or not-at-homes and others said they had had fewer than in previous surveys. This accords with the variation to be expected by chance from one assignment to the next.

The interviewers were classified according to the number of surveys that they had performed previously for NORC. It was found that those who had not performed any previous surveys, and were therefore, for the most part, inexperienced in survey work, had refusal rates about twice as great as the entire group of interviewers; their average seeking and contact times, however, were only slightly higher than the general average. Three interviewers with long experience showed, on the other hand, substantially lower refusal and time rates. Table 12.24 shows the distribution of interviewers' per-

TABLE 12.24

AVERAGE TIME [a] PER COMPLETED INTERVIEW BY NUMBER OF PREVIOUS
SURVEYS DONE BY INTERVIEWER FOR NORC

Minutes per Completed Interview

Previous NORC Surveys	Number of Interviewers	Seeking	Contact	Time Out between Attempts	Total
None	5	9.9	26.4	15.2	51.5
1–3	14	6.9	23.0	17.5	47.4
4–9	15	15.2	27.6	19.2	62.1
10–19	17	9.5	25.2	16.2	50.8
20–29	13	7.0	23.9	19.0	49.9
30–49	7	7.3	26.7	18.6	52.6
50–99	14	8.5	27.0	10.1	45.7
100 or more	3	6.6	19.4	4.3	30.3
All interviewers	88	9.4	25.5	16.0	50.8

[a] Time for all attempts, whether successful or not, is included.

formance, according to the number of surveys they had done for NORC.

The central office staff of NORC regularly rates the interviewers on the quality of the schedules that they return. These ratings run from 1, which is the lowest grade, to 5, which is the highest grade. Table 12.25 shows average times by this rating. The interviewers with the highest grade show the highest average times. It appears that interviewers who do a better job must necessarily take more time to do it; but the averages may vary in response to other factors as well, and hence the apparent relation should be investigated further before conclusions are drawn.

Another rating of the interviewers was made for this study by the central office of NORC. This was based on the examination of their

TABLE 12.25

AVERAGE TIME PER COMPLETED INTERVIEW OR ATTEMPT, BY INTERVIEWER'S
RATING FOR LAST THREE SURVEYS

Minutes per Completed Interview

Average Rating for Last Three Surveys	Number of Interviewers	Seeking	Contact	Time Out between Attempts	Total
–2, lowest	2	26.0	27.2	20.4	73.6
–3	17	13.0	28.7	19.7	61.4
–4	26	7.6	25.0	15.5	48.1
–5	28	8.0	24.3	14.6	47.0
, highest	6	8.1	24.2	13.5	45.9
Not rated	9	9.0	25.0	15.4	49.3
All interviewers	88	9.4	25.5	16.0	50.8

Average Rating for Last Three Surveys	Number of Interviewers	Minutes per Attempt			Total
–2, lowest	2	11.4	11.9	8.9	32.2
–3	17	6.5	14.3	9.9	30.7
–4	26	4.1	13.5	8.3	25.9
–5	28	4.5	13.4	8.1	26.0
, highest	6	3.8	11.4	6.4	21.6
Not rated	9	4.8	13.5	8.3	26.6
All interviewers	88	4.9	13.4	8.4	26.8

time reports in a few previous surveys. It classified them according
to the general impression of expensiveness or inexpensiveness of their
work. The classification was a crude one, since it was not possible
to take into account some factors that might properly be considered
in any attempt to judge the expensiveness or inexpensiveness of an
individual's work. The relation between this rating and the average
time rates is shown in Table 12.26.

The relationship is clear with respect to seven interviewers rated
"expensive" but somewhat weaker in distinguishing "average" from
"inexpensive" interviewers than we might expect if the ratings could
have been made on a highly accurate basis. To make better ratings
we should have an accurate measure of the quality of work to use
with the number of interviews in measuring output. In measuring
output, a simple count of completed interviews, which does not take

TABLE 12.26

AVERAGE TIME PER COMPLETED INTERVIEW OR ATTEMPT, BY JUDGMENT RATING
OF INTERVIEWERS ACCORDING TO EXPENSIVENESS IN PAST SURVEYS

		Minutes per Completed Interview			
Rating	Number of Interviews	Seeking	Contact	Time Out between Attempts	Total
Expensive	97	20.2	36.8	25.5	82.
Average	713	8.2	24.5	16.2	48.
Inexpensive	151	6.3	22.4	16.0	44.
Not rated	262	10.4	25.5	11.7	47.
All interviewers	1223	9.4	25.5	16.0	50.

Rating	Number of Interviewers	Minutes per Attempt			Total
Expensive	7	9.2	16.8	11.6	37.
Average	51	4.4	13.2	8.7	26.
Inexpensive	10	3.3	11.7	8.4	23.
Not rated	20	5.5	13.6	6.2	25.
All interviewers	88	4.9	13.4	8.4	26.

into account the thoroughness of each interview and the accuracy
of the work, would be inaccurate and inequitable.

The study was not undertaken in any attempt to measure the ex
pensiveness or efficiency of individuals. Consequently, no attempt
was made to work out any schéme that might prove more reliable
for judging individual work. Obviously, any organization that main
tains such a system will affect their attrition rates and cost in time
but probably by no more than a change in other features of their
operation would affect them. The whole problem is intimately con
nected with the relation between the interviewers and the home of
fice, a very important phase of survey operations. The purpose of
the log is to emphasize the contribution that interviewers make to
the completion of a survey. It must leave to other studies the prob
lems of relationships with the field staff and the control of the qual
ity of the survey.

12.18 SUGGESTIONS FOR FURTHER RESEARCH

The preceding analysis was a fairly simple study in which the data were obtained from a regular survey, almost as a by-product of ordinary operations. Further studies conducted in this way will add to the results here obtained, but there is a limit to the information that we can get without disturbing the regular procedures. We need to set up experiments designed to separate factors that are entangled in these simpler studies. Such experiments will provide a basis for extending the conclusions to a wider range of new operations and also make it possible to utilize the more general results of the basic sciences, especially psychology. They will knit together research on all forms of field surveys by separating out the phases that are common to other varieties of sampling as well as to quota methods.

The first steps toward experimentation could be taken by assigning more than one interviewer to the same territory, as for example in Mahalanobis' procedure of setting up "interpenetrating samples." This was the procedure used by Moser and Stuart (2). Other steps would be the addition of callbacks to the quota procedure, to discover the biases associated with the loss of interviews at residences where there is no one at home. Research is needed on the effect of assigning quotas by blocks or small areas instead of economic classes. Similar research is needed on the effect of refusals and rejections and on means for reducing the refusal rate.

Studies conducted by the log procedure can be made more valuable by obtaining more information on the way interviewers go about the selection of respondents, their "seeking" operations. These studies would lead to experiments in which the selection of respondents could be related to the location and appearance of their residences, their presence or absence from home at various hours of the day and days of the week, and their other personal characteristics. The selection is undoubtedly related to the attitudes and characteristics of interviewers, each of whom must have a somewhat different set of probabilities of selecting persons within the realm of choice permitted by the quotas and instructions.

The advisability of leaving any choice to the interviewers has been questioned by many survey and research experts and defended by others. It will remain a moot question until accurate and objective evidence replaces the speculation and informal experience that now leads to contradictory advice on this problem of sampling procedure.

REFERENCES

1. Herbert Hyman. *Interviewing in Social Research*, University of Chicago Press, Chicago, 1954.
2. C. A. Moser and A. Stuart, "An Experimental Study of Quota Sampling," *J. Roy. Statist. Soc.*, 116 (1953), 349–405.

CHAPTER 13

Conclusions from Empirical Studies

13.1 WHAT CAN WE LEARN FROM EMPIRICAL STUDIES?

The studies presented in the preceding six chapters are but a few of many that might be made. At this point we should like to consider what can be accomplished by such studies. We also wish to draw some general conclusions in addition to those given in the summary paragraphs at the ends of the chapters.

Empirical studies can contribute a great deal to the improvement of surveys and to the intelligent use of their results. To most people who are not well acquainted with probability theory and sampling doctrine, they are more convincing than theoretical arguments. Hence these studies can be very useful in achieving broad public understanding of the relative accuracy of sampling, though they can disseminate only a modest part of what should be understood by anyone who is dependent on the results of sample survey operations.

Empirical studies are also very valuable in the application of sampling theory since they provide estimates of averages, proportions, variances, correlations, cost rates, and other parameters of the theoretical models that are being applied to particular populations. Such estimates are often not very accurate, but they are much better than mere guesses. They may be quite adequate for practical purposes.

Empirical studies are also very valuable for studying the many phases of the entire survey process that cannot readily be handled by deduction from abstract models.

Such studies are subject, however, to many limitations. A great deal of work is required to produce empirical results that have wide applicability. The implications of the results are not always definite or precise. Also, there are many practical impediments to the completion and publication of such studies. Some of these limitations will be discussed in the sections that follow.

13.2 SOURCES AND FEASIBILITY
OF EMPIRICAL STUDIES

The principal sources of empirical studies are:

1. Published reports of surveys covering variables or populations similar to those in which the survey designer is interested.

2. Results of special studies conducted as integral parts of regular surveys or attached to ordinary survey operations as "free-riders."

3. Special analyses of operating records and data from past surveys to recover "by-products" from information obtained for other purposes.

4. Separate surveys and experiments conducted primarily to obtain data for the empirical studies.

Our experience indicated that comparisons having some value could be made from published reports, but they are not entirely satisfactory. Source 1 will become more fruitful as the use of surveys increases, results accumulate, and reports come to include adequate descriptions and data on the points relevant to empirical studies. Published reports contain too little information on the methods used, insufficient detail in the tabulation, and too many deviations from strict comparability.

The special studies, source 2, are always welcome but are much too rare. They are often weakened by concessions that must be made to keep them from disturbing the regular surveys to which they are attached, but they benefit from both (a) the saving of part of the cost required for an independent survey and (b) the useful collateral information that can be obtained from the regular survey. From the operating point of view there are both advantages and disadvantages in making empirical studies this way rather than as separate surveys.

The possibilities of exploiting information in the files of survey organizations, source 3, we found rather puzzling. Usually we found the organizations friendly and interested but too busy with current work to do any special analyses. Often they had plans to do such studies but they found it necessary to postpone the work again and again. Frequently it was doubtful that the studies would ever be made. Unfortunately it is usually quite difficult for any outsider to come in and make the studies unless, in effect, he becomes a member of the staff for a substantial period of time. This makes a study more costly and less efficient than it would be if it were conducted by members of the staff already well acquainted with the intricacies of the material. The prospective value of the studies may not appear to warrant the cost of analysis from without. The situation is different if adequately prepared graduate students are available for such work. However, many of the possible studies require a level of training and broad experience, at least for the preliminary organization of the work, well above the capabilities of graduate students and members of the ordinary staff. Without a thorough and competent examination of materials and methods from the standpoint of research methodology and theory, many studies of this kind will prove to be invalid or defective. These hazards are much greater with data already collected than they are with data originally collected by the staff that is to analyze them.

Even when the staff of an organization can manage to analyze its operating records and data, the results are often limited because the staff is too close to the operation, has preconceptions and blind spots, sees the defects in competitors' work more readily than in its own, and even has some axe to grind in the field of the study. Often the organization decides not to publish the results or delays publication. All these limitations are natural enough, yet they curb the enthusiasm of students of survey methods who look longingly at the accumulated records of survey organizations and wish they could get at the "gold in them thar hills." There is plenty of rich ore but it takes a good prospector to find it and then it entails some hard digging and difficult refining.

The fourth source is the most potentially productive but is also by far the most expensive. Experiments can be set up in a way that best serves the purposes of an empirical study without compromises to other purposes. If the expense is reduced by experimental designs aimed at spreading the cost over several empirical studies, some compromises will then be necessary among the several research purposes so combined. Hence limitations imposed by the cost of such research tend to restrict the freedom of a research inquiry, even when other entanglements with practical limitations have been removed.

13.3 DIRECT TESTS AGAINST CHECK DATA

The simplest and most satisfactory test of the accuracy of an estimate from a sample survey is a direct comparison of the estimate with the true value of the variable being estimated. This test is also a very valuable source of information about the accuracy of an entire survey process or procedure. When it is repeated under reasonably constant conditions, it provides good estimates of the systematic error or bias of the survey as a whole. Since bias is the most elusive and yet most important aspect of most surveys, direct comparisons of estimates with their true values would seem to be the best possible form of empirical research. Actually they are not, for (a) true values, and even sufficiently accurate approximations to them, are very seldom available, (b) the bias of a survey process is usually combined with large components of random error from which it can be separated only when many repetitions and quite large samples are analyzed, and (c) the bias is usually a function of many situational factors, as well as of various aspects of the survey process, and hence varies somewhat from one repetition of the survey to another.

In spite of these difficulties, comparisons with check data have been quite useful as illustrated in Chapter 7. They have been applied recently in connection with the Post-enumerative Survey conducted to test the 1950 Population Census (1). In various studies they have revealed the existence of large biases and contributed to a sounder understanding of the accuracy of data. Unfortunately they are not always adequate for calibration of the processes and adjustment of the data. Apart from election returns, there is almost a complete lack of check data for the more psychological kind of variables. It would appear unlikely that check data will be obtained for these variables in the near future, and hence other indirect means must be developed and used to test survey results in our major field of interest.

13.4 COMPARISONS OF SEVERAL SURVEYS

In the absence of check data, it is natural to compare one survey with another. This is useful but far from satisfactory. Comparisons demonstrate that different surveys produce different results but they do not show which results are the more accurate or which survey procedures are the more dependable. If comparisons include analyses of the results of many repetitions of the same procedure, they may provide useful estimates of the variance of the results of that procedure but this does not suffice for appraisal of accuracy and dependability. Useful results of this kind were obtained for quota sampling in Chapter 10, and the corresponding methods for probability samples were discussed briefly in Chapter 9.

The greatest limitation of comparisons between surveys is that there are innumerable varieties of surveys and the particular examples that are compared usually differ from each other in many ways. The deductive approach is helpful but leaves many problems of comparability unsolved. In the final analysis of the comparisons, therefore, it is impossible to determine the source of the major part of the difference between their results. The analysis of the 1948 pre-election polls (2) amply demonstrated the impossibility of untangling fully the separate effects of the many basic differences between generally similar surveys.

In spite of this seemingly unsurmountable limitation, we believe that comparisons of surveys can contribute useful results in the future, especially when comparable data on costs can be obtained.

13.5 ANALYSIS OF QUOTA SURVEYS

Since quota surveys have shown major weaknesses on several notable occasions, there is a question about the value of studying them in the future. We found evidence that quota samples do turn out well in some instances but that there is great variation in the performance of quota survey organizations and great difficulty in specifying a quota sampling procedure adequately enough to permit a well-considered judgment of its accuracy. Nevertheless, there are great quantities of data that were obtained by quota methods in the past for which important uses will be found, and it is likely that some kind of quota methods will be used quite frequently in the future. Hence we believe that further study of interviewer-selected samples will be worth while, though the work will be difficult and the results not always conclusive.

13.6 OPERATING PROBLEMS IN PROBABILITY SAMPLING

Empirical studies of surveys conducted according to plans based on probability models have shown relatively large discrepancies between the plans and actual performance. This is especially true for samples of individuals rather than of households. Almost all surveys of psychological variables are necessarily taken from samples of individuals. For such surveys, the planned sample is often completed only to the extent of 80 to 85 per cent. Important questions arise about the effect of omitting 15 to 20 per cent of the sample, since the loss of these individuals may be closely related to variables that are being observed or measured. These questions must be faced anew in each situation. Further research beyond that summarized in Chapter 11 is greatly needed. It should be closely integrated with studies of bias, such as those that are based on check data, and with studies of survey organization and field operations. Studies of interviewing and response bias should also be included, since the efforts to find and obtain the cooperation of potential respondents certainly affect their behavior in the interview.

13.7 ANALYTICAL STUDIES OF THE COMPONENTS OF SURVEY PROCESSES

Survey processes can be analyzed into components that are susceptible of empirical study by themselves. Special experiments can be made on some of them, such as procedures in which interviewers list households, select samples, and interview at the same time. Telephone interviewing can be compared with interviewing by visiting people in their homes. Substitution procedures can be tested. Travel costs deserve further study. People who refuse should be approached again after a lapse of time to determine, if possible, how they differ from people who cooperate. Studies of the internal correlation of clusters and similar population analyses should be extended to provide information for improving sample design. Many other analytical studies are needed to provide working materials for future sampling.

Some of these studies can well be made as part of the quality control and checking operations in regular surveys. Some are a natural extension of ordinary field work. Changes in interviewer assignments that make them equivalent and similar matching of larger parts of survey operations will yield valuable information not now available.

The experience and point of view of the respondent has not been studied but should be. Recording of interviews for analytical study has not been done frequently enough. The original qualifications and training of interviewers that are essential to excellence in interviewing have yet to be determined. The work of interviewers can be made less onerous. Improvements in these and other phases of survey operations can lead to substantial gains in accuracy and repay the cost of finding out how to make them.

13.8 THE OVER-ALL STRATEGY OF EMPIRICAL STUDIES

Anyone who takes a comprehensive view of future research on survey methods must face broad questions of the direction it can best take. How much effort is it wise to devote to empirical studies? How much to deductive and theoretical investigations? How should the effort be apportioned among particular projects in each field? These are all questions that cannot be answered with assurance, but every attempt to make a study implicitly answers some particular question about what is worth doing. To form a helpful background for such decisions, if for no more ambitious purpose, we can speculate about likely answers to the comprehensive question of research strategy. Here are some thoughts about it.

First, some of the past studies tend to suggest that we may be able to progress only a little way in empirical studies before we are stopped by an impenetrable mass of details. It may then be more expensive to seek improvements by such studies than the benefits are worth.

Second, deductive and theoretical investigations may degenerate into relatively trivial explorations after some initial advances. There is no guarantee that their value will increase indefinitely.

Third, the two approaches are more likely to be fruitful if each is brought to bear on the other, raising critical questions, posing problems, and testing each fresh result.

Fourth, results of past studies have illuminated the processes of survey taking, but they have not brought to light the deeper causal processes or more subtle influences that affect the results. In other words, the results are often not definitive, as was observed on page 131. Results may continue to lack definitiveness in the future unless the operations are probed much more deeply, and this probing may tend to disrupt them too much.

Fifth, many of the problems are generic and can best be studied on a broader field of investigation than survey operations. They arise in other forms of organized effort such as administration, field crew operations, communication, and collection of information. These problems rest on common types of human performance and human relations, though they are modified by much that is going on in their broader social context. Therefore, they should be investigated by programs of basic research devoted to the study of these common types of human performance. When these programs are successful, the general scientific theories and general facts which they develop about human behavior can be applied to surveys.

Sixth, any major progress in the improvement of surveys may have to await ingenious innovations and extraordinary discoveries. As in physical science, the germinal ideas may originate in relatively pure research, proceed through testing and development programs, and eventuate in full-scale trials under practical conditions. The last stages must wait on the first.

Seventh, though it may be possible to work ahead step by step, feeling our way, it is very likely that a moderate amount of peering ahead into the darkness will increase our rate of progress. This is a function for research councils and training centers as well as for the critics and philosophers of applied science.

REFERENCES

1. Eli S. Marks, W. Parker Mauldin, and Harold Nisselson, "The Post-Enumeration Survey of the 1950 Census: A Case History in Survey Design," *J. Am. Statist. Assoc.*, 48 (1953), 220–243.
2. Frederick Mosteller, Herbert Hyman, Philip J. McCarthy, Eli S. Marks, and David B. Truman, *The Pre-election Polls of 1948*, Bulletin 60, Social Science Research Council, New York, 1949.

PART III

The Design
of Sample Surveys

CHAPTER 14

The Extent and Nature
of Preparations Vary
with the Circumstances

At this point we turn to another phase of sampling, the planning and preparations that precede the actual operation of a survey. The reader may well ask: "Of what concern is that to me? I'll never make a survey. I'm merely interested in the final results." There is much point to this question. Many readers will find the discussion boring; they are well advised to skip it. Others will find it insufficiently concrete and technical; they should turn to the textbooks and manuals. One of the more ample and profound of these books is Hyman's *Survey Design and Analysis* (1). Another of unusual interest is Ackoff's *The Design of Social Research* (2). Still another excellent treatment of the problem is to be found in Lorie and Roberts' *Basic Methods of Marketing Research* (3). Many readers will find themselves in neither category. They will be interested in the general problems of design that are closely related to what has preceded, especially the complexity of the survey process and the interrelation of the sampling process with other parts of the survey. This is a subject on which relatively little has been written. It has been touched by many of the articles and books that develop techniques in great detail, and it has been discussed in the staff meetings of research organizations, but the problem of design seems to baffle everyone who attempts to discuss it systematically.

An understanding of the problems of design will be of value not only for the actual designing of surveys but also for several other purposes:

1. It may help the user of the final results to understand what was done, removing unnecessary doubts and answering questions that otherwise would make him hesitate to use the results.

2. It will give some users a better opportunity to analyze the results and adapt them for his particular purposes.

3. It will tend to create a demand for information about the designs that are actually followed and thereby encourage research organizations to describe their methods and discuss their operating problems in their reports and elsewhere.

Sampling methods have been improved greatly in the last two decades. As they have advanced, they have tended to require a greater degree of active cooperation by the prospective users in determining as precisely as is feasible the specifications that are to be met and the considerations that should influence technical decisions in the development of the design.

14.1 APPROPRIATE PLANNING AND PREPARATIONS ARE NECESSARY

One of the most important lessons that experienced samplers have learned is that the quality of survey results depends a great deal on the preparations made before the survey is conducted. There is no question about the necessity of making some preparations; the only question is how far to go in planning the details. No simple rule can be found to answer this question, for the extent and the nature of preparations that are worth while depend on many circumstances present in the particular situation. For example, when we set out to study the attitudes of employees in a manufacturing plant, it may only be necessary to obtain a list from the payroll, decide what questions should be asked, and send an interviewer through the plant to talk with every tenth or twentieth employee on the list. In such a simple situation the procedure that comes quickly to mind is frequently a satisfactory one, though not necessarily the best. However, even in apparently simple situations such as this one, an investigation should be made to see that there will be no complications or serious difficulties to interfere with the survey. We should be sure, for example, that the attitudes being studied can be determined accurately by an interview at the employee's bench or desk while he is at work, and that union officials and employees have been consulted and informed in advance of the actual survey operation. The care and detail that may be required in this latter phase of preparation is well illustrated by the experience of the Survey Research Center of the University of Michigan in conducting an attitude survey among the employees of the Caterpillar Tractor Company (4, p. 35). Also, a sample drawn in the manner already noted would not be representative if the employees work in groups of five or ten who are listed together on the payroll with the straw boss or subforeman first. By taking every fifth or tenth name, we should get either all subforemen or none in the sample, instead of an approximation to the correct proportion of 20 per cent.

It is ridiculous, of course, to take unnecessary pains. Often an experienced survey director can proceed without elaborate plans to take a survey that is adequate to the situation and fulfills all essential needs. Surveys should be designed in advance, and procedures worked out and tested, only to the extent that is appropriate in each actual situation. Some survey operators will prefer to go ahead with few preparations, meeting problems only as they arise. Others will try to anticipate possible developments so that they can chart their course in advance. They will look ahead in order to avoid risks whenever possible, and the cost and effort of these precautions is the price they pay for their protection. Further discussion of these two approaches will be found in a paper by Stephan (5).

As complex sampling methods are applied more frequently, and survey results are put to more important uses, planning and preparation become increasingly necessary. Utterly careless or casual operations produce poor results that no analyst, however skillful, can correct or adjust. Even experienced survey organizations frequently produce results subject to unforeseen weaknesses. As an illustration, we present the following quotation from Senf's account of the General Enumerative Surveys of the Bureau of Agricultural Economics (6, p. 115):

Results obtained from the 1947 survey on the value of farm property and on total cash rent paid were not usable. For the United States as a whole the survey obtained a total value figure that was 24 percent above the March 1, 1947, official BAE estimate, and an average value per acre that was 33 percent above the official estimate. Official estimates are derived by applying the percentage change in the BAE index of land values to the 1945 census values. Part of the discrepancy between the survey and the official estimates is probably due to the wording of the questionnaire. The BAE survey asked "How much would the part of this farm (that you own or rent) *sell for?*" In contrast, the census schedule used the term "value," and the schedule to crop reporters, which provides the basis for the BAE index, uses the term "average value." It seems probable that farmers would report the "selling price" higher than they would report "value," particularly in 1947 when land prices were advancing rapidly.

Though it is distressing to discover faults in our own work, it is more discouraging to recognize that work may be subject to additional weaknesses that are hidden. This is likely to happen to those who have a naive approach to survey work. The danger is especially great in undeveloped fields of inquiry for which there are few chances that any other data or subsequent experience will provide a test of results. A study that is merely written up and put on the shelf may never be questioned, but as research is used more intensively as a basis for action, it is subjected to exacting tests that ultimately reveal weaknesses and faulty work. In the field of opinion research, this stage of testing has not yet been fully realized, though it has been approached by the studies of the Committee on the Measurement of Opinion, Attitudes, and Consumer Wants, as well as by various studies conducted by the users and producers of opinion studies in the course of their experience.

No attempt will be made here to settle the question of just how much preparation should be undertaken in particular instances. It is only necessary to recognize that serious attention should be given to preparations before the actual work of conducting the survey is started. The kind of preparation and amount of planning should be determined for each survey as the particular circumstances require.

In many ways designing a major survey is like designing a house or a factory building. The architect starts with a general notion of what his client wants. He then tries to select materials and lay out the rooms and utilities in a way that will suit the desires of the future occupant and meet his needs. In doing this he must take account of the size, shape, and condition of the site and the total amount of money that the owner is willing to pay. The unavailability of certain materials may be a limiting factor. He will incorporate into the plan certain features to protect the occupant against specific hazards. A major change in any one part of the plan may affect other parts. The various parts must fit together properly. No one wants a home in which the stairs come down in the middle of the living room or a factory building in which the shipping platform is located too far from the place where the products are finished and stored. Hence, the planning must proceed all together, though many details can be worked out by themselves. The architect takes advantage of the topography of the site and local materials; he checks the whole plan to avoid possible trouble in using the building when it is completed. The designer of a complex survey procedure works in a similar manner to take advantage of special opportunities and to eliminate avoidable difficulties.

14.2 GENERAL OUTLINE OF PART III

Part III will review the full range of factors that may be considered in designing a complex survey, together with their relationships. Since the extent and detail of the survey design must vary with the circumstances, and the gains in efficiency and economy that can be attained by a complicated design also vary, the reader should not regard Part III as a blueprint to be followed slavishly in every survey. Only so much of it as is appropriate to the particular situation should be considered. Even in designing a very elaborate national sample, some of the subjects discussed in Part III will be unimportant or even irrelevant. Moreover, anyone who has conducted surveys knows that their actual operations have not been arranged as logically or performed as systematically as Part III would suggest. Part III is a reminder of things that may be necessary, a display of the various phases of survey design that may possibly be encountered in one survey or another.

It would take too much space to write a Part III for each type of survey we might encounter. There would be a great deal of repetition. Hence, the whole range of subjects has been set down and their relations, one to another, have been indicated. The reader who is seeking help in planning a particular kind of survey can skip sections that do not seem to apply to the type of survey he intends to make and can modify general statements to adapt them to his particular needs.

Some readers will be disappointed that they are not given a list of exactly what to do in taking a survey. If such instructions were given and a survey turned out badly because of peculiar factors, the reader would surely complain about the advice he had received. It is not difficult to see that a competent expert in sampling procedures shrinks from advising for a situation he has not studied, just as a good physician declines to prescribe for a patient he has not examined. Lacking an opportunity to explore the situation himself, the advisor can only give good advice if the recipient is willing to assume full responsibility for an adequate examination of the factors that may have serious effects on the outcome of the survey. There are a great number of possible factors that he must consider, and some of them may be very elusive. Hence, the reader who attempts to design a sample survey should work out the whole design from the general principles presented here, or if this seems to be beyond his skill, he will do well to seek the services of a competent consultant, or to engage a good research organization to conduct the survey for him.

14.3 SECTION A IS CONCERNED WITH THE PRELIMINARIES TO ACTUAL DESIGNING

This first portion of Part III will be devoted to the assembling of information and materials that can be used in developing the design. Such preparatory work may be done as the design is being developed. However, it is frequently true that more rapid progress will be made by first canvassing thoroughly the job to be done, the materials available for it, and the more important features of the situation in which it must be carried out. The discussion of this preparatory work will run from the crystallizing of the purposes of the survey to the assembling of information about the population to be studied and the attitudes and related variables that will be observed and measured. Thence it will proceed to the material available for measurement and sampling techniques, to the practical possibilities and difficulties that affect their use, and to the marshaling of resources. With these tools and materials assembled, we are then ready to go to work on constructing the final survey design, although sometimes the results of these preliminary steps may be presented to a conference of interested parties in order to determine whether or not the survey should be conducted at all.

It might be noted in closing this introduction that sometimes the preparatory work may involve little more than a simple adaptation of a plan used in some previous survey. After an organization has completed a complex survey, it may repeat the survey at a later date or make surveys of a similar kind without much additional preparation or change in the design. The experience of the previous survey can be used to improve the design, especially if provision has been made to observe and analyze how it worked out. There are many noteworthy examples of this type of work. In particular, we may refer to the Monthly Report on the Labor Force (now the Current Population Survey) of the Bureau of the Census (7, Chap. 12) and the Survey of Consumer Finances of the Survey Research Center of the University of Michigan (8).

Often a design that has been used in one situation can be used in a quite different situation with relatively few changes. However, it is

cessary to check over the special opportunities presented by the new
uation and the additional difficulties it may involve. Also, some of
e difficulties that limited the design in the first situation may not be
esent in the second, and possible features of design that had been
jected may now be reconsidered. How much modification and adap-
tion is warranted is, of course, a matter of judgment since the gains
ay not be worth the additional effort.

REFERENCES

Herbert Hyman, *Survey Design and Analysis,* The Free Press, Glencoe, Ill.,
1955.
Russell L. Ackoff, *The Design of Social Research,* University of Chicago
Press, Chicago, 1953.
James H. Lorie and Harry V. Roberts, *Basic Methods of Marketing Re-
search,* McGraw-Hill Book Company, New York, 1951.
National Industrial Conference Board, *Studies in Personnel Policy: Experi-
ence with Employee Attitude Surveys,* No. 115, New York, 1951.
Frederick F. Stephan, "Sampling," *Am. J. Sociol.,* 45 (1950), 371–375.
Catherine Senf, "A Report on the General Enumerative Surveys," *Agric.
Econ. Res.,* 1 (1949), 105–128.
Morris H. Hansen, William N. Hurwitz, and William G. Madow, *Sample
Survey Methods and Theory,* Vol. I, John Wiley & Sons, New York, 1953.
U.S. Federal Reserve Board, *Fed. Reserve Bull.* (monthly), Washington, D. C.
Various numbers include reports of the annual survey.

CHAPTER 15

A Definite Formulation
of Purposes and Objectives
Is Needed

15.1 THE PURPOSES GUIDE THE DESIGNING

The best starting point for any design is to be found in the aim
purpose that the survey is to fulfill. To produce a satisfactory resu
the designing process must be guided by specific purposes towa
definite goals and objectives. In the end the design will be tested b
the success or failure of the survey in attaining these objectives, i.e
in producing the information that is wanted. The process of design
ing should shape the parts and arrange their relationships to make th
survey as effective as possible in fulfilling its purposes. This can on
be done if the purposes are known; they must be formulated clearl
definitely, and in sufficient detail to provide a working guide for se
ting up the survey.

A number of research directors have reported that their clients ar
superiors usually cannot state their needs adequately, at least n
without being forced to give much more attention to the proble
than they (the clients and supervisors) ordinarily feel is require
When a survey director goes ahead with no more than a general stat
ment of purpose that vaguely but ever so hopefully expresses th
needs and wishes of the prospective client or user, he should expe
to encounter serious difficulty along the way and also when he pr
sents final results. The user may not know what he needs, but he wi
usually have a firm conviction that he has, or has not, received wh
he has a right to expect. This situation has long been recognized b
market research organizations. One instance has come to our atten
tion in which the client is required to fill in a form that asks fo
among other items, the following information:

Study Problem (a concise and brief statement of the situation which has prompted the study. Why is it contemplated? What type of information is required? Generally speaking, in what manner will the study data be used?); *Study Objectives* (. . . listing of the principal fields of investigation which require consideration by the study); *Data Required by Objectives* (. . . a further specification of *each* objective down to the specific data to be obtained from the respondent . . .); *Proposed Questions* (. . . not a questionnaire but rather a means for visualizing the necessary content of the study questionnaire); *Possible Tabulations; Suggested Uses of the Study Data.*

A statement of survey goals and objectives may be especially important and difficult in pure science and in research that is not directly tied to policy and action. There is little that we can say in a general way concerning this, i.e., without discussing specific problems, though Alpert (1) has given the following cogent summary with respect to sociology and sampling:

My basic thesis is that sociological theory must enter into, govern, and guide the development of the appropriate sample design for a given survey purpose. Operationally speaking, survey results are understandable only in terms of the methods by which they are obtained. To the extent that we wish to organize our research in terms of a systematic, theoretical system, we must make method, including sampling method, the servant, and not the master of our theoretical scientific framework. . . . Sampling designs cannot be efficiently and effectively developed for sociological surveys without serious consideration of the relevant theoretical knowledge concerning the subject matter involved in the survey.

Many surveys permit a rather simple exposition of their aims and purposes, even though the sample design required for the efficient achievement of these aims and purposes may be quite complex. As illustrations we may refer to the following:

1. Shaul (2) has described the importance and use of sample surveys in providing demographic statistics in Central Africa. Referring to a specific survey, he states that "The Survey sought to determine: 1. The total indigenous population of each district within a confidence belt of 10 per cent at the 5 per cent level of significance . . ." plus other specified information.

2. In 1944 the Bureau of the Census, at the request of the Committee for Congested Production Areas, conducted special population censuses by the use of sampling methods in specified areas in the United States. This work has been described in a publication of the U.S. Bureau of the Census (3) and this report states: "After deliberations with the Committee on Congested Production Areas, it was decided that samples should be designed with the aim of determining the total

population of each census area within 3 per cent at the outside, whi‹
is to say, with a coefficient of variation of 1 per cent."

3. In May of 1948, eleven leading magazines in Canada formed t‹
Magazine Research Group of Canada and made plans to measu‹
their audiences in a nationwide survey. The results of this study ha›
been described in a report of the Canadian Advertising Resear‹
Foundation (4). This report states (p. 27) that five primary ol
jectives were established for the study, and these are given in tl
report. The principal objective was "To measure the total numb‹
of individuals 15 years and over who look into a typical issue ‹
eleven magazines in Canada."

The statement of goals and objectives for a survey are not alway›
as simple and straightforward as in these examples, particularly ‹
we look at the problem with a view toward an "ideal" use of dat‹
Research and measurement have become regular functions of busines
and government organizations. Surveys are becoming more usef›
and are being applied to increasingly complex problems. With othe
well-established functions, they may share in the responsibility fo
the success of large undertakings in production, distribution, com›
munication, and government. Hence, their part in the division o
labor must be worked out. Their particular objectives must be de
lineated with considerable care in relation to the larger purposes the›
serve. They must be directed toward getting their assigned job done

The complexity noted in the preceding paragraph is exemplified b›
the increasing use of survey methods for research into the behavior o
the individual consumer. For example, Annual Surveys of Consume›
Finances are conducted for the Board of Governors of the Federa‹
Reserve System by the Survey Research Center of the University o‹
Michigan. An article in the *Federal Reserve Bulletin* (5) remark‹
concerning these surveys:

Shifts in consumer spending and saving are important influences upon economic activity. . . . Information on the plans and attitudes of a representative national sample of consumer spending units can be helpful as an indicator of consumer actions over the near-term future. . . . These Surveys are
still largely experimental, but over the past six years Survey findings have
provided helpful clues to prospective tendencies in consumer purchases of
durable goods and houses.*

Further discussion of the use of survey methods in this field will be
found in a book by Katona (6, pp. 302–334) on the psychological

* By permission from *Psychological Analysis of Economic Behavior*, by George
Katona. Copyright, 1951. McGraw-Hill Book Company, Inc.

ialysis of economic behavior. Katona's concluding remarks clearly
t forth the ultimate goals in this type of research, goals that will
low more specific formulation of survey objectives:

Psychological-economic surveys are at the same time a method of basic
search and of research intended to satisfy practical needs. Attempting to
rive at valid generalizations—the aim of basic research—can be combined
ith endeavoring to find solutions to specific problems of the day and even
presents what will ultimately prove the best means of achieving practical
ms. If this function of scientific research is more widely understood, larger
inds than heretofore may be channeled to studies that are basic and there-
re practical.

15.2 EFFECTIVE DESIGNS CAN SERVE MANY PURPOSES
AND COPE WITH UNCERTAINTY ABOUT USE

Survey design would be greatly simplified if each survey had only
ne sharply defined question to answer or purpose to serve. Ordi-
arily, however, there are a number of potential uses for which infor-
iation is sought in a single survey. Such surveys are frequently re-
erred to as "multi-purpose," and they force us to balance effective
esign for one particular purpose against effective design for other
urposes. Also, there may be no certainty about the ultimate impor-
ance of the several purposes when the survey is completed. Inter-
ening events may change their relative importance or induce a sub-
titution of other purposes in their place. It is not possible to antici-
ate all the changes that may occur or all the possible uses to which
he information may be put. But most human activities are quite as
ncertain!

The design of multi-purpose samples becomes simple only when
he fulfilling of requirements for one purpose automatically insures
hat the requirements for the other purposes are satisfied. For ex-
mple, the account of the sample census of congested production areas
3) referred to in the preceding section stated that ". . . a sample ade-
uate to obtain the total population with a coefficient of variation of
 per cent will also provide better precision than is usually required for
he tabulation of most other characteristics of the population such as
he proportions by sex in 5-year age classes, the proportion nonwhite,
r in school, employed, etc., hence the population count was the gov-
rning factor in the sample design." In more complicated situations
t may become necessary to assign relative weights to the various pur-
oses and attempt to arrive at a design that is in some sense "opti-
num" for the weighted aggregate of all purposes. Recent advances

in statistical theory [e.g., statistical decision theory (7)] may aid in finding a practical solution to this problem, but the assignment of precise quantitative weights to the purposes is the first obstacle that must be overcome.

Uncertainty concerning the ultimate use of results to be derived from a survey does not blot out the need for careful preparation or make planning futile. It merely makes it necessary to proceed with care in other situations where risks are involved. All the important possibilities must be considered. Then we decide, from their apparent probabilities and consequences, how to shape our actions so as to reduce both the probability of the unfavorable possibilities and the seriousness of their consequences, should they occur. At the same time, we try to increase the probability and value of favorable possibilities. If we can estimate the probabilities and measure the relative gain or loss from each possible outcome, then we may multiply each gain or loss by its corresponding probability, subtract the results for losses from those for gains, and seek to find a plan of acting that will promise the largest net gain. This is, in essence, the starting point of statistical decision theory referred to in the preceding paragraph and now being presented on a sophisticated level in the statistical literature. Whether or not we can make such a calculation, a similar process of informal judgment can be applied both to uncertainty about the demands that may be made on the survey and to any requirement that the survey fulfill several purposes. The result is a compromise, but it is just as surely directed toward satisfying the multiple purposes and meeting the various contingencies as a simple survey is directed toward its sole objective.

Indefiniteness in the purposes, when it cannot be corrected by further inquiry, creates just such uncertain and multiple aims. But, when the job to be done is not clearly crystallized, we may still develop a specific working statement that sets forth important possibilities; from this statement we can judge the relative weight to be given to each possibility and obtain a basis for shaping the design.

15.3 PURPOSES SHOULD BE TRANSLATED INTO WORKING SPECIFICATIONS

The statement of purposes that provides the starting point for planning the survey should set forth as specifically as possible just what is wanted. If not already in such terms, it should be translated to specify the attitude or opinion variables that are to be measured, the related information that is to be obtained, the estimates and analyses that are to be prepared, the population that is to be represented by the sample, the subgroups of the population for which separate information will be required, the date for delivery of the results, the general requirements for accuracy and detail, and any other particular specifications that the ultimate users may determine. From these specifications the survey director may establish incidental and derived specifications necessary to fulfill the others.

Ordinarily these specifications are worked out in conference with the ultimate users or their representatives. They may be modified for various reasons before the survey is launched. It may be quite satisfactory to leave some of the specifications in general terms. For certain others, any change will greatly alter the problem of design. To avoid drifting away from them when he is occupied with other details of planning and operations, the survey director should review the specifications from time to time and determine whether they will all be satisfied by the survey.

Finally, the specifications will be expressed in a plan of organization for the survey, a time schedule, a budget, and other control devices. An excellent and detailed account of this phase of planning with respect to two nationwide surveys, conducted by the Bureau of Agricultural Economics, has been provided by Brooks (8) and Senf (9).

The preparatory work for various aspects of the survey that spring from the specifications—the population to be studied, the attitudes to be measured, the measurement methods to be used, and the sampling techniques to be employed—will be examined successively in the remainder of Section A.

REFERENCES

1. Harry Alpert, "Some Observations on the Sociology of Sampling," *Soc. Forces*, 31 (1952), 30–33.
2. J. R. H. Shaul, "Sampling Surveys in Central Africa," *J. Am. Statist. Assoc.*, 47 (1952), 239–254.
3. U.S. Bureau of the Census, Sampling Staff, *A Chapter in Population Sampling*, U.S. Government Printing Office, Washington, D. C., 1947.
4. Canadian Advertising Research Foundation, *Audience Study of 11 Magazines in Canada*, Toronto, 1949.
5. U.S. Federal Reserve Board, "Consumer Plans for Spending and Saving," *Fed. Reserve Bull.*, April 1952.

6. By permission from *Psychological Analysis of Economic Behavior*, by George Katona, copyright 1951, McGraw-Hill Book Company, New York.

7. Abraham Wald, *Statistical Decision Functions*, John Wiley & Sons, New York, 1950.

8. Emerson M. Brooks, "The General Enumerative Surveys," *Agric. Econ. Res.*, 1 (1949), 37–48.

9. Catherine Senf, "A Report on the General Enumerative Surveys," *Agric. Econ Res.*, 1 (1949), 105–128.

CHAPTER 16

Essential Information
Should Be Assembled
on the Population, Attitudes,
Opinions, and Related Variables
That Are to Be Studied

16.1 INTRODUCTION

Starting from the working statement of purpose and specifications, it is necessary to complete the determination of just what group of people is to be covered by the survey and to assemble pertinent information and materials about the population and its subgroups that will be useful in drawing and getting in touch with the sample. These steps are particularly important since the general character of the population and many of its particular characteristics will limit the opportunities available for survey design. The costs and efficiencies of different procedures vary from population to population. The relative feasibility and efficiency of various kinds of sampling units and means of contact with the sample are closely related to the characteristics of the population. The actual measurements and observations, and the materials available for use in estimation and analysis, are also affected by the specification of the population. Hence, not only is it necessary to specify precisely the population to be studied but it is also necessary to assemble and review information about the population as carefully as other preparatory material on which the design must rest.

Questions about the population are customarily regarded as part of the sampling phase of a survey. Questions about the attitudes, on the other hand, are customarily regarded not as part of the sampling but as a wholly separate and almost unrelated phase of the survey.

This view of psychological variables, when pushed to the extreme, is quite distorted since ultimately it is the distribution of possible values of the variable that are actually being sampled. The population elements, if the attitudes can be separated from them, are merely vehicles and intermediaries in the process of obtaining a sample of attitudes (or measurements of them). The identities of individual persons do not appear in the final product of the survey, and the non-attitudinal characteristics of the population are ignored or are subordinated to the attitudes. The results would be much the same if the attitude data were transmitted directly to the analyst by some imperfect process of telepathy without any information on which value of the attitude belonged to which person. Now, of course, the attitudinal results are referred back to, or are associated with, the population. Other characteristics of the population elements, such as geographic location, etc., are brought into the analysis and are related to the attitudinal variables.

The somewhat novel position just expressed is an enlightening alternative to the natural but unduly restrictive assumption that the sampling can only have a very remote relation to the measured attitudes in contrast to its close relation to the people who are studied. A more sophisticated scientific point of view regards the process as one of drawing a sample from a set or collection of attitude measurements rather than from a population of human biological organisms. There is no fundamental conflict between these points of view; they are both simplifications and idealizations of the actual sampling operations (1, pp. 8–12).

Information must be assembled on the attitude variables and other variables that are to be studied because they affect profoundly most of the phases of the entire survey design. They constitute, first of all, the basic criteria of the representativeness of the sample. They determine very largely the kinds of measurement operations that are feasible (e.g., mail questionnaire, personal interview, direct observation, laboratory test, etc.) and therefore influence the choice of methods of getting in touch with the population. Through these direct effects, the attitude and other variables in turn play an important role in determining cost factors, cooperation and refusal rates, the techniques of drawing the sample, and the effectiveness of the survey in satisfying the purposes for which it is taken. The effectiveness of various possible control variables, the detail of subdivision of the results in the analysis, the size of the sample, and the ultimate accuracy of the survey depend on the attitudinal variables as well as on the general purposes and the population.

A very clear-cut illustration of these points concerning the variables to be measured can be obtained by considering a situation in which we have to choose between using a mail questionnaire and a personal interview. Assume that a complete listing of the population exists, that the population elements are spread over a wide geographic area, and that there is no non-response problem. If the attitudes and related variables can be adequately measured by means of a mail questionnaire, then the sampling problem is much simpler than for a personal-interview survey, and we avoid the cost of training and paying interviewers. On the other hand, if personal interviews are required to obtain information on the attitude variables, the sample must be drawn so as to avoid excessive travel costs and we must face the problems of hiring, training, and supervising interviewers. Admittedly this is an oversimplified example, since mail surveys are always plagued by the problem of non-response, yet it makes clear the manner in which the attitude variables influence the entire survey operation.

We shall now consider some aspects of the population specification and of the assembly of information on the variables to be measured that may be dealt with during the preparatory stages of design. These will be treated in succession.

16.2 SPECIFICATION OF THE POPULATION MUST BE COMPLETED

A. Population Specification May Alter Survey Results Considerably

The working statement of purposes, discussed in the preceding chapter, will probably indicate the general type of population that is to be sampled. It may be the entire nation or some regional or local population. It may be restricted to certain age groups, to men or to women, to racial, religious, occupational, or economic groups within the total population. However, unless the survey director is willing to take the risk of confusion during the survey and uncertainty about the results after it is completed, he must decide, in advance, certain additional questions of detail about the population that is to be studied.

The specifications may be partly functional and partly a matter of status; the distinction is not always sharp. For example, they may call for participants in certain activities, members of certain organizations, users of certain products, readers of certain publications, voters, radio listeners, movie patrons, labor leaders, business executives, downtown office workers, farmers, soldiers, veterans, college students, or migrants. Each group has its own special characteristics by which it may be defined, and members of the group are distinguished from persons who do not fall in the same classification.

The distinguishing characteristics of a special group are usually not simple. Frequently a person can move into or out of the group and, hence, questions arise about persons who are in transition or whose cases are peculiar in some respect or who are marginal to the central concept of what a member of the group should be. These atypical and transitional persons may be excluded or included in the population.

Decisions to exclude or include particular subgroups, in effect, redefine the population. If the persons affected by such decisions are relatively numerous, or if they differ greatly from the rest of the population in variables that are to be studied, then the redefinition of the population may have important effects on the results of the survey. When the ultimate user receives the results, he will need to know just what population was sampled so that he can know just what the results mean.

B. Some Groups Are Included or Excluded for Substantive Reasons

The decisions about the marginal groups will usually be made for one of two general reasons, substantive or operational. The substantive reasons are those concerned with determining the kind of people who will be appropriate for the general purposes of the study. For example, the 1935 urban Study of Consumer Purchases was largely confined to native white families with both husband and wife present. The Census excludes from its monthly survey of the labor force the military and institutional groups, i.e., members of the armed forces, inmates of penal and mental institutions and of homes for the aged, infirm, and needy. For some other purposes these groups are included. Similarly, some pre-election polls specifically attempt to exclude all classes of non-voters. These are matters of trimming the population to fit the concepts that will be used in the analysis and interpretation.

Many surveys of radio listening, magazine reading, and opinion on public issues are limited to persons who have passed their eighteenth or twentieth or twenty-first birthdays. This procedure is in part a trimming to the concept of what is wanted, but it is somewhat arbitrary since there are few sudden changes that occur on these birthdays. Even those who are twenty at the time of the survey may be twenty-one and eligible to vote when election day comes around. And surely the needs, wants, and reactions of the younger part of the population are as real and important as those of the adults. Ordinarily, this conventional division is satisfactory because the results would be changed only slightly if some other minimum age had been set. In any case, the decision would be arbitrary to a large degree.

C. Some Groups Are Included or Excluded for Operational Reasons

The operational decisions about the definition of the population flow not from what will be desired in the analysis, but from what is feasible or what is convenient in conducting the survey. This was true of the exclusion of members of the armed forces from election polls during World War II. Military regulations prohibited the polling of officers and enlisted men by unofficial agencies. The loss of this segment of the voting population was a great embarrassment to analysis but was dictated by stern necessity.

Difficulties in canvassing the "floating populations" and inmates of institutions probably led to their exclusion from the Census labor force statistics. Hence, at least this part of the group excluded from the population is trimmed off for operational if not for substantive reasons. A similar operational reason led the Federal Reserve Board's Survey of Consumer Finances, conducted by the Survey Research Center of the University of Michigan, to exclude about eight million persons living outside the United States, at military reservations, in institutions, or in hotels, large boarding houses, and tourist camps.

The exclusion of certain groups for operational reasons is a matter of definition of the population to be studied. It should not be confused with the omission of certain kinds of people from the sample because they are not found at home or refuse to cooperate, and it should not be ignored in the analysis and interpretation of the results of the survey.

D. The Geographic Location of the Population Must Usually Be Defined

Most populations that are subjected to study are confined to some definite geographic area. The national boundaries, regional divisions, city limits, or other lines mark off an area, and persons who live within the area, and meet other qualifications as well, are regarded as members of the population. All who live without are excluded. Sometimes the populations consist of persons who work in the area regardless of where they live or who are present within it (a *de facto* population) as residents or visitors at a specified date.

Substantive and operational reasons apply to the geographic phase of the definition of the population. The decision is not always obvious. For example, in a survey of the labor force in St. Paul, Minnesota, it was decided for various reasons to use the city boundaries of St. Paul as the area, excluding Minneapolis, which borders St. Paul so closely that the Twin Cities are really Siamese twins joined by a steady flow of traffic and by many common interests. The suburbs and nearby towns and farms were also excluded. But this decision to include only part of the larger community left another study to be made. The concepts of a labor market and a labor force could be defined in terms of either where workers worked or where they lived. The distinction was complicated in the case of workers who take temporary quarters near their work or look for work in several areas. The solution was to use both definitions and, hence, to work with two populations that largely overlapped (2, pp. 75–76):

In the design of the sampling survey, several questions had to be answered. The first problem was one of defining where the labor force being measured was located. The answer seemed rather obvious, for this was to be a study of the St. Paul labor force. But did the term "St. Paul labor force" mean persons residing in St. Paul, persons working in St. Paul, or persons both residing and working in St. Paul? Thus, was a man who lived in Minneapolis but worked in St. Paul a part of the St. Paul labor force? Further, was it not artificial to survey the labor force within the city, a political unit, and ignore the labor force in the remainder of the metropolitan area, a geographical unit? These and many similar considerations led to the establishment of two different criteria: One classification included those employed in St. Paul regardless of where residence was maintained, and the other classification involved residence in St. Paul without regard to place of employment. This latter classification appeared essential if comparisons were to be made between relief status and labor force status, because relief is granted on the basis of political, not economic, units, and on the basis of residence rather than place of work.

E. Special Problems of Definition Arise for the Farm Population

Somewhat similar problems arise in rural surveys. Should the farm population include farmers who live in town and travel out to work on their farms? Should it include farmers who live on the farm only during the active seasons of planting and harvest, such as the "suitcase farmers" in some of the Northern wheat country? Should sons and daughters who are away at work or at college be included? And how much of an agricultural operation is a tree nursery, an idle farm,

r an institution with a farm worked by the inmates? Both sub-
tantive and operational reasons will influence the decision. Usually
hese questions have an insignificant effect on the survey; occasion-
lly they are of the greatest importance, not only for factual surveys
ut for studies of rural opinion and rural wants.

These difficulties with the definition of a farm have been summarized
y Leon E. Truesdell of the U.S. Bureau of the Census (3):

In some types of statistics the nature of the fundamental unit is such that
ccuracy of count is absolutely impossible, at least in the present stage of the
evelopment and application of definitions. The census count of farms, for
xample, is subject to tremendous variation growing out of differing interpre-
ation of the traditional definition of a farm. Fortunately the count of
arms is among the less important of the results of the farm census and
ariations in it do not affect materially such more significant items as the
creage of major crops or even the numbers of livestock, since the marginal
racts of land subject to being included or excluded in accordance with the
him of the enumerator contain negligible acreages of crops and have negliible numbers of livestock.

As previously noted, this same problem of definition of a farm must
e faced in attitude and opinion surveys. For example, a sample surey to obtain farm operators' opinions on selected agricultural poli-
ies and programs was carried out in New York State in the summer
f 1951 (4). This report states: "All of the 1500 farmers interviewed
eceived one-half or more of their income from the operation of a
arm [survey definition of a farmer]. Practically all of them were
ommercial farmers. The findings therefore represent the opinions of
his group of farmers. They do not represent the opinions of all farm-
rs meeting the census definition of a farmer, that is, a person living
n three or more acres of land or one who produces products worth
250 or more." It was pointed out in the report of the study that
his change in definition of a farm operator had an especially marked
ffect on the distribution of size of farm. ("Fewer than 1 out of 10 of
he farmers interviewed had less than 50 acres, and fewer than one-
ourth had less than 100 acres. About 3 out of 10 census farms had
ess than 50 acres, and almost half had less than 100 acres.") We
night well expect that a farmer's attitudes and opinions concerning
gricultural policies and programs would be related to the size of the
arm he operates, and that the survey results would therefore vary
vith the definition of the population.

F. The Population May Be Defined by Official Lists or Form
Status

Some populations can be defined and located by their membershi
in an organization such as the Bar Association, American Medic:
Association, or trade unions. Others are defined by subscription list:
business accounts, or tax records or license files. The Veterans Ad
ministration has conducted surveys by drawing samples from its of
ficial records, thereby defining the population as the men and wome
who were in the records, less those who had died. Selective Servic
sampled its registrants from official lists, excluding, of course, thos
men who had volunteered before their age group was called to registe
and who had not yet returned to civilian life.

The use of a list is often convenient and may also be a very appro
priate definition of the population for substantive reasons, though i
may present problems of persons in transition or unsettled statu:
An excellent account of the use of lists in the study of income fro:
independent professional practice, the professions being medicin:
dentistry, law, certified public accountancy, and consulting engineer
ing, has been given by Friedman and Kuznets (5, Appendix A).

16.3 SURVEYS FOR SCIENTIFIC PURPOSES PRESENT DIFFICULT PROBLEMS OF DEFINING THE POPULATION

Surveys that are made for purposes of immediate application t
practical problems usually seek information about a population as i
exists on some date or in some brief period and at a particular plac:
In this sense they share with historical studies a time and place par
ticularity that science seeks to escape in its generalizations and law:
True, the practical surveys are often projected into the future and t
other places, but they are not generalized to the same extent as scien
tific findings (6).

Though large-scale surveys of attitudes for purely scientific pur
poses are not numerous, they may be expected to become increasingl:
common in the future. One example of such surveys is the Milban!
Memorial Fund's Study of Social and Psychological Factors Affect:
ing Fertility (7). This survey was planned in 1938 by a group o
scientists who wanted to make at least a beginning in the scientifi
study of individual motivations regarding the size of family. A lis:
of twenty-three hypotheses was prepared, no one of which was limite:
explicitly to any period of time or specified population. Several o
the hypotheses involved attitude variables explicitly. It is probabl:
that the demographers who drafted the list of hypotheses would re

strict them, however, to populations living under Western civilization following the industrial revolution and probably even more narrowly to the population of western Europe and North America during the past fifty years or so, excluding populations located in, or recently removed from, relatively isolated rural areas. Perhaps the group did not face the question of delimiting the population to which these hypotheses should apply, but it clearly did not intend to confine them to Indianapolis, the city that was selected for the Study.

Anyone approaching the scientific problem of testing these hypotheses by observation or experiment will encounter the question of the scope of their application, including the kinds of populations in which they may be found valid. This raises the question of defining the population to be sampled in any particular test. Should it be broadly representative of the American people as they are today? Obviously, this is not necessary or even desirable, since there are many special factors at work and many unique conditions in the national population. A scientific study is aimed at testing relatively simple general statements that may summarize known facts. It would be useless to set up a separate statement for each combination of factors and every particular situation.

Scientific laws and hypotheses ordinarily imply that if a certain set of factors is present, then a certain set of results will follow, *except for the effects of any additional factors that may be present to modify those results.* The effects of these other possible factors are described by other hypotheses or laws, applicable to many or all the situations in which these factors may be found. Hence, to test a hypothesis, we shall attempt to eliminate the effect of these factors that are not included in the statement or else take account of their effects by using hypotheses and laws that are not being tested at the moment.

Considerations of this kind, in combination with a limited research budget, may have led the Milbank Memorial Fund to restrict the population to be studied to couples married in 1927, 1928, or 1929, who were native white and Protestant, educated at least through elementary school, with husband under forty and wife under thirty at marriage, neither having married previously, and resident in a large city most of the time since they were married. These restrictions influenced the choice of Indianapolis as the city in which such couples would be studied. Thus, the population was narrowed down by the elimination of a number of possible disturbing factors to "a group sufficiently homogeneous that in the final analyses of socio-economic and psychological factors affecting fertility it would not be necessary to subdivide the couples by such factors as color, nativity, religion, type of community of residence since marriage, duration of marriage, and age. With a limited number of couples such subdivisions could easily result in groups too small to yield valid results" (7).

Undoubtedly many other disturbing factors were not controlled or eliminated. If they were all relatively small and not very highly correlated, then their net effect would be similar to that of random disturbances commonly referred to in other scientific work as "experimental error." The important point in this example is that sampling for scientific research may require the definition of quite different populations than those that are defined for most opinion polling and market research. Furthermore, the problems of generalization in studies of this kind are difficult in that they cannot be firmly based on statistical theory. For example, assume that the Study of Social and Psychological Factors Affecting Fertility was ultimately interested in the population of all couples (satisfying the requirements set forth in the preceding paragraph) residing in the United States. Clearly the portion of this population residing in Indianapolis, or any randomly drawn sample thereof, does not constitute a random sample from this population. We might choose to regard it as a cluster sample (each city constituting a cluster), but a sample consisting of only one city permits no estimate of the between-cluster variability. Also the selection was not at random. Accordingly, it is not possible to attach estimates of sampling error to the results of the survey, regarding these results as estimates for the entire population. This does not mean that generalizations cannot be made, only that such generalizations must be based on non-statistical arguments, possibly along the lines suggested in the preceding paragraphs of this section.

The foregoing difficulties concerning population definition and generalization of results arise in much of current social science research. Nationwide public opinion surveys include the "informed" and the "uninformed." Studies are made in particular universities, in particular communities, in particular plants, and with readily available subjects; and yet generalizations are usually desired for a much larger population than the one the samples can be said to "represent" on firm statistical grounds. These are problems that have been somewhat neglected in social science research, in spite of their great and fundamental importance. (See the discussions in 8, pp. 100–112; 9, pp. 9–14; and 10.)

16.4 MATERIALS RELATING TO THE POPULATION SHOULD BE ASSEMBLED FOR USE IN THE DESIGNING

A. Information about Important Groupings and Relationships within the Population Is Needed

Entirely apart from the definition of the population, information about various subgroups and about relations between various parts of the population, as well as practical knowledge about people and how they live, is helpful in designing efficient survey procedures. The geographic concentrations and patterns of distribution of the population may be known well enough to permit large gains in the sampling and estimating if full advantage is taken of such knowledge. In less obvious ways, knowledge of the behavior and psychology of various types of people may be utilized to good effect. Information about variables that are highly correlated with those being studied can be used to increase the precision of the results and test representativeness. This information will also be very helpful to the ultimate user in applying the results of the survey to his problems. Hence, a systematic review of available information about the population is an important part of the preparatory work.

The kinds of information that are generally available include previous surveys, the Population Census, data in official records or data produced as a by-product of commercial as well as government functions, knowledge of the population accumulated by persons whose work keeps them in close touch with it, and various kinds of maps, directories, and other special materials.

As sampling surveys become more frequent and important, information about the population accumulates. Accumulated data and experiences should be made more readily available to the community of research workers, and procedures for obtaining additional information should be incorporated in regular surveys whenever that can be done without difficulty. Examples of the kinds of information about the population that will be very useful for sample design, and yet readily obtainable if an attempt is made to get it, are given throughout Part II and in the remainder of Part III.

B. Information about Means of Making Contact with the Elements of the Population Should Be Assembled

Since surveys involve expenses that can be reduced by good planning and management, it is highly important that we review what is known about ways of getting in touch with the population or, more specifically, with the sample of persons who will represent it. Among the possible means of reaching the sample are the mails, telephone, and personal interview at home, at work, or elsewhere. It may not be necessary to make a separate approach to each person if the channels and regular contacts of other organizations can be used to provide the contact. Thus, in a radio-listening survey during the war when gasoline was rationed, the school system was used; questionnaires were distributed to the pupils to take home and bring back after their parents had filled them out. Questionnaires about preferences and reactions to a product may be included in the package and the consumer asked to mail his reply when he opens the package. Insurance agents, police, mail carriers, and public health nurses have been asked to obtain data for a survey while making their rounds. In all such instances, the activity with which the survey is connected may have very important effects on the kinds of sample and answers obtained, but the expediency of utilizing existing mechanism of contact with the public is clear. Discussions of and references on these points may be found in Chapters III and IV of Parten's book on survey methods (11).

When the survey is to be conducted by field interviewers, information about the possible means of travel will be useful. There has been very little progress on the problem of laying out the most efficient travel routes. This is a problem of both the effect on the sample and the effect on costs of modifying the interviewers' work assignments and travel arrangements. A great deal of very practical work should be done on this problem, and some preliminary investigations have been reported in an unpublished doctoral thesis by McCreary (12), dealing with surveys carried out in Iowa. Some preliminary inquiries about the practices of milk delivery, meter reading, and similar field activities proved fruitless, but there may be some value in investigating the corresponding problem outside the field of sample surveys.

In considering the means of contact with the sample, it is necessary to take account of the problem of identifying a person who is to be included in the sample and distinguishing such persons from those who are to be excluded. Though the latter may be removed from the sample later, if some of them get into it, good methods of making this identification and of checking to see that the separating has been done effectively are of first importance.

16.5 PRELIMINARY LIST OF ATTITUDE QUESTIONS SHOULD BE MADE UP AND STUDIED IN PREPARATION FOR THE ACTUAL DESIGNING OF THE SURVEY

Drawing up the final list of attitude questions may take more time than can be spared in the preparatory period, but it should be possible to derive from the purposes of the survey a reasonable approximation to the final set of questions. This preliminary list will guide the initial assembling of information about the attitudes to which the questions relate. Sometimes the survey personnel will be sufficiently familiar with these attitudes to require no further preparation before they proceed to design the survey, but it is easy to be overly optimistic on this point.

Attitudes differ greatly in their amenability to observation and measurement. Attitudes on the details of sexual behavior covered by Kinsey, Pomeroy, and Martin (13) cannot be surveyed as readily as political opinions. Racial and religious prejudices are more elusive than preferences among nationally advertised products. Yet even in the easier subjects, prestige factors and ignorance or indifference complicate the measurement process. Knowledge of such factors is useful since the design might include devices to minimize their effect. These devices are not limited to the wording of questions and approach to the respondent but may include the method of assigning respondents to interviewers, the choice of time and place for the interview, the use of secret questionnaires or other means of attaining anonymity, and many other procedures. The experience of research organizations in dealing with these problems should be made generally available. Good beginnings have been reported in several conferences and in the scientific periodicals, and detailed references to these will be found in the following chapter, which deals more intensively with the measurement process.

16.6 ADDITIONAL VARIABLES MAY BE VALUABLE

The list of questions should be supplemented by a list of additional variables that are correlated with the attitude variables and will be used for control and design of the sampling, interpretation of the responses, and improvement of the estimates. Such variables are of special value in the first and last of these uses if their distribution in the population is available from other sources such as a census. Information about these variables and the approximate degree of relationship to the attitude variables will be needed. Some of the data may be available beforehand; some must be obtained in the survey or by other means. Though age, sex, occupation, education, economic level, and geographic location have been most generally useful, many other variables may be applicable in particular surveys. It would be neither possible nor desirable to attempt to give here an exhaustive listing of potentially useful variables. These can come only from an analysis of a specific survey situation. However, solely as illustrations, we may note the following variables that have been used in general attitude surveys: marital status, owned or rented home, number of children in various age groups, religion, racial background, occupation of father, membership in social and fraternal groups, veteran status, political affiliations, etc.

16.7 INFORMATION ABOUT DISTRIBUTION AND CHANGE SHOULD BE ASSEMBLED

The manner in which attitudes and related variables are spatially distributed over a population or change with time directly affects the relative efficiency of possible sampling procedures. The relative similarity of attitudes between members of the same household provides guidance on whether only one or more than one person should be taken for the sample in each household. Surprisingly, almost no attempt has been made to determine the degree of similarity for representative attitudes. Also, almost nothing has been done to measure the relative similarity among neighbors within blocks or other small areas. The application of such information to cluster sampling is direct and valuable, and it is easily demonstrated that the proper use of this information leads to large gains in efficiency. Some investigations on this are described in Chapter 9, and the appropriate theory has been given by Cochran (14, Section 6.9 and Chapters 9, 10, and 11). As more information of this kind becomes available, it will contribute to efficient survey design.

The study of change of attitudes over time raises other problems for statistical analysis and for sample design. Increasing use is being made of the "panel" technique in which the same individuals are interviewed at a number of points in time (15; 16, Chap. 18). Dropouts from such panels become a source of great concern. Also, sample design must be related to the manner in which attitudes change, and statistical analyses must take into account the correlation introduced by repeated interviewing of the same individuals. The Committee on the Measurement of Opinions, Attitudes and Consumer Wants has sponsored a study on the use of panels, paralleling the Study of Sampling. Consequently, the Study of Sampling has not placed any particular emphasis on the special problems raised by the use of the panel technique.

16.8 A PRELIMINARY OUTLINE OF THE ANALYSIS OF THE SURVEY RESULTS SHOULD BE PREPARED

The size of sample, method of drawing it, and other features of the survey design will be affected by the kind of analysis to be made of the results. If the sample is to be subdivided into many classifications, or if the analysis will be in proportions, regressions, correlations, factor analysis, measures of change from preceding surveys, or other derivative estimates instead of in simple frequencies of the observed variables, then the sample may need to be much larger or perhaps it can be smaller. The kind of sample that will be most effective will generally be different in each case.

Here again, a composite or weighted judgment may be necessary to determine a satisfactory design. There may be some one variable or one estimate that transcends the others in importance and provides a single criterion for the design, but even its implications for the design will depend on the manner in which the results are to be analyzed, interpreted, and used. A recent paper by Deming (6) illustrates these points in making a distinction between "enumerative" and "analytic" surveys. Thus Deming says:

. . . we shall see that it is often impossible to design a survey that will supply economically information for both enumerative and analytic purposes. For example, in a marketing survey, the best design for an estimate of the number of people who prefer to use ground coffee at home, rather than soluble coffee, requires, for greatest economy, one type of sample design; whereas, a study of the *reasons*, or even of the difference in the two proportions, requires another design. One must be prepared to make some sacrifices in precision, as it may not be economical to satisfy both aims simultaneously.

The analysis may be largely one of cross-tabulations with interpretative comment, a form of analysis illustrated in Zeisel's book (17). For this, the table forms should be prepared with titles, stubs, and captions in nearly final form and guessed or dummy numbers in the cells. The latter are important for the calculation of sample size and forecasts of what comparisons can be made effectively among subgroups within the "breakdown." For more complex analysis the processes should be rehearsed and examined to determine as definitely as possible what will be required from the sample to permit the analytical studies that are to be made. The results of such pretests of the analysis are often very informative and sometimes startlingly so. They may lead to a complete revision of purposes or even abandonment of the survey as not feasible. Early discovery of the futility of a proposal is a great gain. More often, pretests will lead to the selection of techniques and designs that are most efficient for the satisfaction of the ultimate purposes.

No claim is made here that the results of surveys can be anticipated very accurately when new problems are being investigated. Very rough guesses may suffice, however, for effective incorporation of knowledge about the attitude variables in the shaping of the survey procedures.

REFERENCES

1. William Feller, *An Introduction to Probability Theory and Its Applications,* Vol. 1, second edition, John Wiley & Sons, New York, 1957.
2. University of Minnesota Industrial Relations Center, *Local Labor Market Research,* University of Minnesota Press, Minneapolis, 1948.
3. Leon E. Truesdell, "The Problem of Quality in Census Data," paper read before the Population Association of America, Philadelphia, May 23, 1948.
4. Edward O. Moe, *New York Farmers' Opinions on Agricultural Programs,* Cornell Extension Bulletin 864, Ithaca, N. Y., 1952.
5. Milton Friedman and Simon Kuznets, *Income from Independent Professional Practice,* National Bureau of Economic Research, New York, 1945.
6. W. Edwards Deming, "On the Distinction between Enumerative and Analytic Surveys," *J. Am. Statist. Assoc.,* 48 (1953), 244–255.
7. Clyde V. Kiser and P. K. Whelpton, "Social and Psychological Factors Affecting Fertility. IV. Developing the Schedules and Choosing the Type of Couples and the Area to be Studied," *Milbank Mem. Fd. Quart. Bull.,* 23 (1945), 386–409.
8. Proceedings of the Second International Conference on Public Opinion Research, Williamstown, Mass., 1947.
9. Leonard S. Cottrell, Jr., and Sylvia Eberhardt, *American Opinion on World Affairs in the Atomic Age,* Princeton University Press, Princeton, 1948.
10. Quinn McNemar, "Opinion-Attitude Methodology," *Psychol. Bull.,* 43 (1946), 289–374.
11. Mildred Parten, *Surveys, Polls, and Samples,* Harper & Brothers, New York, 1950.
12. Garnet E. McCreary, Cost Functions for Sample Surveys, unpublished Ph.D. thesis, Iowa State College, Ames, Iowa, 1950.
13. Alfred C. Kinsey, Wardell B. Pomeroy, and Clyde E. Martin, *Sexual Behavior in the Human Male,* W. B. Saunders Company, Philadelphia, 1948.
14. William G. Cochran, *Sampling Techniques,* John Wiley & Sons, New York, 1953.

5. Paul F. Lazarsfeld, Bernard R. Berelson, and Hazel Gaudet, *The Peoples Choice,* second edition, Columbia University Press, New York, 1948.
6. Marie Jahoda, Morton Deutsch, and Stuart W. Cook, *Research Methods in Social Relations,* The Dryden Press, New York, 1951.
7. Hans Zeisel, *Say It with Figures,* Harper & Brothers, New York, 1947.

CHAPTER 17

Initial Knowledge
of the Measurement Process

17.1 THE INTERDEPENDENCE OF SAMPLING
AND MEASUREMENT

The preceding chapter discussed in general terms the manner in which the opinion and attitude variables and the additional variable required for their analysis and interpretation influence the entir survey design. These influences become evident when we conside the various measurement procedures that may be used to obtain th data required to investigate and define any given attitudinal situa tion. Circumstances such as the following may arise:

1. The measurement process used to obtain a specific type of data is known to be subject to a standard error of 5 per cent, yet an at tempt is being made to develop a sample design that will yield re sults within 3 per cent of those that would be obtained if perfec measurements were made on the entire population.

2. A certain portion of the available resources is free to be allo cated either to a refinement of the measurement process or to a re finement of the sample design.

3. A sample that is spread widely (i.e., geographically or other wise) may be obtained only if relatively simple interviewing tech niques are used; the use of more accurate intensive techniques will re quire much concentration or grouping of interviews into clusters read ily accessible to the interviewers who are competent to use the tech niques.

Even though a consideration of the problems and questions arisin in connection with the measurement process cannot properly be re garded as a primary assignment for a study of sampling methods, th circumstances just listed make it apparent that sample design canno be divorced from their treatment. For this reason it is important tha an initial examination of the measurement process be made before we proceed to details of the sample and survey design. A summary discussion of the steps that might be followed in this initial examination will now be given. It should be noted that it is not our purpose to give here an extended treatment of attitude definition and measurement. Rather, we wish only to present such materials as are necessary to illustrate the points made. The interested reader may wish to refer to books such as those by Jahoda, Deutsch and Cook (1, Chaps. 4, 5, 6, and 7), Festinger and Katz (2), Guilford (3), Lindzey (4), and Stouffer et al. (5).

17.2 FEASIBILITY OF VARIOUS METHODS OF MEASUREMENT

In the context of this chapter the term "measurement instrument" will be used to refer primarily to any technique that can be applied to obtain information about a variable from an individual element of the population. It will not necessarily refer to the procedures that are used to combine related pieces of information into a final attitude measurement such as an average or the results of scaling, latent structure analysis, discriminant function analysis, and index construction. This restriction is imposed in order to avoid the difficulties involved in discussing procedures so complex that sampling theory has not yet been adequately developed for them. However, as research proceeds it may become increasingly possible to tie the sample and survey design directly to the end product rather than to the pieces that go to make up the end product.

A wide variety of procedures are potentially available for consideration in choosing a measuring instrument. Many of these involve the presentation of a question to a population element, an answer to which will either allow the respondent to be classified into one of a number of categories or furnish a numerical quantity to be associated with the respondent. The variations arise from the manner in which the question is presented to the respondent, the form in which the question is asked, and the use made of the obtained answer. Some of the well-known possibilities are as follows:

1. Manner of presentation—by United States mail, by telephone, by an interviewer, or in a group situation in which each individual fills out his own questionnaire (e.g., an employee attitude survey).

2. Form of the question—answer must fall into a set of predetermined categories, a free answer is allowed, or respondent is urged by

probing to amplify his answer until he has thoroughly expressed hi
opinion.

3. Use of response—a simple tabulation of the number of response
falling into a particular category, the combination of this respons
with others similar to it (coding) in categories determined after a
inspection of the answers, or a verbatim reporting of free answers.

Actually, any such listing cannot be kept up to date without consid
erable effort since new techniques are continually appearing and be
ing tested. For example, Sanford (6) reports on the use of a simpl
cartoon-like projective device which he states may ". . . make pos
sible the use of large and representative samples in testing personalit
hypotheses heretofore based and tested by the study of small number
of people drawn from the 'captive' undergraduate population."

In addition to methods involving some form of questioning, ther
are others that depend on observation of the population element'
words or actions. This observation may be made either by a human
observer or by a mechanical device and may occur in either a planne
or unplanned situation. Thus "Mass Observation," as developed i
England and described by Willcock (7), is an attempt to apply the
techniques of anthropology "to the study of the habits, lives, and
beliefs of the British." Mass Observation has asked volunteer observers to keep day-to-day diaries of everything that happened to
them, has used special investigators to keep careful records of al
that was said and done in specific situations, and has supplemented
these materials by information derived through the use of more standard polling devices. The A. C. Nielson Company studies radio and
television audiences by means of a mechanical device which is attached to the sets in a sample of homes and which furnishes a continuous record of the periods the set is turned on and the station to
which it is tuned. Many other examples of observational approaches
to measurement can be found in the literature, and references to some
of these will be found in Parten (8, pp. 82–84).

Past experience has built up a large body of general knowledge concerning these and other methods of measurement, and this knowledge
can be immediately brought to bear when a specific problem of measurement is first examined in broad outline. This information will be
most helpful in providing answers to the following two questions:

1. Which measurement procedures are feasible for obtaining the
required information? *This question is, of course, asked with reference to the attitudes and opinions under study and the population
to be surveyed.*

2. Which of those procedures not eliminated under the first question can be integrated into the survey design so as to obtain the specified information and keep within the survey budget?

At this stage in the planning, a consideration of measurement procedures in the light of these two questions will be most helpful in
discarding many of the methods on reasonably obvious grounds without examining them in great detail. This does not mean that every
time a sample survey is designed a complete list of every available
measurement method should be made, with subsequent elimination of
all but a few of them. Rather, it means that this preliminary investigation and elimination proceed almost automatically; it is instructive
merely to recognize that certain measurement methods have been
ruled out, even though no formal work is necessary for their elimination.

The elimination of measurement procedures for obvious failure to
conform to the restrictions implied by the two questions is quite
straightforward. It is not feasible, even if it were ethical, to assign
agents to shadow selected individuals and listen in on their private
conversations until each spontaneously expresses to someone his intention to vote for a particular candidate. Nor can we find out by
observation, without violating the secrecy of the ballot, how a person
voted in a United States presidential election. Mail surveys cannot
be used unless an address listing is available for the population elements. Telephone interviews would not be feasible for obtaining detailed personal information regarding economic or social behavior. A
variety of reasons for elimination immediately come to mind in a specific situation and, at this stage, most of the decisions are made without any special study or delay.

17.3 ASSEMBLING MATERIALS ON PERFORMANCE

As a result of the preliminary investigation and elimination, only
a few measurement procedures will be left for detailed consideration.
In order to prepare for a final choice of one of them and for its inte-
gration into the survey design, we investigate next, in some detail,
previous uses or special studies that have been made of each remaining instrument. Accounts of these previous uses and studies are very
likely to present data describing the method with respect to practical
application and accuracy, and perhaps even to present a discussion
of other aspects of its performance, such as difficulties encountered
and their treatment, personnel needed, costs, etc. Sometimes it may

even be necessary to conduct special presurvey studies in order to decide between two or more competing procedures. One excellent account of such research has been provided by Marks and Mauldin (9) in connection with the preparation for the 1950 Census, the purpose of this piece of research being to obtain information on the relative merits of the following: (a) self-enumeration household schedule, (b) self-enumeration individual schedule, (c) direct interview, and (d) direct interview plus leaving the schedule with some additional questions for the respondent to fill out.

There are many different facets on which we may wish to check a measurement procedure (and the results it produces) or compare two or more competing procedures. Moreover, a wealth of material on many of these items already exists in the literature, and undoubtedly much more can be found in the files of any research agency. As examples, we present the following list of questions, together with some illustrative references where partial answers may be found:

1. How accurately will respondents provide information of a factual nature (e.g., age, occupation, income, rent, etc.)? A large amount of data on this topic has been provided by the Milbank Memorial Fund's Study of Social and Psychological Factors Affecting Fertility, in which the answers given to a short questionnaire for a number of factual items were compared with those given to more detailed questioning for a selected group of people [see Section 7.3, Table 7.5, and (10)]. Similar data for labor market studies have been given by Gladys Palmer (11, 12). Also, Parry and Crossley (13) have reported on the validity of responses to survey questions with respect to such "prestige" items as voting participation, community chest contributions, and the like.

2. How closely will answers to a fixed question agree as between two or more points in time?

3. How well can individuals predict what their actions will be under a future set of circumstances? A general discussion of this problem has been given in a paper by Dollard (14) entitled "Under What Conditions Do Opinions Predict Behavior?" The specific problem of predicting the postwar plans of soldiers has been treated by Clausen (5, Chaps. 15 and 16), and Katona (15) has discussed the economic behavior of consumers on the basis of survey data.

4. How much change will be introduced into answers by a change in the wording of a question or by asking an entirely different question on the same subject? An intensive investigation of knowledge concerning the "single" question has been given by Payne (16).

5. What differences exist when the questions are asked by an inter-

viewer, when they are presented by means of a mail questionnaire, and when they are presented in a group interview situation? Some research on this problem has been performed. For example, Kahn (17) has described an experiment, conducted in connection with an employee attitude survey, in which "The aim of the research was to test the relative effectiveness of the intensive, 'open-ended' interview and the fixed-alternative, written questionnaire for collecting data from a population of industrial workers regarding various aspects of the work situation." A similar experiment was conducted by the Research Branch of the Information and Education Division in the War Department in World War II (5, pp: 718–719) to compare classroom administration of questionnaires with the results obtained through personal interviewing.

6. What proportion of respondents will fail to return a mail questionnaire, what proportion will refuse to cooperate with an interview, and what proportion cannot be economically contacted for a personal interview? How do these rates vary with the population and with the subject of investigation? The use of mail questionnaires in the Veterans Administration has been described by Clausen and Ford (18); a report on some pertinent research and references to other materials in the literature are also given. The problems of nonresponse in personal interview surveys and the literature on this subject have been reviewed in Chapter 11.

7. What are the relative costs of alternative methods of administering a specified instrument?

8. How accurately can responses to free-answer questions be coded for purposes of analysis? Woodward and Franzen (19) and Durbin and Stuart (20) have reported some research directed at this problem.

All these points, plus many more which have not been mentioned, will make their effects felt throughout the entire survey design and, in particular, will condition the sample design. Past experience will aid in the final selection of a measurement instrument by pointing out the errors associated with each (to be considered in conjunction with the specifications and needs), the relative costs (to be considered jointly with needs and resources), and the operational characteristics (to be considered in relation to the sampling and field work). Here again it is not always necessary that this collection of data and evidence on performance be carried out in any formal manner. Many of the decisions are more or less obvious, particularly when they have been made many times before in similar situations. However, it must be clear that these steps are implied, so that when entirely new situations arise, planning can be carried out in the most efficient manner.

17.4 EXISTENCE, EXTENT, AND CONTROL
OF MALFUNCTIONING

There are circumstances under which any particular measurement process can be expected to malfunction. Some of these circumstances will be recognized from past experience as dangerous, but under a new set of circumstances it may not be certain that the instrument will perform according to expectations. Before a choice of methods is made, it is extremely important that the risk of failure be assessed for each method with respect to the conditions under which it is to be applied. This assessment may be made on the basis of past experience and reasonable inference about the future. If the risk of malfunctioning cannot be avoided, then the next step is to try to evaluate the consequences of partial and complete failure in terms of the specifications and needs, and to attempt to find ways of eliminating, reducing, or controlling these consequences.

Typical varieties of malfunction are those in which the attempt to measure leads to refusal to be interviewed, declining to answer some questions, evasive answers, claiming to be "undecided" or "don't know," misunderstanding the question, incomplete and erroneous responses, etc. Many of these possibilities for malfunctioning depend on a limited number of very well-known factors such as time at which the measurement process is carried out, organization (i.e., government, private research, educational, etc.) under whose name the measurement instrument is presented to the population element, multiple meanings and ambiguity of words, interpretation which the respondent places on the question asked, existence (in the mind of the subject) of extraneous threats and rewards in the measurement situation, and the physical location and state of mind of the subject at the time of the measurement.

Unlike the previous two steps outlined for the examination of measurement methods, this step probably should be carried out in a formal way for every study. Because the usual reasons for malfunctioning are so well known and are so amply illustrated in the literature, it is easy to say "This can't happen to me," with the result that it does happen. A list of many forms of malfunctioning has been given by Deming (21, Chap. 2), and a consideration of such a list for each proposed method of measurement will be well worth the effort. As illustrations of the more "unthought" of difficulties, we present the following two quotations from Deming:

Advance publicity by press and radio is very helpful, even if the publicity is unfavorable. A whispering campaign, however, may be disastrous. One

terviewer that I talked with recently had sat three hours in the kitchen
a respondent listening to family history and reasons why she was not
ing to give answers to the interviewer's questions, all as a result of coach-
g by neighbors who perhaps justifiably thought that their area was being
ersurveyed [p. 34].

The measurement of total annual sales, total annual postal traffic in vari-
s classes, telephone, telegraph, freight, passenger, or air traffic, or move-
ent of some particular commodity, consumption of foods of various kinds,
the pattern of consumption or service rendered, and a host of other prob-
ms which require totals or averages over a year or some other period, pre-
nt difficult problems because of heavy weekly or seasonal variability. Ac-
ally, in such problems it is necessary to recognize the fact that a good
mple of time is as imperative as a good sample of areas, business establish-
ents, families, or anything else [pp. 43–44].

If it is impossible to eliminate all sources of malfunctioning, then
t least it may be feasible to include methods of evaluating their ef-
ects in the survey design. For example, when it is felt that the
easurement process will be conditioned by the auspices under which
he measurement process is presented, but the magnitudes of the ef-
ect are not known for the various possible auspices, a portion of the
easurement may be performed under one set of conditions, another
ortion under a different set, etc. All such considerations will clearly
nfluence the sample design.

17.5 THE HUMAN ELEMENT IN MEASUREMENT

The measurement process is further complicated by the fact that
ll or part of it will usually involve the use of interviewers and other
uman agents. Not only will the behavior of these agents in carry-
ng out instructions influence the results of the measurement, but their
ersonal characteristics, as they react with the personalities of their
ubjects, may also have an effect on the measurement. It is appro-
riate to examine each of the possible measurement methods with re-
pect to the effects that can be expected to follow from the use of ob-
ervers and interviewers in place of instruments less affected by the
'human factor," and to see whether any predictions can be made in
advance concerning the existence and magnitude of such effects.
Under many circumstances it may be possible to institute a degree of
control over the performance of the human agents in the survey procedure, or at least to provide a means of assessing their influence.
These comments are, of course, not meant to imply that mechanical
measurement devices, such as the Nielson audiometer, have no characteristic malfunctionings and weaknesses. They obviously do. The

use of these means of observation is extremely specialized, howeve
and will not concern us here.

The types of information that we look for in relation to this top
are varied and involve such items as deliberate cheating, failure to u
derstand or follow instructions, conscious or unconscious efforts to i
fluence the respondent to the interviewer's point of view, errors
writing down replies given by subjects and errors in classifying the
replies into categories (either as they are given to the interviewer
as the questionnaire was coded in the office). Material on many
these items is now available in the literature. For example, one
the earliest studies was Stuart Rice's "Contagious Bias in the Inte
view" (22). It was found that in a social study of 2000 destitu
men, the reasons given by them for being down and out carried
strong flavor of the interviewer. Results recorded by a prohibitioni
showed a strong tendency among the men he interviewed to ascrit
their sorry existence to drink; those interviewed by a man with sc
cialist leanings showed a strong tendency to blame their plight on ir
dustrial causes. The men may have been glad to please anyone wh
showed an interest in them. An extensive review of the literature o
this subject of interviewer effect has been provided by Moser (23)
and a very comprehensive study has been published by Hyman and hi
associates (24).

All the evidence gathered in this exploratory stage will be valuabl
not only in the choice of a measurement method but also in indicatin
some of the features that must later be incorporated into the sampl
and survey design. At the very least, tentative decisions can b
made on the preparation of instructions for the interviewers and othe
field workers, on the degree of supervision they should receive i
carrying out their work, on the amount of checking that is necessary
and on whether or not it is desirable to incorporate check features int
the sample design itself. The process of setting up a sample desig
in such a manner that some measure is obtained of the performanc
of the human agents has been used extensively by Mahalanobis (25
26).

17.6 COLLECTION OF ADDITIONAL DATA
FOR EVALUATION OF MEASUREMENT

Even after all sources of existing data on the performance charac
teristics of one or more methods of measurement have been tapped
there may still remain areas of uncertainty that must be filled in be
fore a final choice can be made or before the final results of the con-

emplated survey may be used. If uncertainty remains about the measurement instrument, then the only solution is to conduct a pilot study. This is, of course, a well-known operation in settling on the final form that one or more questions are to take, or on the final form f an entire questionnaire. However, there is not any reason why the function of a pilot study could not be made broader. For example, f we are undecided on whether or not to use a mail questionnaire on population for which no data on response rate are available, we may send the questionnaire to a small portion of the population to obtain an estimate of the unknown response rate. If a preliminary survey is needed to obtain data for the design of the final sampling procedure, additional information may be obtained on the measurement process at the same time.

Also, there always exists the possibility of integrating into the final survey design the means for obtaining data on all aspects of the measurement process set forth in this chapter. For example, it is frequently desirable to divide the sample into two comparable halves and use one form of question on one half and another form on the other half. This is a procedure sometimes used by the American Institute of Public Opinion to provide a basis for the comparison of alternative wordings of a question. Mahalanobis also uses essentially this procedure in trying to detect the effects human agents may have on the survey process. He calls this the system of *interpenetrating samples* (26, p. 432). The sample for the survey was made up of two interpenetrating subsamples, each distributed over much of the same area. The sample units were taken in pairs, and one of each pair was allotted at random to the first subsample and the other unit to the second subsample. Information was collected independently for each subsample by a different set of investigators. To avoid any chance of records being compared, the two sets of investigators did not work in the same area at the same time.

Techniques similar to those mentioned in the preceding paragraph may sometimes prove helpful in trying to estimate the sampling error of a particular sample design or in trying to gain data that will be useful for the planning of future surveys (see Chapters 8, 9, and 10). Consequently, even though a measurement process must be decided on before the survey goes into the field, there still remains the opportunity of discovering something about its performance characteristics during the operation of the survey, *if this eventuality is planned for in advance.*

17.7 RELATION OF MEASUREMENT TO OTHER PHASES OF THE SURVEY PROCESS

Though some indication has been given throughout this treatment of the relation between measurement and the other phases of survey design, it seems important to summarize briefly some of the major points that arise. Perhaps the most important of these is concerned with the accuracy of the measurement process. If the survey is not measuring what it should measure within, say, five percentage points, it is obviously inefficient to use a sample design that will yield results within one percentage point of those that would be obtained from perfect measurement of the entire population. Moreover, it is frequently necessary to decide on the proper allocation of resources between the sampling and the measurement process. Specifically, we might have to choose between a refinement of the measurement process (e.g., an increase in the training and supervision of interviewers) and an increase in the sample size, which will lead to a decrease in the sampling error.

In addition to these considerations, it has also been pointed out that a suitable sample design may allow some evaluation of the measurement process (including the effects of having human agents and respondents) during the actual carrying out of the survey. Moreover, the nature of the sample affects the use and accuracy of the measurement procedure, and this procedure affects the degree to which a given sample can be realized in the field. All these facts lead to one conclusion, namely that the sampling and measurement are quite interdependent and that it is usually unwise to divorce completely the consideration of one from the other.

REFERENCES

1. Marie Jahoda, Morton Deutsch, and Stuart W. Cook, *Resarch Methods in Social Relations,* The Dryden Press, New York, 1951.
2. Leon Festinger and Daniel Katz, editors, *Research Methods in the Behavioral Sciences,* The Dryden Press, New York, 1953.
3. J. P. Guilford, *Psychometric Methods,* second edition, McGraw-Hill Book Company, New York, 1954.
4. Gardner Lindzey, editor, *Handbook of Social Psychology,* Vol. I, Addison-Wesley Publishing Company, Cambridge, Mass., 1954.
5. Samuel A. Stouffer, Louis Guttman, Edward A. Suchman, Paul F. Lazarsfeld, Shirley A. Star, and John A. Clausen, *Measurement and Prediction,* Vol. IV of Studies in Social Psychology in World War II, Princeton University Press, Princeton, 1950.

3. Fillmore H. Sanford, "The Use of a Projective Device in Attitude Survey-ing," *Publ. Op. Quart.*, 14 (1950), 697–709.

7. H. D. Willcock, "Mass Observation," *Am. J. Sociol.*, 48 (1943), 445–456.

8. Mildred Parten, *Surveys, Polls, and Samples,* Harper & Brothers, New York, 1950.

9. Eli S. Marks and W. Parker Mauldin, "Response Errors in Census Re-search," *J. Am. Statist. Assoc.*, 45 (1950), 424–438.

10. Clyde V. Kiser and P. K. Whelpton, "Social and Psychological Factors Af-fecting Fertility. III. The Completeness and Accuracy of the Household Survey of Indianapolis," *Milbank Mem. Fd. Quart. Bull.*, 23 (1945), 254–296.

1. Gladys L. Palmer, *The Reliability of Response in Labor Market Inquiries,* Technical paper 22, U.S. Bureau of the Budget, Washington, D. C., 1942.

2. Gladys L. Palmer, "Factors in the Variability of Response in Enumerative Studies," *J. Am. Statist. Assoc.*, 38 (1943), 143–152.

3. Hugh J. Parry and Helen M. Crossley, "Validity of Responses to Survey Questions," *Publ. Op. Quart.*, 14 (1950), 61–80.

4. John Dollard, "Under What Conditions Do Opinions Predict Behavior," *Publ. Op. Quart.*, 12 (1948), 623–633.

5. George Katona, *Psychological Analysis of Economic Behavior,* McGraw-Hill Book Company, New York, 1951.

6. Stanley L. Payne, *The Art of Asking Questions,* Princeton University Press, Princeton, 1951.

17. Robert L. Kahn, "A Comparison of Two Methods of Collecting Data for Social Research: The Fixed-alternative Questionnaire and the Open-ended Interview," a paper presented at the Annual Meeting of the American Sta-tistical Association, 1952.

18. John A. Clausen and Robert N. Ford, "Controlling Bias in Mail Question-naires," *J. Am. Statist. Assoc.*, 42 (1947), 497–511.

19. Julian L. Woodward and Raymond Franzen, "A Study of Coding Reliabil-ity," *Publ. Op. Quart.*, 12 (1948), 253–257.

20. J. Durbin and A. Stuart, "An Experimental Comparison Between Coders," *J. Marketing,* 11 (1954), 54–67.

21. W. Edwards Deming, *Some Theory of Sampling,* John Wiley & Sons, New York, 1950.

22. Stuart A. Rice, "Contagious Bias in the Interview," *Am. J. Sociol.*, 35 (1929), 420–423.

23. C. A. Moser, "Interviewer Bias," *Rev. Int. Statist. Inst.*, 19 (1951), 1–13.

24. Herbert Hyman, *Interviewing in Social Research,* University of Chicago Press, Chicago, 1954.

25. P. C. Mahalanobis, "On Large-Scale Sample Surveys," *Phil. Trans. Roy. Soc.,* B231 (1944), 239–451.

26. P. C. Mahalanobis, "Recent Experiments in Statistical Sampling in the Indian Statistical Institute," *J. Roy. Statist. Soc.,* 109 (1946), 325–370.

CHAPTER 18

Review of Possible
Sampling Procedures
and Assembly of Information
about Their Performance

18.1 BROAD DECISIONS ON SAMPLE DESIGN MUST BE MADE DURING PREPARATORY STAGE

Just as for the measurement process, it is frequently desirable to undertake a broad preliminary examination of the various alternatives from among which a sampling procedure must ultimately be selected. In speaking of sampling procedures at this stage in the preparatory work, we must of necessity limit our attention solely to their general outlines, leaving the determination of most of the specific details until designing has progressed further. The primary purposes in conducting this examination are to discover procedures that may be well suited to the problem, to eliminate from further consideration those procedures that are obviously not applicable because of the specifications and needs of the problem at hand, and to pull together any readily available materials that may facilitate further examination or be of aid in the detailed planning to follow.

Under the "ideal" circumstances ordinarily assumed for the presentation of the theory of random sampling, such an investigation would not be necessary. We have only to take as large a sample as resources allow, or take a sample that will provide the required precision, or balance cost and precision against the costs of making certain decisions on the basis of the obtained data. Unfortunately, questions of costs and practicability usually preclude the use of simple random sampling. Thus lists are rarely available from which to make the selection; the elements in a random sample will be widely scattered geographically, and interviewing costs will be excessive unless he sample elements can be "clustered" into compact geographic groups; a random sample may not provide a sufficient number of people who are to be the object of special study; and the use of certain known characteristics of the population may enable us to obtain the desired degree of accuracy with a smaller sample than would be required by simple random sampling. All these required departures from simple random sampling increase the number of possibilities that may be considered in sample design; each should be examined carefully to insure that "good" or "best" procedures are not overlooked. These possibilities also increase the opportunities to make serious mistakes or blunders.

Even in apparently simple situations, preliminary work may show that simple random sampling cannot be used and that more complex sampling procedures must be considered. For example, the director of a study to be carried out in a medium-sized Eastern city (population approximately 125,000) attempted to obtain a listing of the names of parents of school children from which to draw a sample for the purpose of comparing those who participated in the activities of the Parent Teachers Association with those who did not. It was found that such a list could not be released by the school authorities under a policy established several years earlier after an unfortunate experience with the misuse of such a list. Consequently it was necessary to take a regular population sample and then eliminate families that did not have school children. The sample could have been drawn in any one of several ways, including systematic selection from a city directory of names or addresses, or random selection of a sample of blocks followed by systematic selection of a sample of households from a complete listing of the chosen blocks. A decision had to be made among these possibilities.

The purpose of this chapter is to set forth some of the considerations that may enter into the preparation of a preliminary list of possible sampling procedures, recognizing that it is not always necessary or desirable to carry out this activity in any formal manner. For example, if a poll is to be conducted in a college student body for which a directory exists, there is no need for first choosing a sample of rooming houses and dormitories. A systematic selection from the directory will prove quite adequate for most purposes.

18.2 MANY TYPES OF DESIGN MAY BE ELIMINATED ON THE BASIS OF GENERAL CONSIDERATIONS

In listing sampling procedures for further consideration, perhap the most important question to ask is "Do I need a design which wi produce results free from bias (or with measurable bias), assumin that it is executed according to plan, and which will furnish a measur of sampling error?" Presumably the answer to this question is cor tained in the specifications for the survey, but it is well to re-examin the whole matter at this stage in the design process. It is all too eas to sidestep the question by choosing a sampling method that is cus tomary in a particular field of investigation or by imitating som successful study made under different conditions. Then we mis opportunities to obtain more appropriate and valuable results b tailoring the sampling procedure to fit the actual condition unde which it will be used and to fit the purposes of the survey. Fo example, this uncritical use of a prevailing method characterized th vogue of quota sampling in the period that ended in 1948 and ha characterized the use of probability sampling since. Some of th considerations to be weighed in this phase of survey design are dis cussed in Chapters 4, 9, and 10 of this book, by Deming (1, Chap. 1) and by Hansen, Hurwitz, and Madow (2, pp. 4–10, 52–53, 68–74) An example of the danger of ignoring important possibilities b merely imitating common practice in the selection of a samplin procedure is described by Alpert in a recent paper (3), from whicl the following quotation is taken:

Finally, I should like to comment briefly on the role of fashion in the de velopment of sampling theory. Fashion is the enemy of the utilitarian. I the field of techniques, it operates especially to increase the incidence o faulty applications of the particular technique which happens to be in fash ion. At this moment, probability sampling is very much *the* fashion. It i not surprising, therefore, to find surveys being designed to include probability samples even though such a design is not the most effective or most relevan to the situation being studied.

Brief mention may be made of one example. In the plan for a study involving investigating people's reactions, behavior patterns, and attitude under conditions of disorganization and disruption, it was proposed to use a probability sample. . . . What I should like to question is whether a probability sample is at all appropriate under the particular circumstances My hunch is that such a sample was proposed chiefly because it is the pre vailing fashion among survey specialists. Interviewing the individuals par ticularly affected, and especially those who are most severely affected, would appear to be of paramount importance. (It was not the purpose of the survey to estimate magnitudes for the disruption but to determine psychological reactions.) On a probability sample design of the type proposed, the interviewers might well have missed the very individuals who would be in the most propitious situation to provide the data sought. . . .

Of course, it is also true that the tendency to imitate what is "fashionable" may be the first step toward introducing superior methods of sampling into a new field. For example, anthropologists have long used such techniques as interviewing informants, studying life histories, and observing participants and have rarely attempted the application of survey methods. One break with tradition, described by Streib (4), was a survey-type study conducted among the Navaho Indians. Though he used a "convenience" method of selecting a sample of respondents, Streib expresses the opinion that more precise sampling methods could easily have been utilized in this particular population.

An excellent example of a sampling survey that was not restricted by adherence to a single variety of sampling procedure, but combined in a fruitful manner probability sampling of individuals with purposive sampling of communities, is to be found in a study of public reactions to the presence of atomic energy plants in a number of communities; the study was conducted by the Survey Research Center of the University of Michigan. In an unpublished paper Metzner commented that:

. . . the departure from a probability design is unusual at the Survey Research Center. . . . Yet in the case of this study we somewhat unhappily but with forethought used a purposive design! . . . The answer to this anomaly lies in the kind of problem and its implications. The basic problem was to discover the degree and extent of fear, if any, related to the presence of a nearby atomic energy installation. . . . Our basic approach was to match communities lying within twenty-five miles of a major atomic energy installation with similar communities at a distance from such installations. . . .

The manner in which purposive and probability designs that appeared to be required by the objectives of the study were combined was described in the report of the study (5) as follows:

A principal objective of the study of public responses to atomic energy uses was to determine what influences, if any, the presence of a nearby major atomic energy activity had on the information, opinions and reactions of people living in the surrounding area. In accordance with this objective, people living in seven areas were first selected for interview. They all lived within twenty-five miles of one of seven major atomic energy installations. For the people in each of these areas, people in two other areas having no installation were selected for interview. These two "matching areas" were chosen for their similarity to one of the "installation areas" in size, industrial composition, average rent, and section of the United States. . . .

Within each community, individuals were selected for interview by objective probability techniques. . . .

This quasi-experimental design allows a comparison of people from presumably similar areas, except for the presence or absence of an installation nearby. The "matching areas" are as similar to their respective "installation areas" as was permitted by the *natural* variety of American communities and their characteristics, the accuracy and amount of available information on the two types of communities, and the judgment and information of the research staff of the more subtle sociological factors involved in the matching. Obviously, it was impossible either to *create* (or "contrive") matching communities or to select them at random. Since no randomization procedure could be introduced in the selection of the matching or "control" areas (although selection of respondents within areas was random), statistical procedures dependent on the assumption of independent selections are invalid. However, the selection of *two* matched areas for every installation area does give an estimate of the residual variation after matching.

It is very difficult to establish general rules for determining when it is necessary or desirable to use probability model sampling procedures. Even in certain specific instances the problem may be hard to solve in a realistic manner. It is still more difficult to establish such rules for other types of sampling. However, we definitely can recognize gradations in situations, some examples of which might be as follows:

1. There are circumstances under which the use of something other than a probability model sample is clearly indicated. Thus the sample used for pretesting a questionnaire does not ordinarily have to be chosen in any objective fashion. We merely wish to insure that certain "extremes" in the population are represented in the sample, and we are in no way concerned, for example, with the relative proportion of such extremes in the population. (See the account by Marks and Mauldin (6) of the October 1948 pretest of Census procedures.) Similarly, if we wish to make a community study in several United States cities belonging to a specified size class, a judgment or purposive procedure may be the best way to select them. It is highly questionable whether we could make precise statistical statements about the entire population of such cities from a study of several randomly selected cities. The procedure used would therefore take into account as many as possible of those considerations judged to have an important influence on the study. The discussion of the Study of Social and Psychological Factors Affecting Fertility in Chapter 16 provides a case in point.

2. There are other situations in which we may desire a probability model sample but find that certain practical considerations will not permit the use of the necessary procedures. At the same time, it may

appear that data derived from other types of samples will be quite valuable. For example, consider the case of a doctor who wishes to determine the effect of a new treatment or drug on persons suffering from a specific disease. For *convenience,* he may choose to experiment only on those patients who are available in a single hospital. It is quite clear that this sample of patients may not be representative of all patients having the disease. The mere fact that they are in the hospital may mean that they are more seriously ill than those who are not in the hospital, that they come from a higher social class, etc. However, this lack of representativeness might not be at all important if the doctor's sole purpose in conducting the study were to obtain some preliminary notions concerning the action of the treatment or drug. Further work would then be required to generalize the results to the entire population of persons suffering from the disease. Problems of this kind are illustrated by the opening of the Clinical Center of the National Institutes of Health.

3. Finally, we can distinguish cases in which the importance of obtaining objective estimates of sampling error and eliminating sampling bias is unequivocal. Many of the surveys conducted by the various agencies of the U.S. Government for the guidance of administration provide illustrations of such situations. The data must stand by themselves and are to be used by many different individuals for many different public purposes. Even here, however, it is not always possible to state exactly what degree of accuracy is needed, and a recent paper by Hansen, Hurwitz, and Pritzker (7) stresses the point that increased attention should be given to the determination of needed accuracy.

The variety of situations that range between these three examples leads to the only general rule for selecting a sampling procedure: Do not choose blindly when selecting a sampling procedure but try honestly to determine what is "best" for each particular situation according to realistic and adequate standards.

18.3 FURTHER CONSIDERATION OF DETAIL MAY PERMIT THE ELIMINATION OF OTHER TYPES OF DESIGN

After we have chosen the general classes of sampling designs that are to be retained for more detailed consideration, we can turn to material that has been assembled on the population and on the measurement process and examine the lessons that past experience has taught with respect to the use of these designs. It would be impossible to attempt to cover here all aspects of a design that might be investigated during this preparatory stage, but we shall indicate some of the major points to which attention might be devoted. Most, if not all, the comments will be directed toward probability model designs.

The points to be covered can most easily be set forth as answers to the following questions. First: *Can a design be executed according to plan or will losses due to "not at homes" and "refusals" negate the value of having a probability model sample?* The amount of loss that we may anticipate through "non-response," "not at homes," and "refusals" is conditioned primarily by such factors as the attitudes being measured, the population being surveyed, the form of approach to the sample elements, and the training interviewers have received, not by the method used in selecting the sample (see Chapters 16, 17, and especially Chapter 11). Nevertheless, if such losses are expected to be so large as to introduce a potential bias of serious magnitude and we still wish to conduct the survey, we may consider the possibility of changing from a probability model procedure to some other type of selection process. An argument against doing this is, of course, that probability model procedures will permit the incorporation of a maximum bias figure for these losses into a mean square error, whereas such biases are hidden and non-assessable in other types of procedure—except as indicated by calibration through past experience. There is no clear answer to this problem at the present time, and we simply raise it for consideration.

Are there alternative ways of identifying and contacting population elements, and what are their effects on the choice of a sampling procedure? There are many situations in which population elements may be identified and contacted in alternative ways, each possible approach having its own unique effects on the selection of a sampling procedure. Sometimes the choice between alternatives is clearly dictated by the specifications, but in other instances each of them may have to be carried along until a consideration of possible sampling procedures determines the final approach to be used.

A few examples will serve to illustrate this particular point. In one instance a group was interested in studying, among other things, the serviceability of women's blouses in relation to price paid, fabric, manufacturer's instructions on care, amount of wear, and purchaser's care. Two alternative approaches were possible. A direct sample of dwelling units could have been chosen in the defined geographic area and the study based on the experience of women living in these dwelling units; or, a sample of stores could have been selected in the area and all purchases of blouses noted during a selected time period by direct observation, the purchasers to be contacted for later interviews. The sampling problem is quite different in these alternatives; one involves a sample of dwelling units, and the other involved choosing a sample of stores and a sample of time periods for observation (with possible loss of mail order purchases and out-of-area shopping). There were, however, other considerations to be taken into account. The dwelling unit sample would have dealt mainly with purchases made in the past. Prices might have been forgotten and manufacturer's instructions on care lost. On the other hand, the store sample would require the cooperation of the stores for the placing of observers, the time periods for observation would have to be chosen carefully, and the volume of purchases would have to satisfy needs as to sample size. Thus the two alternative approaches raised quite different problems with respect to sampling and other considerations.

Problems raised by alternative ways of identifying and contacting population elements become especially acute and important when we are dealing with populations whose members are "thinly" dispersed throughout a much larger population (e.g., the population of "aged," of users of special equipment, of newsstand purchasers of certain magazines, etc.). Some discussion of sampling the "aged" population in the United States has been given by Stephan (8). In general, it will be necessary to search for special ways of getting at such groups in order to avoid the high costs of having to contact a large number of individuals in order to find one member of the population. For example, we may utilize other surveys, distribute questionnaires with each purchase of a particular type of equipment, or obtain lists as a by-product of other operations that reach the people we seek.

Are there available the necessary materials for determining the design details? The available information that has been assembled with respect to the population will have a marked effect on the types of sample design to be considered. Lack or inadequacy of lists means that simple random or systematic selection of individual elements from the population cannot be used; lack of data may mean that all quota and purposive selection procedures must be discarded, that stratification (at least with disproportionate allocation) cannot be used, that certain efficient forms of estimation (e.g., ratio and regression) are not possible, and that special forms of selection (e.g., probability proportionate to size) are not applicable; lack of detailed maps may mean that the primary sampling units cannot be made an "optimum" size or that it will be necessary to engage in more extensive and costly field operations than would otherwise have been required. Factors such as these become especially noticeable when survey work is being conducted in certain areas of the world, for example, in many regions of Asia and Africa. It is therefore instructive to read and study accounts of such surveys, for example, Shaul's description of survey work in Central Africa (9) or that by Jessen, Blythe, Kempthorne, and Deming, of a survey in Greece (10).

Are there special considerations (e.g., travel facilities, location of an existing field staff, etc.) that will eliminate certain types of design? Frequently it is necessary to pay particular attention to the location of an existing field staff, as well as to transportation facilities. One illustration is provided by the Monthly Report on the Labor Force, which originated in the Work Projects Administration as a monthly sample survey of the population with the primary purpose of measuring the number of unemployed (11). In 1942 this activity was transferred to the Bureau of the Census and developed as the Current Population Survey. It was broadened to provide additional information needed by the Selective Service System, the War Manpower Commission, and other agencies in formulating policies on manpower. In setting up the sample design for MRLF, a number of administrative requirements were imposed, in addition to the requirement that a probability model sample be used, and these are set forth in the following quotation taken from a paper by Hansen and Hurwitz (12):

> It (the sample) should provide for a closely-knit field organization of the type that had been operating formerly (i.e., in the WPA sample) and which had demonstrated its ability to produce field results of high quality in rapid time schedules. The existing field organization involved the use of about 60 full-time local supervisors, each of them having 5 to 15 part-time enumerators working under his immediate supervision. The maintenance of this type of organization meant that the total number of different areas to be included in the revised sample could not be expanded greatly beyond the 60 included in the original sample, at least not for the immediate future.

In addition,

> It shall retain as many of the counties in the old sample as possible, to avoid moving all the field offices into new counties at one time. The design of the sample shall provide for a minimum of travel relative to total amount of information obtained, and should provide for continued operation on rapid time schedules.

Similar requirements are imposed by most survey organizations maintaining a permanent field staff, and these have to be taken into account in judging procedures.

What can past experience teach about the choice of possible sampling procedures? It is always valuable to examine reports and other

:cords of similar surveys. The primary contribution to be made
:nsists in finding out such information as: (*a*) How was the sampling
:ctually carried out? (*b*) What was needed in the way of prepara-
:on? (*c*) What personnel were used in conducting the survey? (*d*)
Iow long did it take? (*e*) What difficulties were encountered in the
:peration? (*f*) How well did the sample turn out as measured by
:mparison with known population characteristics and by computation
f sampling errors and biases? (*g*) What recommendations were made
:n the conduct of future surveys of a similar nature? The experience
:etailed in Part II of this book should be helpful in evaluating this
/pe of information.

On the basis of this information, it may then be desirable to make
:rther eliminations from the list of possibilities, *making sure that the
:vidence used arises out of a situation similar in all important re-
:pects to the one for which a sampling plan is needed.* In the event a
:ethod is not eliminated, the data and information collected will prove
:seful in the more detailed planning and comparison stages. The data
:ay also suggest further information that should be obtained, either
:y preliminary surveys or in conjunction with the actual survey.
"here are many reasons for elimination which may arise during this
:ork, and only a few will be cited for illustrative purposes. Thus
:ne plan might require too much time for designing the sample; an-
:ther might require materials that are badly out of date at the present
:me (e.g., census material, city directories); and a third might show
:n extreme deviation from the population value for a variable which
: highly correlated with the present subject of investigation and for
:hich no controls can be incorporated into the sample.

The great value of past experience for these preliminary investiga-
:ions and frequent disappointments with the inadequate descriptions
:f sampling given in the analysis of many surveys stress the need for
:lways preparing a careful account of sampling procedures to accom-
:any, or to be incorporated in, the final report. A brief summary of
:he reasons for eliminating many of the possible methods that could
:ave been used would often be very beneficial to the future develop-
:ent and use of sampling.

18.4 PREPARING FINAL LIST OF POSSIBLE PROCEDURES AND ASSEMBLING MATERIALS TO AID IN DETERMINING THEIR DETAILS

The final result of this preliminary examination of sampling pro-
:edures will be a list of all those that have not been eliminated, to-

gether with the known advantages and limitations of each. These procedures will be considered in the detailed planning operations to be outlined in succeeding chapters.

Advantages and limitations will relate to the items discussed in this chapter, and to any others that may seem pertinent because of the peculiar circumstances of the intended application. Thus we should have information on: (a) the materials and personnel needed for designing the sample (maps, data, lists, technical advisors, etc.), (b) the special training which field personnel may require, and estimate of its cost, (c) the relationship of the sampling procedure to the method of approach to be used, (d) time and resources required for preliminary operations (pretesting, prelisting, special surveys to obtain additional data), (e) time and resources needed for carrying out the measurement process on the designated sample and their relation to the survey specifications and needs, (f) advance predictions of the bias and variance of the estimates to be expected from the sample, either on the basis of sampling theory or on the basis of prior experience and (g) possibility of evaluating the sampling procedure after one or more samples have been drawn, including resources needed for this phase of the operation.

The net result of this preliminary work may be a decision that a sampling advisor is needed, particularly if three or four possible procedures still remain and the individual carrying out the survey does not feel competent to assess them on the theoretical grounds of sampling theory. Although theoretical consideration would not necessarily eliminate any of them, at least it would build a firmer foundation for carrying all of them along into the detailed stages of planning so that the final choice of the one to be used will be made more efficiently.

The principal point to be made in this respect is that the sampling advisor must be given an adequate opportunity to exercise his knowledge and experience. He must be brought into the discussion of specifications and needs for the study and, if he feels it necessary for the planning, must be allowed to expend resources in preliminary survey or in the analysis of previous surveys. He can frequently suggest special materials to be used in the planning and will be in an advantageous position to judge the degree of similarity between the present problem and past experience.

The preparatory work of assembling information and materials for use in the actual drafting of the design leads up to conferences at which major decisions are to be made. Preparations usually do not terminate abruptly but may continue into the active work of shaping the survey procedures. There need be no sharp separation of the two phases of survey design. Usually one grows naturally out of the other. Some anticipation of the later stages of designing is inevitable during the preparations. New problems for exploration will arise during the actual designing. Moreover, the time schedule may make it necessary to start designing well before the preparations are completed. The preparatory work is not an end in itself; it merely gives initial direction to the designing and simplifies it, making it more efficient and less expensive. The experience of many survey directors is that good preparation is the distinguishing mark of competent survey operations.

REFERENCES

1. W. Edwards Deming, *Some Theory of Sampling,* John Wiley & Sons, New York, 1950.
2. Morris H. Hansen, William N. Hurwitz, and William G. Madow, *Sample Survey Methods and Theory,* Vol. I, John Wiley & Sons, New York, 1953.
3. Harry Alpert, "Some Observations on the Sociology of Sampling," *Soc. Forces,* 31 (1952), 30–33.
4. Gordon F. Streib, "The Use of Survey Methods Among the Navaho," *Am. Anthropologist,* 54 (1952), 30–40.
5. Burton R. Fisher, Charles A. Metzner, and Benjamin J. Darsky, *Public Response to Peacetime Uses of Atomic Energy,* Vol. I, Survey Research Center, Institute for Social Research, University of Michigan, Ann Arbor, 1951, pp. 2, 5, Appendix B.
6. Eli S. Marks and W. Parker Mauldin, "Response Errors in Census Research," *J. Am. Statist. Assoc.,* 45 (1950), 424–438.
7. Morris H. Hansen, William N. Hurwitz, and Leon Pritzker, "The Accuracy of Census Results," *Am. Sociol. Rev.,* 18 (1953), 416–423.
8. Frederick F. Stephan, "Sampling for Old-Age Research," Appendix in Otto Pollak, *Social Adjustment in Old Age: A Research Planning Report,* Bulletin 59, Social Science Research Council, New York, 1948.
9. J. R. H. Shaul, "Sampling Surveys in Central Africa," *J. Am. Statist. Assoc.,* 47 (1952), 239–254.
10. R. J. Jessen, Richard H. Blythe, Jr., Oscar Kempthorne, and W. Edwards Deming, "On a Population Sample for Greece," *J. Am. Statist. Assoc.,* 42 (1947), 357–384.
11. Lester R. Frankel and J. Stevens Stock, "On the Sample Survey of Unemployment," *J. Am. Statist. Assoc.,* 37 (1942), 77–80.
12. Morris H. Hansen and William N. Hurwitz, *A New Sample of the Population,* U.S. Bureau of the Census, Washington, D. C., 1944.

CHAPTER 19

A Broad View
of the Drafting of Designs

19.1 INTRODUCTION

The preparatory stages in survey design, described in Chapters 14 through 18, sharpen the specifications and assemble working materials relating to the various components of the ultimate design; these stages may possibly force decisions on the broad outlines or specific details of some of these components. Once this work has been done, it becomes necessary to focus attention on the survey as a complete and integrated plan of operation and to think in terms of actually applying techniques that may have previously been considered only in broad generality. Though our discussion will proceed as though only one design were being prepared, it is important to recognize that in practice we may be developing two or more alternative plans simultaneously, with a final choice to be made only after each plan is complete. Such plans may, of course, have many features in common, even to the extent that one is only a modification of the other. Nevertheless, even a minor modification in one component may make its effects felt throughout the entire design, and thus the final choice should be made only on the basis of over-all performance.

Drawing up a plan for a complex survey is a complicated process requiring the well-coordinated development of all parts in relation to the whole. Only in rare instances will it be possible to work out the various details of measurement and sampling in a piecemeal fashion and arrive at a design that is in any sense "best" or "optimum." This fact has been illustrated in the account of the Post-Enumeration Survey of the 1950 Census provided by Marks, Mauldin, and Nisselson (1):

> Although the Post-Enumeration Survey possessed certain distinctive features peculiar to its purposes and content, its history illustrates the complex decision process involved in survey planning. This decision process requires the balancing of cost and accuracy considerations both within a particular aspect of a survey and among different aspects. The over-all cost of a survey and the accuracy of the estimates the survey provides are resultants of the costs and accuracies associated with a variety of processes—sample selection, interviewing, coding, tabulating, etc. Since most surveys have a fixed over-all expenditure limit, increased expense to improve the accuracy of one phase will usually require curtailed expenditure and lowered accuracy in some other phase —for example, may require a decrease in sample size with consequent increase in variance. Between survey processes a balance is necessary; to achieve a reduction of 5 per cent in coding error at the expense of decreased checking of the interviewing with an attendant increase of 10 per cent in interviewing error is hardly desirable, particularly if the interviewing error is already larger than the coding error. Within a given survey process—for example, interviewing—an increased expenditure for training must be balanced against an increase in expenditures for supervision, interviewer salaries, more detailed questions, etc., in terms of the relative effect of the various expenditures on the quality of the interview output.

Actually, the goal implied in this statement is one that cannot be achieved, except by intuitive means, with the present state of knowledge concerning certain aspects of the survey process. Recognition of this fact prompted the preparation of the paper from which this quotation was taken, as evidenced by the introductory comment:

> This paper points up the extent to which decisions on survey design have to be reached on the basis of intuition and opinion, due to the absence of any satisfactory objective data. Such analysis directs attention to the major gaps in the knowledge of survey technique—the spots where there is need to superceed art with science—and may raise as subjects for investigation, points which have been implicitly accepted as axiomatic.

Not only is it true that we may miss "best" or "optimum" designs by treating the components of a survey in a piecemeal fashion, but we may also become involved in a circular process which might preclude the possibility of ever arriving at a feasible plan, or at least of arriving at such a plan in a reasonable length of time. A simple illustration of the type of dilemma that might arise is given by the following example. Suppose the design is concerned primarily with the number

of sampling points (or primary sampling units), the number of interviews to be obtained at each point, the length of the interview, the
accuracy and detail required by the contemplated uses of the survey
results, and the costs of carrying out the survey. Let us suppose
further that the details of each of the first three items are determined
in isolation. Then we might find ourselves going through a series of
successive revisions such as those shown in Table 19.1.

TABLE 19.1

SUCCESSIVE REVISIONS OF SURVEY DESIGN

Order of Consid- eration	Component of the Design	First Draft	Second Draft	Third Draft	Fourth Draft
A	Number of sampling points	Fixed [a]	Decreased from 1	Same as 2	Same as 3
B	Number of interviews at each sampling point	Fixed [a]	Increased from 1	Decreased from 2	Same as 3
C	Length of interview	Fixed [a]	Same as 1	Same as 2	Decreased from 3
D	Accuracy and detail for contemplated uses	Adequate	Inadequate	Adequate	Inadequate
E	Costs	Too high	Satisfactory	Too high	Satisfactory

[a] Starting numbers, possibly based on judgment or prior survey experience.

Thus, in the process of adjusting the first three components in isolation, the entire design process is essentially going in circles without
settling down rapidly to the desired result. Such situations, and others
of equal importance, can be expected to occur in highly articulated
systems. The most desirable solution is to determine all adjustable
phases of the system simultaneously and to avoid the fixing of details
at too early a stage in the design process. In some instances it may
even be necessary to defer decision on certain aspects of the design
until the survey is almost ready to go into the field, and the Post-
Enumeration Survey provides a case in point.

It is not to be expected that the remaining chapters will tell exactly
how to carry out this balancing operation and when to fix certain
details. Nevertheless, the mere recognition that we are engaged in

this type of process will have a tendency to improve the ultimate design and will possibly spur survey operators to obtain information, not otherwise needed, that will be of assistance in future design problems.

19.2 PREPARING A ROUGH SKETCH OF THE ENTIRE DESIGN

Building on the preparatory work, it should now be possible to sketch the principal features of one or more feasible designs. This sketch, for one particular design, will serve as a guide for the detailed planning to follow and, since it will emphasize the over-all aspects of the survey, as a test of decisions already made. In any specific situation, great variability will undoubtedly exist between different aspects of the design with respect to the detail that can be set down in these sketches—and even greater variability may arise between different survey situations. Thus some portions of a design may already be fixed down to minute detail, others may be final only in broad outline, and still other vital and far-reaching portions of the design may be in such a state of flux that two or more alternative plans must be carried on until a final decision can be made between them. For example, the survey specifications and preliminary work will ordinarily define the population in unequivocal terms. On the other hand, the mere decision to use blocks as primary sampling units in a probability model design does not determine whether or not the blocks are to be stratified and whether they are to be selected randomly or according to some other model. Finally, if decisions have not already been made on such matters as a mail questionnaire versus a personal interview, or a probability model sampling procedure as opposed to some judgment procedure, then it will almost certainly be necessary to prepare and work with more than one sketch.

The purpose of this sketch is to indicate where detail must yet be determined, to warn early if special pilot or exploratory studies are needed before final decisions can be made (either for the specific determination of a particular design or for the final choice among designs), and especially to emphasize the interrelationships between the various components of a design. It is also at this point that we must budget a portion of the total resources to planning work such as writing and pretesting questionnaires and designing the sample and to the hiring and training of interviewers. Following the preparation of complete sketches, it is next necessary to make each plan complete by filling in the detail—at least until we can decide to exclude plans from further consideration. Filling in these details will naturally overlap and merge in a variety of ways into the preparatory stages of planning and into the field operation of the final plan.

19.3 SPECIFIC DETAILS OF SAMPLING TO BE DETERMINED

We shall now present a brief discussion of some of the points that must be treated in the development of detailed plans. The list that follows is not meant to be exhaustive, only indicative. Moreover, the ordering is not necessarily the order in which actual work would proceed. As a matter of fact, work will ordinarily proceed on all points simultaneously.

Since a major emphasis of this book is on the choice of sampling procedures, it seems natural to consider first various aspects of sampling that must be fixed in some detail in order to provide a complete plan. Following this discussion of points directly related to sample design, a rather brief outline will be given of other items that must be included in filling out a survey design. The amount of space accorded to these latter is not to be taken as a measure of their relative importance but only as another indication of the already noted emphasis on sampling.

A. The Sample Size

Though later revisions may be made and probably will have to be made as the design develops, the preparatory consideration of specifications and needs (resources, accuracy, desired subtabulations, etc.) will frequently permit a workable approximation to the ultimate sample size. Such an early estimate, no matter how rough it may be, will facilitate the planning of phases of the survey other than sampling, particularly with respect to the allocation of resources in time and money among the various items in the budget. For example, a size estimate will give some indication of the cost and time that will be required for carrying out interviewing and other field work and for processing the completed schedules.

The accuracy of a first approximation to the final sample size will depend on a large number of factors, including the manner in which the specifications are phrased, the knowledge already assembled concerning population characteristics, and the type of sample design that will be used. This may be especially easy if simple random sampling is to be applied to estimate a single population quantity and the specifications are stated in terms of "fixed amount of money to spend" or "fixed degree of sampling variability to be attained." However, even here, difficulties such as the following may arise: the cost of a single interview is not known, there are differential costs between various members of the population, or the population variance of the variable being studied cannot be estimated accurately.

Any deviation from simple random sampling (e.g., stratified random sampling, multi-stage sampling, and systematic sampling), or from a simple statement of specification (e.g., where we attempt to assign loss values to errors of estimate or to design a multi-purpose sample), will make the final sample size more and more dependent on unknown quantities and on other phases of the survey design. It may make the accuracy of the first approximation to sample size quite inaccurate. For example, city samples of dwelling units are frequently obtained by first choosing a simple random sample of blocks and then selecting a constant proportion of dwelling units from each of the sample blocks. Since blocks vary in size, the number of dwelling units in the final sample is itself a variable whose actual value in any particular situation cannot be known until after the sample has been drawn, although the average size of such a sample would ordinarily be known in advance.

Finally, we note that the concept of the sample size as a random variable becomes of paramount importance when we attempt to apply the theory of sequential analysis. [See Wald (2).] A truly sequential test of a statistical hypothesis calls for examining randomly selected elements one by one, studying the total available sample after each element has been added, and continuing this process until some specific criterion is satisfied. It would be difficult from a practical point of view to use this procedure in studying human populations that are dispersed over a wide geographic area, but some appropriate modification of the truly sequential test procedure (e.g., selecting successive *groups* of people rather than individuals) may find extensive future use in the field of sample surveys. Some results on this are already available in the field of acceptance sampling (3) and in some surveys.

B. Distribution of the Sample over the Population

Having made a rough determination of sample size, we may examine and state the requirements concerning the distribution of the sample over the population. By problems of distribution, we refer to the manner in which the sample is to be spread over the geographic area occupied by the population or over the various population subgroups. The decisions which ultimately must be made on these points will relate primarily to (a) the availability of personnel (e.g., the location of interviewers at definite points throughout the population area), (*b*) the degree of supervision that must be maintained over field operations, (*c*) the definition of special groups for which analyses may be required in the final report or whose proper representation in the sample will enhance its precision, (*d*) the cost of travel within and between interviewing points, and (*e*) the relative sampling variability as between a widely "dispersed" sample and a "concentrated" sample.

Points *a* and *b* involve broad administrative considerations on which tentative decisions can and should be made early, since they are concerned with the organizational structure of a research operation and since they markedly affect all aspects of the survey design. It would be extremely difficult if not impossible to offer recommendations on such items outside of a specific setting. The type of requirement that may be imposed in this connection has already been well illustrated in Section 18.3 with respect to the original design of the sample for the Monthly Report on the Labor Force.

Point *c* refers specifically to the techniques of stratification whereby the population is divided into subgroups and a sample is selected from *each* such subgroup, either for the purpose of improving the over-all precision of the sample or to provide a sufficient number of cases in special groups for purposes of analysis. The determination of these subgroups or *strata* and the proportion of the sample which is to come from each is directly related to problems of precision and costs, and specific recommendations and techniques will be discussed in the following chapter.

Finally, we note that points *d* and *e* are particularly concerned with the choice of sampling units (see following reference to units of sampling and Chapter 20).

C. Determination of the Units of Sampling

It is not necessary that the ultimate population element be the actual unit in terms of which the sample is drawn. Frequently it is advantageous, or even necessary, to carry out the sampling in several stages. That is, the population is first separated into a set of mutually exclusive and exhaustive groups of elements; a *sample of these groups* is selected; and finally, all or a sample of elements from the chosen groups constitute the total sample to be studied. The reasons for considering groups or clusters of elements as sampling units stem from considerations of cost, practicability, and administration. Thus groups can frequently be identified when the individual population elements cannot, and clustering the sample elements on a geographic basis will reduce the amount of travel required for obtain-

g personal interviews. On the other hand, the use of cluster sam-
ling units often leads to a loss in precision over that which would be
btained by a simple random sample of the same size.

The detailed outline being prepared can probably state the units
at are feasible under a particular set of circumstances, based on a
nowledge of the population and on an examination of past experience
similar surveys. However, the final choice of units and their inte-
ration into the survey design may have to be left until a more de-
iled examination can be carried out, and even in apparently simple
tuations data may be lacking on which to base precise and objective
ecisions concerning sample design. For example, households consti-
te clusters of individuals from which one or more respondents may
e chosen for the sample, and yet little or nothing is known concern-
g the homogeneity of members of the same household with respect
attitude and opinion variables. More detailed treatment of these
nd other points relating to units of sampling will be found in Chap-
r 20.

. Fixing the Techniques to Be Used in the Selection Process

On the basis of the preparatory work, one or more selection pro-
dures must now be fitted into each of the designs being developed.
he reason for using the phrase "one or more" is that most complex
esigns will include more than one stage of sampling, and there is
o need to employ the same broad procedure at each stage. The
ctual incorporation of a sampling procedure into a design plan will
sually follow a logical sequence of events. The details of the pro-
dure will be worked out so as to fit the particular situation, and
lans will be prepared for the more or less mechanical task of carrying
ut the final selection of sample elements. Finally, an evaluation
f the sample design will be made in relation to the cost and perform-
nce of the total survey.

If a probability model procedure is to be used, there is now avail-
ble a large body of theory and experience which can be consulted
nd drawn upon (4, 5, 6, 7, 8). For example, we can find the ap-
ropriate manner in which to allocate a sample of fixed size among
rata in order to secure maximum precision; or, if a fixed amount of
oney is available and the cost of an interview varies between strata,
rmulations are given of the "best" possible way to spread the re-
urces among the strata as measured in terms of the precision of
mple estimates. The theory and quoted experience is also extended
handle the problem of multi-stage sample designs, either when the
mpling units are equal in size or when each sampling unit contains

a different number of individual population elements. Many way
of integrating cost considerations into different designs are discusse
Little of the experience quoted in these books deals directly with th
opinion and attitude surveys, but many parallels can be drawn an
the appropriate use of such theory will tend in time to produce a bod
of experience for this field. (See Chapter 9 for several illustration
and a further discussion of this point.)

Even after the choice of a probability model design and the dete
mination of the pertinent theory, other details must be fixed. F
example, provision must be made for deciding where some of the san
pling operations are to be carried out—in the office or in the field. Thr
if a unit has been chosen for the sample, the final choice of residenc
within the unit may be carried out in the office after a prelisting h
been completed, or it may be accomplished in the field by interviewer
A choice between these alternatives must ultimately be based on th
cost of carrying out the prelisting, on the cost of training the inte
viewers to carry out the additional work in the field, and on the exter
to which it is felt that the interviewers can be trusted with this pha
of the operation. One point at which the selection is usually entruste
to the interviewer is in choosing an individual at random from
designated household. Procedures have been devised for accomplish
ing this without making the interviewers' burden too heavy (9).

Finally, we note that designs other than those based on probabilit
models also require the specification of many details. For exampl
quota sampling procedures demand particular care with respect to th
form in which quotas and instructions are given to the interviewer
Are the quotas going to be on marginals alone, or are they going t
be on cross breaks between the control variables? Are they going t
restrict the interviewer to working in a small area, or will they perm
him to travel over a large area? Is the interviewer allowed to pick u
respondents in public buildings and in the street, or must all of h
interviewing be confined to the home? The two or more alternativ
decisions that can be made on many of these questions run at cro
purposes to one another. If the restrictions on the interviewer are in
creased, then his personal judgment is clearly less likely to influenc
the final selection of respondents. On the other hand, additional re
strictions will make it harder for him to fill his quotas and may lea
to increased costs, cheating, and the like. A great number of thes
problems are intimately related to the way in which the quota inter
viewer actually fills his assigned quotas. Many aspects of this pro
cedure were discussed in Chapter 12.

E. Choosing an Estimation Procedure

The end result of any survey operation is the preparation of one or more numerical quantities, *estimates*, which will serve as approximations for unknown population characteristics. The estimation procedure specifies the manner in which information derived from the sample is to be combined with information obtained from other sources into the final estimate; ordinarily the estimation procedure must be determined jointly with the sample design. In simple surveys, this poses no particular problems. Thus if we take a random sample, we may only need to compute averages and proportions in a straightforward manner. However, in a complex survey design many possible forms of estimate may exist, and it may be necessary to investigate each of them in order to arrive at a final decision. Advance planning is particularly necessary if the same end result can be attained either by incorporating special features into the sample design or by appropriate choice of an estimation procedure. For example, if the population can be divided into a number of groups of known size, is it better to stratify the groups by size and pick out a random sample from each stratum or to choose the groups with probability proportionate to size? Similarly, it may be necessary to decide between stratifying the population in advance and choosing a random sample from each stratum, choosing a random sample from the population and attempting to gain the effects of stratification after the sample has been drawn, or choosing a random sample and using the known information about the population for making a ratio or regression estimate.

The theoretical and computational details of many estimation procedures, such as ratio and regression methods, are too extensive to be covered here, but an adequate treatment will be found in the books mentioned earlier (4, 5, 6, 7, 8). Nevertheless, it is not difficult to set forth the considerations that enter into the choice in any specific situation. These are:

1. It is generally desirable, all other things being equal, that an estimate be easy to compute. Complicated computational procedures may require a great deal of time, may require the use of expensive computing machines, and will probably require the services of highly trained and experienced personnel. It is for this reason that there is a strong preference for *self-weighting* sample designs, that is, sample designs which admit estimates requiring no weighting computations.

2. It is necessary that the requisite data be available. Thus one form of estimate for a particular quantity may require a knowledge of the total number of elements in the population. If this is **not**

known, then this form of estimate is valueless. These additional data may come from either of two sources, namely, previous studies or censuses of the population or special questions asked during the course of the survey. Usually, it is a combination of the two which is important. For example, if we know the proportions of the population that fall into subgroups defined by a variable known to be highly correlated with the variable under investigation, then information on this auxiliary variable can be obtained from each sample element. The resulting data can be used to compute regression estimates or ratio estimates and to make weighted estimates similar to those used in conjunction with stratification.

3. Though these questions of convenience, cost, and availability of needed data play an important role in the final determination of a procedure of estimation, they must always be considered in conjunction with the sampling distribution of the estimate. In other words, the reason we are willing to pay more for one estimate than for another is that the preferred estimate has a smaller variance or mean square error. This may at first glance seem to throw the choice between estimates back to a choice between sampling methods. However, it is frequently possible to have more than one method of estimation associated with a given sampling procedure, and the problem of optimum design must take both the sampling procedure and the form of estimation into account.

F. Providing for the Estimation of Sampling Error

As a final point, we note that any sample design should be arranged so that an estimate of sampling error can be obtained from the chosen sample. This may involve the incorporation of special features into the sample design, the obtaining of special information from respondents, the use of special systems for identifying respondents, and the like. These points are discussed and illustrated in Chapters 9 and 10.

19.4 OTHER DETAILS OF COMPLETE SURVEY PLAN

Over and above providing for the details of sample design, each plan that is now being carried along must make provision for all other aspects of a complete survey design—for example, the questionnaire, the survey personnel, the field application of the sample design, the actual interviewing process, and the analysis of the survey results. Certain of these components may have at least as important an influence on the final results as does the sample design. For example, though Hansen, Hurwitz, and Madow present a mathematical model for response errors, they emphasize the foregoing situation in the following quotation (6, Vol. II, p. 280):

The paucity of dependable data on response errors is unquestionably the greatest present obstacle to sound survey design. In survey after survey, losses in sampling efficiency are taken on the basis of quite dubious assumptions about the magnitudes and distributions of response errors. Frequently, this point is obscured by the implicit (practically "unconscious") nature of survey designers' assumptions regarding response error.

It can be expected that an increasing amount of experience and data will become available relating to response errors, interviewer effects, and the like, and that ways of integrating these into "survey performance functions" will be found. However, until this stage is reached, decisions must still be made on the basis of the best available information. Listed below are some of the major points that must be included in the survey plan.

A. Drafting and Testing the Questionnaire

The method of approach to the sample elements (i.e., mail, personal interview, group interview, etc.) and the data required by the specifications will ordinarily have been fairly well determined during the preparatory work. These two factors will influence and to a large extent fix the general form and content of the questionnaire that is to be used, and it is now necessary to translate these influences into an actual draft questionnaire. In carrying out this work, we should keep in mind that the questionnaire will, in turn, have effects throughout the survey design. Thus the use of open-end interviewing techniques will frequently require more intensive training of interviewers than will fixed-answer questions; the length of the questionnaire will be reflected in the interviewing costs and the degree of cooperation to be expected of respondents; the organization, form, and content of questions will also affect this cooperation; and the proper framing of questions may do much to reduce or increase response error. Assistance in framing questions, in integrating questions so as to provide scales and indices, in placing questions in their "best" relationship within the questionnaire, and in finding references to appropriate experience in the literature can be found in such books as Payne (10), Parten (11), Jahoda, Deutsch, and Cook (12), and in Festinger and Katz (13).

The detailed plan should also provide for the pretesting of the questionnaire. The pretest will give information on the average length of the interview, on the devices needed to secure and maintain the respondents' cooperation, on the types of answers to be expected, and on the extent to which these answers provide the desired data. It can forewarn of difficulties to be encountered and may indicate ways of overcoming them.

B. Recruiting, Training, and Supervising the Field Staff

If the time schedule is to move smoothly, plans must be instituted early to insure an adequate and well-trained staff for conducting the survey in the field. If an organization maintains a permanent group of field workers, this may involve nothing more than proper scheduling. However, if this is not the case, we must face the problems of recruiting, training, and supervising those individuals upon whom depends to a large extent the entire success of the undertaking. Careless execution can invalidate the results obtained from even the soundest of survey designs. Moreover, this phase of the operation must receive very serious attention in the budget since training and supervisory expenses can be expanded or contracted almost at will, and there are no very concrete principles to guide us.

Field work will ordinarily involve two quite different types of operation. One relates to the sample design whereas the other is part of the measurement process. Certain sample designs call for the preparation of a list of dwelling units in specified areas, from which the actual sample elements are chosen by an office staff; others require the interviewer to make this listing and at the same time select the potential respondents according to stated rules; and still others may demand interviews with every element in an assigned area. The field work related to the measurement process is of course the actual interviewing.

It is well known that errors can occur in the performance of such tasks as listing and that an interviewer may also introduce error through cheating, inaccurate entries, conscious or unconscious influencing of the respondent, and the like. Unfortunately, most of the experience in the literature merely reports incidents where these effects have been shown to exist rather than indicating definite ways to eliminate them or to provide for them in the survey model. For example, Mannheimer and Hyman (14) have treated some of these errors (in listing dwelling units, in the selection of the sample of dwelling units, and in the selection of individuals within dwelling units) in connection with a city sample; Durbin and Stuart (15) have investigated the differences in response rates between experienced and inexperienced interviewers with the conclusion that "Though the inquiry has demonstrated the inferiority of the students in obtaining interviews when compared with professional interviewers, it tells us nothing of the causes of the differences and whether they can be easily remedied"; the U.S. Bureau of the Census has conducted work on response errors (16, 17); Sheatsley has prepared a series of papers giving an analysis of interviewer characteristics and their relationship to performance (18, 19); and a comprehensive study has been reported by Hyman on interviewer effects in opinion and attitude measurement (20).

The foregoing points and references make it clear that serious attention must be given to recruiting, training, and supervision of field workers and that errors and differences can be detected by appropriate supervision and design (see Chapter 21). However, it is frequently not at all clear how far we should go in training and supervision, or what steps we should take if differences between interviewers arise in the analysis of a particular survey.

C. Obtaining and Maintaining the Cooperation of Respondents

It is essential that the sample elements cooperate thoroughly during the measurement process. This is important not only for assuring the quality of response but also for securing a response at all. There are many devices that may be applied in this respect. Thus we consider such items as the following: preparation of a statement to be given if a respondent asks why he has been chosen for the sample, the offering of prizes and premiums, presentation of the survey under favorable auspices, the use of airmail and special delivery stamps on mail questionnaires, etc. Under some circumstances it may be desirable to obtain a local sponsoring organization. Thus the Milbank Memorial Fund Study of the Social and Psychological Factors Affecting Fertility (see Chapter 16) obtained the local Council of Social Agencies as sponsor. This Council issued credentials to each interviewer, authorized newspaper publicity, and answered the queries of persons who were suspicious of the interviewers or their questions. Some groups will require special approaches in order to obtain their cooperation. For example, physicians should undoubtedly be asked for appointments and, unless they prefer other arrangements, the interviewing should be conducted in their offices. Individuals who refuse to be interviewed may require attempts by supervisory personnel or formal letters asking for cooperation.

Once original acceptance has been obtained by the interviewer, the maintenance of cooperation ordinarily depends on such items as the length of questionnaire and the proper placing of embarrassing questions. However, panel studies may raise particularly vexing problems in this respect since repeated calls have to be made on the members of the selected sample. A very detailed report of experience in the establishment and maintenance of a panel has been given by Industrial

Surveys, based on work carried out for the Bureau of Agricultural Economics in New York City (21).

D. Timing the Survey

The point in time at which a survey is put into the field and the length of time it takes to complete the interviewing may have a pronounced effect on the survey results. Here again it is difficult to make general recommendations, but some of the things that should be kept in mind during the planning stage are as follows: (a) Is the time period in any way unrepresentative with respect to the generalizations that are desired? For example, would an employee-attitude survey conducted during the course of a strike provide information on "normal" attitudes toward the company? Or should studies of food habits be carried out during a holiday or religious seasons when "normal" dietary habits are modified to some extent? (b) Will particular circumstances such as impassable roads in winter weather and vacations during the summer make it difficult or impossible to obtain a high rate of response? (c) Will unexpected events occur during the course of the survey so that some respondents are interviewed later on in a different context than that of others who were interviewed early? (d) Are questions being asked about events that happened so far in the past that there may be serious recall errors coloring the responses?

Some of these considerations may even have to be taken into account in the sample design. For example, the effect of unexpected and interfering events may be guarded against by designing a number of independent samples that are to be interviewed in succession. Further comments on this possibility will be found in Chapter 21.

E. Processing the Results

Finally, it is desirable to check on plans for actually processing the results, particularly in respect to the manner in which they will influence the design of the sample and the questionnaire. For example, all or part of the questionnaire may be precoded so that IBM cards can be punched directly from them. Detailed consideration is given to many pertinent parts of this operation in Chapters XIII, XIV, and XV of Parten's book (11).

19.5 SOME GENERAL PRINCIPLES OF STRATEGY

While we are working out the foregoing or related details, there are a number of general principles that should be kept in mind. These apply to almost any phase of the design process, and consequently

me of them have already been mentioned in a particular context.
hey are being reiterated here for special emphasis. Most, if not all,
: them are simple matters of common sense, but it is for this very
ason that they may be forgotten or ignored if not set directly be-
re us. These principles are given in the succeeding paragraphs.

*Continual checks should be made for feasibility, consistency, com-
leteness, and possible malfunctioning.* As the drafting proceeds, every
ffort should be put forth to check the developing plan with respect
• these requirements. Serious operational trouble may be prevented
y exercising care in drafting and preparing field materials. Thus
ailure to provide an adequate operational definition of the population
an lead to the necessity for many spot decisions on the part of field
orkers or supervisors. These spot decisions may be only time con-
uming if adequate records are maintained, but they may cause an
nknown amount of non-comparability among sample elements if not
ecorded. Similarly, if sample areas are to be recorded on a map,
iey must be regarded from the point of view of the person who must
ocate and identify them in the field. Much time and accuracy may
e lost if they are not located according to easily discernible natural
oundaries.

*Adequate provision should be made for control of operations and for
hanges during the course of operations.* The survey design should
ontain provision for an adequate system of supervision and control.
t should be alert to meet emergencies that arise and to order neces-
ary changes that will not invalidate but will conserve the effectiveness
f the plan. Since no plan is final until it has been carried out, changes
uring the field work should be regarded as changes in the actual plan
nd taken into account in the analysis.

The extent to which we may be able to provide such a system of
ontrol depends on the way in which we are able to resolve the con-
ict between flexibility and generality of plans with the specificity
esulting from the choice of particular plans. High efficiency may be
ttained only at the cost of inflexibility. This may make the plan
ail badly if any part goes wrong. Like biological evolution, undue
pecialization may prove adverse to survival when conditions change.
'he plan should be prepared with such contingencies in mind, but of
ourse some risks must be taken.

The exhaustion of funds or time provides one illustration of the type
f major contingency that may arise and that can be allowed for in
esign. Thus it may be desirable to draw two or more independent
amples and to conduct the field work in such a way that it can be cut
hort at a number of different points in time, with the assurance that

a "decent" sample will always be available for analysis. This related to the subsample method of estimating sampling variabili discussed in Chapter 9.

Systematic records should be maintained. For control purposes a use in the analysis, it is highly desirable to maintain a systemat record of decisions, reasons for them, troubles that arose, special co ditions observed, and operating experience. This means that efficie and convenient record systems should be prepared in the drafti process for use in the conduct of field operations. These record sy tems should be designed to account for all changes in original desig time and performance records, unsuccessful attempts, and the lik Not only may this material modify the subsequent treatment of sche ules and their analysis, but it will also help to fill in some of the ga in knowledge concerning the survey process. Many of the resul reported in Part II of this book were made possible only becau special records were kept during the course of surveys. (See especial the discussions in Chapter 11 on callbacks and refusals and in Chapt 12 on interviewer performance with a quota sample design.)

There are, of course, serious psychological tendencies to be ove come in maintaining many of these records since people think th this paper work is useless or less important than the completed sche ule. This attitude can be overcome in part by specific compensation personnel for completion and quality of records.

Sample design and field work may require outside contracto Finally, it may be necessary or desirable to deal with outside adviso or contractors if it is apparent that our own organization is not pr pared to complete a design or carry it through the field operation This might be because of lack of skilled personnel to design the detai of a sample, because of the lack of an adequate field staff, or becau of the lack of needed physical resources (maps, data, etc.). In su cases, it will be necessary to obtain bids and time estimates for t required work in order to see whether such action will fit into t budget and time schedule of the survey. There are many universi research centers, market research organizations, and some governme agencies that are prepared to offer such services.

This chapter has briefly discussed many of the items that must consired in designing a sample survey for use in the measurement opinions, attitudes, and consumer wants and has set forth some of t general principles that may be applied in the design process. Eac of these could be successively examined in great detail, but this ha not been possible in the present book because of lack of space, becau

the emphasis on sampling, and because of the attention that many
them have received in the quoted references. Consequently, in
e next three chapters we shall restrict ourselves to a number of spe-
al topics which are closely related to sample design and to which
sufficient attention has been given in other work or for which data
eculiar to the measurement of opinions, attitudes, and consumer
ants are required before full use can be made of the existing theory
sample design.

REFERENCES

1. Eli S. Marks, W. Parker Mauldin, and Harold Nisselson, "The Post-Enumeration Survey of the 1950 Census: A Case History in Survey Design," *J. Am.
Statist. Assoc.*, 48 (1953), 220–243.

2. Abraham Wald, *Sequential Analysis*, John Wiley & Sons, New York, 1947.

3. Harold F. Dodge and Harry G. Romig, *Sampling Inspection Tables*, John
Wiley & Sons, New York, 1944.

4. William G. Cochran, *Sampling Techniques*, John Wiley & Sons, New York,
1953.

5. W. Edwards Deming, *Some Theory of Sampling*, John Wiley & Sons, New
York, 1950.

6. Morris H. Hansen, William N. Hurwitz, and William G. Madow, *Sample
Survey Methods and Theory*, Vols. I and II, John Wiley & Sons, New York,
1953.

7. P. V. Sukhatme, *Sampling Theory of Surveys with Applications*, Iowa State
College Press, Ames, 1954.

8. Frank Yates, *Sampling Methods for Censuses and Surveys*, second edition,
Charles Griffin & Company, London, 1953.

9. Leslie Kish, "A Procedure for Objective Respondent Selection within the
Household," *J. Am. Statist. Assoc.*, 44 (1949), 380–387.

10. Stanley L. Payne, *The Art of Asking Questions*, Princeton Universtiy Press,
Princeton, 1951.

11. Mildred Parten, *Surveys, Polls, and Samples*, Harper & Brothers, New York,
1950.

12. Marie Jahoda, Morton Deutsch, and Stuart W. Cook, *Research Methods in
Social Relations*, The Dryden Press, New York, 1951.

13. Leon Festinger and Daniel Katz, editors, *Research Methods in the Behavioral Sciences*, The Dryden Press, New York, 1953.

14. Dean Mannheimer and Herbert Hyman, "Interviewer Performance in Area
Sampling," *Publ. Op. Quart.*, 13 (1949), 83–92.

15. J. Durbin and A. Stuart, "Differences in Response Rates of Experienced and
Inexperienced Interviewers," *J. Roy. Statist. Soc.*, 114 (1951), 164–206.

16. Morris H. Hansen, William N. Hurwitz, and Leon Pritzker, "The Accuracy
of Census Results," *Am. Sociol. Rev.*, 18 (1953), 416–423.

17. Eli S. Marks and W. Parker Mauldin, "Response Errors in Census Research,"
J. Am. Statist. Assoc., 45 (1950), 424–438.

18. Paul B. Sheatsley, "An Analysis of Interviewer Characteristics and Their Relationship to Performance: Part I," *Int. J. Op. Att. Res.*, 4 (1950), 473–498.

19. Paul B. Sheatsley, "An Analysis of Interviewer Characteristics and The Relationship to Performance: Part II," *Int. J. Op. Att. Res.*, 5 (1951), 79–9

20. Herbert Hyman, *Interviewing in Social Research*, University of Chicag Press, Chicago, 1954.

21. U.S. Bureau of Agricultural Economics, *Problems of Establishing a Con sumer Panel*, Marketing Research Report 8, Washington, D. C., 1952.

CHAPTER 20

The Use of Population
Subgroups in Sample Design

20.1 INTRODUCTION

There are usually many ways in which any particular population may be divided into subpopulations or subgroups. Sometimes these groups can be defined before a sample is actually drawn, and their appropriate use in sample design may mean the difference between doing or not doing a study, may decrease the sampling variability of survey estimates, may allow us to attain a specified level of precision at decreased cost, or may insure that desired analyses can be made with the sample that is finally selected. Under other circumstances it may be impossible to delimit the groups until certain information has been obtained from the individual sample elements, but even here proper planning and design may make the use of such groups almost as beneficial as if they had been designated in advance. Much of modern sampling theory relates to the utilization of these population subgroups for the improvement of sample design, but all too frequently we lack the data and experience necessary to apply the theory effectively to the measurement of opinions, attitudes, and consumer wants. This chapter will review several problems that arise in the use of population subgroups.

The succeeding discussion will distinguish between *stratification* and *multi-stage sampling*, the two techniques most definitely associated with the use of groups of population elements. Stratification ordinarily involves the following steps: (*a*) The population is divided into mutually exclusive and exhaustive subpopulations or *strata*. (*b*) A sample of elements is selected from *each* stratum. (*c*) An estimate is made for *each* stratum. (*d*) These estimates are combined to provide an estimate for the entire population. (The strata estimates may also be compared with each other.) Multi-stage sampling, though it

also makes use of groups, is quite different in approach and purpos
as will be shown in some detail at a later point in this chapter. The
a two-stage sample would be drawn as follows: (a) The populatio
is separated into a set of mutually exclusive and exhaustive groups
elements. (b) A *sample of these groups* is selected. (c) Either a
the elements from the chosen groups or a sample of them constitut
the final sample to be studied. If the chosen groups are subsample
the subsampling can be done by a further application of some variet
of multi-stage sampling or by a single-stage sampling procedure
partly by one and partly by the other method.

The groups of individual population elements that are to be use
as sampling units are frequently referred to as "clusters." It shoul
be noted that two-stage cluster sampling becomes stratified samplin
if a 100 per cent sample of clusters is chosen and a sample is draw
from each cluster. Also it is apparent that the two techniques ma
easily be combined. For example, in drawing a two-stage sample o
households in a city (blocks and households), the blocks may b
stratified with respect to some such variable as average rental valu
of dwelling units before the actual selection of blocks and household
is made.

20.2 DETERMINATION OF USEFUL AND FEASIBLE POPULATION SUBGROUPS

There is usually a wide variety of considerations upon which w
may base the definition of one or more systems of groups within an
given population. We shall now describe some of the more commo
considerations, keeping in mind that they may be overlapping. I
each situation many specific considerations will also influence th
choice of subgroups.

A. General Bases of Grouping

1. Physical Proximity. When a population is dispersed over a wid
geographic area and each individual element can be associated eithe
with a point in the area or with a certain portion of the area, it is al
ways natural to consider the possibility of forming groups withi
which the members are in some sense physically "close" to one an
other. Such groups are usually defined on the basis of political an
natural physical boundaries that are commonly used in reportin
Census figures or in drawing maps; they may of course vary greatl
in size. Thus it is possible to consider such units as broad geographi
regions, states, counties, cities, townships, wards, election precincts

ocks, and the like. Sometimes slightly different but related ap-
oaches are possible. For example, a household or a group of say
ve households (a segment) marked off on a map or on a directory
reet listing may form a cluster. Also, organizations have selected
mples of business establishments (scattered over the United States)
y first choosing a sample from a list and then associating with each
osen establishment the three or four additional establishments that
e geographically closest or have, for instance, treated magazine sub-
ription lists in a similar manner. The former of these is discussed
an unpublished paper by Frankel (1).

The groups defined on the basis of physical proximity may be used
ther for stratification or for multi-stage sampling. The distinction
ill be discussed in what follows.

2. Organizational Attachment. Most individuals are members of
ne or more identifiable and formally organized groups and may pos-
bly be contacted through these groups. Such groups have many dif-
erent reasons for existence—social, professional, service, and so on.
hus we can consider members of the same household or spending unit,
roups of workers having common employers, groups of workers em-
loyed in separate plants, groups of workers belonging to different
bor unions (or the locals thereof), doctors in the same county med-
al society, lawyers in the same firm, farmers belonging to different
nits of the Grange, businessmen who are members of different Rotary
lubs, college students belonging to fraternities or sororities, etc.
pecial groups in the population may frequently be economically con-
acted only through such organizations. For a study on the mobility
f tool and die makers, conducted by the Bureau of Labor Statistics,
sample was obtained by first selecting a sample of firms (each firm
onstituting a group of tool and die makers) which employed such
orkers. The major problem is to make sure that the groups used
ontain all the members of the desired population, or at least such a
igh proportion that the remainder can be safely ignored.

3. Similarity in Size. It will frequently happen that information is
vailable from which the elements in a population may be classified
ccording to their size, or some closely related variable, and that
tratification or grouping according to size will be beneficial in in-
reasing precision of sample estimates. Counties or cities may be
rouped according to their total population; companies may be
rouped according to the total number of workers they employ; and
usiness establishments may be grouped on the basis of total sales or
otal amount of product produced.

4. Demographic Characteristics and Functional Similarity. Eac individual possesses many personal characteristics which may b directly observable or for which information may be readily obtaine these characteristics are often of interest in opinion and attitud studies. Groups may be formed on the basis of these characteristic Thus we have men and women, old people and young people, individ uals with a large amount of formal education and individuals wit a relatively small amount of information, persons registered to vot families with relatives in foreign countries or in the armed forces an families without such relatives, and so on almost indefinitely. Fre quently these are the types of groupings for which individual class fication will be impossible until the sample has been drawn, and s their use will be restricted to sample analysis and estimation rathe than to sample design.

Even the preceding systems of groups do not exhaust the possibilitie that should be kept in mind when designing a sample. For exampl positional characteristics are frequently of great importance, as evi denced by the use of systematic sampling procedures. If an ordere list of population elements exists or can be constructed, we can procee through such a list taking, say, every tenth element. This define groups on the basis of relative position on the list; if a random start ing point is chosen, one of these groups is thereby randomly selecte for the sample.

B. Questions of Feasibility

During the preparatory work, as outlined in the earlier chapters o Part III, data and other materials will have been collected relatin to the population, attitude and opinion variables, and sampling pro cedures. These materials, in conjunction with the survey specification for time, cost, and purpose, must now be used to integrate a system o systems of subgroups into the sample design. Specific discussion of th points to be considered in stratification and in multi-stage samplin will be given in the next two sections of this chapter, but here we ma set forth some of the general questions that may be raised concernin the feasibility of various systems of groups.

1. Have the groups that we wish to use already been defined, o must the survey budget bear the cost of such definition? For example when *areas* are to be used as the ultimate sampling units, it is usuall desirable that they be small in size. In some instances this definin process is a large and expensive operation, as has been evidenced ir the preparation of materials for the Master Sample of Agriculture

(2); in others it may be relatively simple (e.g., use of a street map of a small city to mark out blocks). A possible mitigating feature of the cost of defining *areal* segments or clusters is that their use and therefore their cost may be spread over a large number of surveys.

2. Are the maps and other materials required for giving group identification to field workers available, or must the cost of this operation also be borne by the survey? For example, small areal segments marked out on one master map must ordinarily be transferred to smaller maps that can then be given to interviewers.

3. If organizational groups are to be used for drawing and contacting a sample, will these groups release membership lists and cooperate with the survey—at least to the extent of not actively opposing it?

4. If groups are to be defined on the basis of interview results, can the appropriate materials be obtained accurately enough by the ordinary methods, i.e., through the questionnaire or through the observations of an interviewer?

5. Is previous experience available for assessing the effects of various systems of groups on the cost and precision of the survey, and can additional experience be obtained from *this* survey through appropriate design?

20.3 STRATIFICATION

A. The Goals of Stratification

There is probably no technique that is more widely used in sample design than stratification. This is undoubtedly partly because a stratified sample has a certain intuitive appeal. That is, the subgroups or strata are each represented in the sample in their proper proportions. It is tempting to argue intuitively that such a sample must therefore be "representative" of the population from which it is drawn. Actually, this intuitive appeal can be justified in an objective fashion, *provided the samples within each of the strata are selected in an appropriate manner.* It is not enough to state that a sample is "representative" and therefore "good" *simply* because of controlled representation of various subgroups in the population. In the succeeding portions of this section it will be assumed, unless otherwise stated, that a random sample is chosen from within each stratum.

The goals that can be realized, to a greater or lesser extent depending on the circumstances, through the use of stratification are the following:

1. Given a specified level of sampling variability that must be attained for a population estimate, an appropriately chosen stratified

random sample will require a smaller total sample size than will an unrestricted random sample. Sometimes the gain will be so small that it is not worth the trouble, but this matter will require further amplification.

2. If the survey is to provide some results for each of the strata as such, rather than for the population as a whole, stratification will permit adjusting the size of sample taken from each stratum so that it is adequate for this purpose.

3. Certain types of cost and administrative requirements may sometimes be best met through the use of stratification. For example, the cost of obtaining data may vary greatly from stratum to stratum, as between urban strata and rural strata. Reduction in the size of the sample in a high-cost stratum may permit a large enough increase in the size of sample in low-cost strata to reduce the sampling variability of population estimates.

The actual use of stratification techniques demands answers to three questions: (a) How are strata to be defined and how many of them shall there be? (b) How should the sample be allocated among the strata? (c) What gains can be expected through the use of stratification in any particular situation? We shall now briefly discuss these points in connection with opinion and attitude measurement.

B. Stratified Random Sampling with Proportional Allocation

Much of the work in the measurement of opinions, attitudes, and consumer wants is concerned with the estimation of the proportion of individuals in a population who fall in a certain attribute category— for example, the proportion of adults approving a stated governmental program, the proportion of families planning to buy a new car in a specified interval of time, the proportion of adults who may be classified as having a certain degree of "racial prejudice," etc. Though the expanding use of various types of scales and indices may shift some of the current emphasis on proportions toward a consideration of quantitative variables, we shall here restrict ourselves to dealing with the case of proportions.

The primary effects of stratification on the sampling variability of a population estimate can be most easily set forth if we consider *proportional allocation*. That is, a system of strata is given, the total size of sample is fixed, and this sample is then allocated among the strata in accordance with the number of elements they contain. Thus a stratum containing 10 per cent of the population will also receive 10 per cent of the sample. Such a procedure of allocation may be termed *self-weighting* since an unbiased estimate of a population proportion may be obtained by analyzing the sample as a whole rather than stratum by stratum and then weighting these results.

The sampling variability of an unbiased estimate is measured in terms of the variance of that estimate. It is a relatively simple matter to determine the variance of an estimate of the type that is here being treated. Appropriate derivations and formulae may be found in books on sampling theory and practice, for example, in Cochran (3, pp. 90–93), but for our purposes we need only one simple result. The variance of estimates of a proportion made from stratified samples of size n with proportional allocation is smaller than the variance of the same estimate made from simple random samples of size n by

$$R = \Sigma W_i (p_i - p)^2 / n$$

where W_i is the proportion of the population in the ith stratum, p_i is the true proportion of individuals in the ith stratum falling in the specified attribute category, and p is the true proportion of such individuals in the population (i.e., the quantity being estimated). It is also assumed that the size of the total population is large and that the sample size is small in comparison with the population size.

From this relationship we see that the effectiveness of stratification in reducing the variance of estimates is measured by the term $R = \Sigma W_i (p_i - p)^2 / n$. This in turn depends on the size of the differences that exist between strata. If there are no differences between strata, then from the point of view of sampling variability stratified random sampling is no better than simple random sampling. From this statement we can, of course, also infer the opposite. If—and this is purely theoretical—strata can be formed so that all elements within each group are identical, then we need take only one element from each stratum and the resulting estimate properly weighted will have no sampling variability. Thus we obtain one of the guiding principles for defining strata: *The individuals within any one stratum should be as much alike as possible with respect to the variable under investigation and the strata should differ as much as possible.*

The foregoing points have long been recognized. However, what has not been so generally realized is that stratification frequently offers only slight gains in sampling variability over simple random sampling when it is applied in opinion and attitude measurement. This can be illustrated by some computations presented in Table 20.1, which shows the relative efficiency of stratified random sampling for situations in which there are three strata, each of equal size, and for varying degrees of differentiation between strata. It will be seen that substantial gains are realized only for populations C, H, I, J, K, and L, and that these gains require much greater differentiation between strata than is ordinarily possible. An actual example illustrating this fact is given in Section 9.3C and Hansen, Hurwitz, and Madow (4, p. 53) make the same point.

The examples in Table 20.1 are simplified so that they refer to three strata of equal size. Variation in strata sizes would also influence the results. Nevertheless, they indicate that we may not be obtaining as much gain from stratification as might be intuitively expected. These remarks are not meant to imply that stratification should be avoided when only negligible gains in sampling variability can be achieved. It is still necessary to consider cost and administrative advantage and to obtain adequate sample sizes within the separate strata. Also, it may be that some form of allocation other than proportional will be worth while.

C. Optimum Allocations

It is a truism in sample design that the more we know about a population and about the costs of conducting a survey, the better we can design a sample to achieve stated objectives. In our discussion of stratified sampling up to this point, we have really used only the known relative sizes of the strata to allocate the sample. Even this information was not strictly necessary since exactly the same goal can be attained (proportional allocation) by using a *constant sampling rate* in each stratum. Other information was applied in evaluating the variance of the estimate, but this task frequently can be deferred until a sample has been chosen and is not essential for the allocation.

Assuming that strata are already defined, we may now ask what other types of information might be available for design purposes, what specific uses could be made of this information, and what would be the effects on the cost and sampling variability of survey estimates. For the conditions under which we are discussing stratification—estimation of a proportion and simple random sampling in each stratum—sampling theory recognizes two additional pieces of information that can be used in design. These are: (a) from past experience it may be possible to assign differential dollar costs per interview to the separate strata, and (b) some reasonable estimates of the true proportion of individuals falling in the desired attribute category for each stratum may be known from other sources such as past surveys. The second point really refers to the variance of the variable under study, but for proportions this variance depends directly on the proportion.

TABLE 20.1

SIZE OF STRATIFIED RANDOM SAMPLE WITH PROPORTIONAL ALLOCATION
REQUIRED TO GIVE THE SAME VARIANCE OF ESTIMATE AS A SIMPLE
RANDOM SAMPLE OF SIZE 100 [a]

(Three strata, each including one-third of the total population)

	Proportion of Individuals Falling in Specified Attribute Category in Stratum			Size of Stratified Random Sample
Population	1	2	3	
A	0.50	0.40	0.30	97
B	0.50	0.20	0.10	85
C	0.50	0.10	0.01	72
D	0.20	0.10	0.05	96
E	0.20	0.10	0.01	94
F	0.10	0.01	0.001	94
G	0.70	0.50	0.30	89
H	0.80	0.50	0.20	76
I	0.80	0.20	0.10	57
J	0.90	0.50	0.10	57
K	0.99	0.50	0.01	36
L	0.90	0.20	0.10	47

[a] If the sample is some other size, the stratified random samples will have sizes that are in the same proportion as those shown in the table.

Various formulations of the "optimum" problem can now be set up and solved. These formulations are as follows:

1. *For a fixed total sample size and with no attention paid to variable costs between strata,* what allocation of the sample among strata provides the estimate with smallest sampling variability?

2. *For a fixed total amount of money to spend and with variable costs between strata,* what allocation of the sample among strata provides the estimate with smallest sampling variability?

3. *For fixed sampling variability and with variable costs between strata,* what allocation of the sample among strata requires the smallest total cost?

In each instance the estimate is an appropriately weighted average of the proportions found for the strata in the sample. Formal solutions to these three problems are presented in any of the modern books on sampling theory, and so we shall not reproduce them here. In general, they give quantitative results that can be characterized in the following way:

1. The larger the relative size of a stratum, the greater should be the sample size from that stratum.

2. The closer the true proportion in the attribute category in a stratum is to one-half, the larger should be the sample taken from that stratum.

3. The greater the cost per interview in a stratum, the smaller should be the sample size from that stratum.

4. These relationships may work in opposing directions, producing a net effect on the sample that is not as great as any one of them would have by itself.

The effects of the foregoing may perhaps best be illustrated by means of a simple example. Suppose we consider a population divided into three strata having the following characteristics.

Stratum	Relative Sizes of Strata	Fraction of Individuals Having Specified Attribute	Dollar Cost per Interview
1	¼	0.50	$2
2	¼	0.20	5
3	½	0.10	9

There are many ways in which we can draw samples from this population and make comparisons between the variances of resulting estimates. However, we choose to do this in two steps, the first ignoring costs and the second taking costs into account.

For the first step, let us assume that we can draw a random sample of size 100, a stratified sample with proportional allocation that will give the same variance of estimate, or a stratified sample that will give the same variance of estimate and will use the smallest possible total sample. It can be easily shown from the previously quoted theory that the following results hold.

	Fraction of Sample to Be Taken from Stratum			Total Sample Size
	1	2	3	
Proportional allocation	¼	¼	½	85
Allocation giving smallest total sample size	⅓	⁴⁄₁₅	⅖	81

'rom these values we see that the "optimum" allocation has, in comparison with proportional allocation, shifted sample elements out of
the third stratum into the first and second strata. This is in line with
our earlier comment that the closer the true proportion in the attribute
category in a stratum is to one-half, the larger should be the sample
taken from that stratum. However, we also note that this shift has
not really given us much of a gain in sample size—only from 85 to 81.
This is a fairly typical result, as has been pointed out by Cochran
[3, p. 92). In other words, "optimum" allocation seems seldom to
give appreciable gains over proportional allocation when we are estimating population proportions; and, since it is "disproportionate," it
adds to the complications of analysis. Survey results must be obtained stratum by stratum and then weighted in order to obtain unbiased population estimates.

As a second step, cost considerations can be introduced into this picture. Suppose we have approximately $1000 to spend and wish to
compare different allocations with respect to the sampling variability
of their respective estimates. The following four cases will be distinguished.

1. A simple random sample of size 160 can be drawn. The average
cost of such a sample will be $1000. The cost of any given sample
would deviate from this average, depending on the chance division
of the sample among strata.

2. A stratified random sample with proportional allocation can be
drawn. In this case a sample of 160 will give a total cost of exactly
$1000.

3. Using the allocation, obtained previously, that gave the smallest
variance for fixed sample size, we can adjust the sample size so that
the total cost will be approximately $1000.

4. Finally, fixing the total cost at $1000, we can simultaneously
adjust the total sample size and the allocation among strata so as to
obtain the sample size and allocation that will give the smallest possible variance.

The pertinent numerical details for these four cases are shown in
Table 20.2.

All four of these cases give a total cost of approximately $1000, but
the respective sample sizes and estimates of variance are quite different. The third case perhaps requires a special word of comment.
The allocation obtained in the preceding discussion, where costs were
ignored, was used and the sample size was then adjusted until a total
cost of approximately $1000 was obtained. This procedure applied

TABLE 20.2

SIZE AND VARIANCE FOR A RANDOM SAMPLE AND THREE STRATIFIED SAMPLES

Type of Sample	Fraction of Sample to Be Taken from Stratum			Total Sample Size	Total Cost	Variance of Estimate [a]
	1	2	3			
1				160	$1000	100
2	¼	¼	½	160	1000	85
3	⅓	4/15	⅖	179	1003	72
4	0.48	0.25	0.27	216	1004	66

[a] The variances for cases 2, 3, and 4 are given as percentages of the variance for simple random sampling. The simple random sampling variance is therefore 100.

very well to this situation, but such results are not to be expected in general. It so happened that the original strata characteristics associated low-strata cost with high-strata variability. If the reverse had been true, the third procedure would not fare well at all in comparison with the fourth. The allocation in the fourth procedure is worth noting. The first stratum has low dollar cost per interview and has a true proportion near one-half. Therefore almost half the sample has been placed in this stratum, even though it contains only one-quarter of the population. That this shifting was worth while is evidenced by the variances of the second and fourth procedures—85 and 66.

The foregoing discussion and examples illustrate not only the effects of stratification but also the line of reasoning that runs through much of present-day sampling theory. This line of reasoning is exemplified in the following quotation from Deming (5, p. 3).

The statistician's aim in designing surveys and experiments is to meet a desired degree of reliability at the lowest possible cost under existing budgetary, administrative, and physical limitations within which the work must be conducted. In other words, the aim is *efficiency*—the most information (smallest error) for the money. These aims accord with Fisher's principles of modern design of experiment.

D. Defining the Strata

The preceding treatment of stratified random sampling not only has demonstrated how to evaluate the effectiveness of such a procedure in connection with the estimation of a population proportion, but in so doing has also provided guideposts to use in the original definition of strata. Basically, the approach is to break up the population into groups in such a manner that the proportion of individuals falling in the specified attribute category *within a group* is as close as possible to either zero or one—that is, the groups are to be *internally homogeneous* and the strata are *heterogeneous from group to group.*

Under ideal circumstances this goal would be accomplished by using only two strata, one containing all individuals in the specified attribute category and the other containing the remainder of the population. This is clearly an impossibility since one of the major aims of most opinion and attitude studies is not only to obtain an estimate of a population proportion but also to discover the characteristics of the population elements which will distinguish to some extent between these two groups. Consequently, we are led to consider the use of variables which judgment and past experience have shown to be most highly correlated with the subject of investigation, for which data are available to determine allocations and to make estimates, and which will fit into the survey design from a practical point of view. Thus religion would be a characteristic that would quickly come to mind in studying birth control. Yet there is usually no accurate classification of a population by religion, and so data and other information would not be available for allocating and selecting the sample and for making estimates. This lack of data frequently forces us to use only the more easily observable and more commonly reported variables (e.g., geographic location and characteristics of the geographic unit as reported in the Census) for stratification.

The definition and use of strata become particularly difficult tasks if a survey is multi-purpose in character. A system of strata advantageous for one variable may not be worth while for another, and vice versa. As yet there is no comprehensive theory for designing multi-purpose samples, although Hagood and Bernert (6) have explored the possibility of employing component indices, derived by using the methods of factor analysis as a basis for stratification.

In designing samples for opinion and attitude surveys, the foregoing problems are not nearly as crucial as they frequently become in surveys dealing with quantitative variables, especially where the variables have highly skewed distributions. Thus it has been demonstrated that proportional allocation gives substantial gains over random sampling only when there is quite marked differentiation between strata, and that optimum allocation (where differential costs are ignored) offers only slight gains over proportional allocation. Moreover, for all practical purposes, estimates obtained from stratified random samples with proportional allocation cannot have greater sampling variability than would be given by random samples of the same size.

This is not necessarily true for other forms of allocation, suggesting that stratification should be tied to the variable of paramount importance in a survey, and that proportional allocation should be used unless there is clear evidence that something of importance will be gained by using a different allocation.

E. Miscellaneous Considerations in the Use of Stratification

The preceding treatment of stratified random sampling has emphasized its relation to the sampling variability of a population proportion and has introduced costs in the form of differential unit costs between strata. As noted in the introduction to this section, there are a number of other aspects of stratification that need to be mentioned briefly.

Strata can be used as administrative units. Large-scale surveys extending over a wide geographic area may require a number of field offices to handle the hiring, training, and supervision of interviewers. Under such circumstances, it is usually possible to define geographic strata in such a way that a field office can be located in each stratum, and to it can be assigned the responsibility for the field work conducted in that stratum. This type of design is particularly essential when a permanent field organization is required for repeated surveys. One example of such a design is provided by the Current Population Survey of the Bureau of the Census, described by Hansen, Hurwitz, and Madow (4, Chap. 12).

Strata can be subpopulations that are to be compared or studied as separate populations. Frequently opinion and attitude surveys are not so much concerned with obtaining an estimate for the entire population as they are in making comparisons of the separate strata. Stratification then serves as a technique for assuring that an adequate sample size will be obtained from each stratum, the term adequate usually being defined in terms of the ability to detect any desired degree of difference between strata. Employee attitude surveys provide a simple illustration of this point since results are almost certainly to be required separately for professional staff, supervisory personnel, and production workers, and proportional allocation of a sample would ordinarily provide very few interviews from the professional and supervisory groups.

Another requirement of the allocation may be that the separate strata estimates have equal precision. If finite population correction factors cannot be ignored, this becomes a more complex theoretical problem, a discussion of which has been given by Frankel and Stock (7).

Stratification gains can be realized after a random sample has been drawn. There are many situations in which a system of strata may be defined, the relative sizes of the strata are known from a census or other source, and yet the sampling from strata cannot be controlled because an individual element cannot be classified into a stratum until after he has been interviewed. Education, for example, provides a case in point. Though the U.S. Census provides the relative sizes of the various educational classes, this information cannot be used directly in sample design, even though education may be a most effective variable for stratification. However, a random sample may be drawn, the sample elements classified into their appropriate strata on the basis of interviews, the analyses performed stratum by stratum, and the results weighted in accordance with the known relative sizes of the strata. It can be shown (3, p. 104) that the sampling variability of such an estimate will be very close to that of the estimate made from a stratified sample with proportional allocation, except under unusual circumstances.

Systematic sampling procedures can provide automatic stratification with proportional allocation. If systematic sampling is used in conjunction with a list in which elements are arranged by groups, or strata, then proportional allocation among the groups is obtained.

In conclusion, we revert to our opening remarks in this section concerning the intuitive appeal of stratification and point out that, even though we know that only small gains are being realized through the use of stratification, this intuitive appeal may help to reassure potential users who have an unreasoned aversion to results obtained from samples and who fear that the sample will not represent properly the principal subgroups in the population.

20.4 MULTI-STAGE SAMPLING

A. The Goals of Multi-stage Sampling

The principal reasons for using groups of population elements as sampling units are quite apparent and have been set forth briefly in Section 19.3. First, groups of elements can often be identified when individual elements cannot; and second, the clustering of sample elements on a geographic basis can lead to economies in travel and other types of field work. Unfortunately, these gains are usually accompanied by increases in the variance of estimates obtained from samples of the same size. Therefore we are faced with the problem of balancing costs and precision against each other. This section will set forth some of the considerations that must enter into this balancing operation.

In contrast with the preceding treatment of stratification, we shall find it somewhat difficult to give a reasonably comprehensive treatment of multi-stage sampling, even though the possible effects of these techniques on the costs and precision of survey results are likely to be much more pronounced than those of stratification. These difficulties arise in the following ways.

1. Intuition will not ordinarily be as reliable for assessing the effects of multi-stage sampling as it is for stratification.

2. The theoretical treatment of stratification is relatively straightforward, whereas the theory needed for most *practical* applications of multi-stage sampling tends to be complex, with many possible alternatives available at various points in the sample design.

3. The customary analyses performed for almost any opinion or attitude survey will provide data that can be used to evaluate the effects of stratification on the precision of survey estimates, whereas rather complicated special analyses are required to evaluate the similar effects of multi-stage sampling. In addition, very little in the way of post-survey analyses are available in this respect.

4. Since costs are extremely important for cluster sampling, much more careful and detailed field records of costs and time must be maintained to aid the process of future design than is true for stratification.

Recognizing these limitations, we shall now review briefly the use of multi-stage sampling in the measurement of opinions, attitudes, and consumer wants. Actually, our discussion will be limited to two-stage sampling since this illustrates the major points to be made.

B. General Effects of Cluster Sampling on the Precision of Estimates of Population Proportions

As a starting point, let us assume a very simple situation and draw on intuition to the fullest possible extent. Suppose we have a population that can be divided into subgroups or clusters, *each cluster having the same number of population elements.* If we wish to talk in terms of a concrete example, it is possible to think of the adult population in a city as living on blocks, each block having the same number of adults. A two-stage sample would be obtained in this situation by first selecting a *random sample of blocks* (where not all blocks are included in the sample) and then taking all or a sample of individuals from each of the chosen blocks.

To illustrate the effects of cluster sampling, let us furthermore assume that we can arbitrarily assign individual elements to clusters, and let us distinguish between the three methods of assignment shown

TABLE 20.3

THREE POSSIBLE METHODS OF ASSIGNMENT

Method	Description
A	Individuals are randomly assigned to clusters without regard to any of their characteristics.
B	Individuals are assigned to clusters in such a way that *within any cluster* either all individuals fall in the specified attribute category or no individuals fall in it. That is, there is no variation between elements within a cluster.
C	Each cluster contains exactly the same proportion of individuals in the specified attribute category as does the entire population.

in Table 20.3. As we shall see shortly, methods B and C represent two extremes and A falls in between the two.

With this as background we can now argue concerning the effects of cluster sampling on the precision of an estimate of a population proportion. Finite population correction factors will be ignored in this approach.

If the assignment to clusters is according to method A, then any random selection of clusters and any random selection of individuals from within the chosen clusters will be equivalent to drawing a random sample of the same total size from the entire population, without regard to clustering. That is, no gain or loss in precision, in comparison with random sampling, is effected in this situation.

Now consider the assignment to clusters according to method B. As far as precision is concerned, it does no good to take more than one element from each sample cluster since all elements within the cluster are identical. Therefore, if more than one element is taken from each sample cluster, the sampling variability of the resulting estimate will be larger than what would be given by a number of elements drawn at random without clusters. The extreme loss is of course attained when all elements are taken from each sample cluster.

Finally, if method C is used, we need study only one cluster since this is equivalent to studying the entire population. Our estimate will have no sampling variability if we take a 100 per cent sample from the selected cluster.

The foregoing observations can be stated a bit more formally in the following manner. The more homogeneous are the elements within a cluster—and therefore the greater the differences between clusters —the greater will be the sampling variability of an estimate made from a cluster sample in comparison with a simple random sample of the same size. Conversely, the more heterogeneous are the elements within a cluster—and therefore the smaller the differences between clusters—the smaller will be the sampling variability of an estimate made from a cluster sample in comparison with a simple random sample of the same size. At some point between the two extremes, equality of precision for cluster samples and simple random samples of the same size will be attained.

The implications of the foregoing remarks in regard to the definition of clusters for use in multi-stage sampling are in direct contrast to the criteria developed for the definition of strata. That is, elements within clusters should be as unlike as possible whereas elements within a stratum should be as like as possible. Sometimes the relative degree of homogeneity or heterogeneity is measured in terms of the *intraclass correlation coefficient*, but it is not necessary to define a particular measure of homogeneity for our present discussion.

The theory required to deal with two-stage cluster sampling, where the clusters are of the same size, is readily available in recent books on sampling—Cochran (3, Chap. 10), Deming (5, Chap. 5) and Hansen, Hurwitz, and Madow (4, Chap. 6), and special attention has been given to the estimation of proportions by McCarthy (8). However, at the present time there is very little experience available concerning the relative efficiency of cluster sampling when it is used in the measurement of opinions and attitudes, unless we are willing to attempt to transfer the experience of the Bureau of the Census with respect to population, housing, and agricultural investigations [see, for example, Hansen, Hurwitz, and Madow (4, pp. 588–617)]. Further discussion of this point, together with some illustrative material, may be found in Chapter 9 of this book.

In spite of this lack of quantitative experience with which to design cluster samples, it seems intuitively clear that the sampling of geographic clusters will be less efficient than the sampling of individuals for most attitude and opinion variables. People living in close geographic proximity will tend to be alike on such characteristics as race, education, and income, thus introducing homogeneity within clusters for attitude and opinion variables. The need at the moment is to build up a body of experience to make this intuitive observation more precise.

In concluding this subsection, we note that the "natural" clusters (dwelling units, blocks, cities, counties, etc.) ordinarily used in the design of sample surveys will not all contain the same number of population elements. This introduces a large number of complications in theory, and we may introduce different sampling procedures and different estimation procedures and must frequently use biased estimates which have to be evaluated in terms of mean square error. For example, the clusters can be sampled so that larger clusters have a greater chance of being included in the sample (probability proportionate to size), the proportion that is subsampled varying from cluster to cluster inversely to its original chance of being included in the sample. The pertinent theory for these various possibilities will be found in the previously cited books on sampling; some of the theory is illustrated in Chapter 9. Also, a very usable "how to do it" account of the procedures for obtaining a two-stage sample in a city has been given by Kish (9).

C. General Effects of Cluster Sampling on the Cost of a Survey

As stated previously in this section, the manner in which a cluster sample is set up can have a marked effect on total survey costs. A few observations will perhaps help to make this clear.

1. The larger the clusters that are to be used in the design, the less chance there is that a survey will have to bear the cost of defining them, of preparing special maps for field work, etc. Thus in a city sample, the use of units smaller than those that can be defined with streets as boundaries usually entails rather tedious and costly preparatory work.

2. The fewer the clusters drawn into a sample, the smaller will be certain types of cost. For example, if city blocks are the units and if the chosen blocks are to be prelisted, then the cost of prelisting is the same whether one or ten dwelling units is finally chosen from this prelisting, and the total cost of prelisting for a fixed number of dwelling units will vary inversely with the size of cluster.

3. The travel time required for interviewers to go from one potential respondent to another is less when respondents are grouped in a small area than it is when they are spread over a large area. This means that the units should be efficient for the work of interviewers, i.e., no more time should be spent in unproductive travel between units than is required by considerations of precision.

It is extremely difficult to discuss the problem of costs apart from a specified setting, but a number of individuals and organizations have attempted to set up general cost functions to serve as rough guides in cluster sample design. An extended discussion of the development of several such cost functions is given by Hansen, Hurwitz, and Madow (4, pp. 270–284), one of these cost functions being

$$C = C_0\sqrt{m} + C_1 m + C_2 m\bar{n}$$

In this expression, C stands for the total variable cost of the survey (i.e., excluding fixed costs), m is the number of clusters included in the sample, $\bar{n}$ is the average number of interviews per cluster, $C_0\sqrt{m}$ is the cost of travel between clusters, C_1 is the cost of adding a cluster to the sample (including cost of selection, cost of prelisting, etc.), and C_2 is the cost of an interview (cost of interviewing, reviewing questionnaire, etc.).

Whether or not the foregoing cost function and the constants that appear in it (C_0, C_1, and C_2) provide an *accurate* representation of variable costs of a survey is not too important. It is only necessary that they give a good approximation to the relation between total cost and m, the number of clusters drawn into the sample and $\bar{n}$, and the average number of sample elements per cluster. This question is, of course, a matter that must be decided for each survey organization. If it does provide this rough approximation, then we see that the sample features that make for low costs, i.e., few clusters in the sample and many elements per cluster, will ordinarily make for large sampling variability. It is therefore necessary to use the same balancing operation that was applied to stratification where there were differential costs between strata. That is, we fix the total amount of money to be spent and try to choose m and $\bar{n}$ so that the resulting estimates will have the greatest precision, or we fix the desired degree of precision and try to find the values of m and $\bar{n}$ that will give the smallest total cost. Such solutions are referred to as "optimum" solutions.

There are very few published accounts of the costs associated with surveys. An interesting analysis of the costs of some of its surveys was provided by the Bureau of the Census in an unpublished memorandum. Moreover, this Study of Sampling was not even able to ascertain whether cost records were available within other survey organizations with which to approximate these indicated analyses. Certainly more experience should be assembled and made available on this phase of sample design. Of course we may argue that such a "fixed" approach to sample design is not realistic under the present conditions of opinion and attitude measurement because of inability to quantify errors of measurement, inability to state precisely the uses to which survey results are to be put, and the like. Nevertheless, no matter what approach may ultimately turn out to be "best," some form of cost specification will be necessary.

The contents of this chapter, in conjunction with quoted references and other portions of this book, have attempted to show (a) that the systems of subgroups incorporated into a survey may have far-reaching effects on the cost and performance of the plan; (b) that there presently exists a large body of theory relating to "fixed cost, maximum precision" and "fixed precision, minimum cost" solutions; (c) that stratification does not carry the same weight in opinion-attitude studies that it does in some quantitative investigations, unless the strata are important for administrative or analysis purposes or unless there are extreme differential costs between strata; (d) that cluster sampling is frequently necessary for practical reasons and that it can have marked effects on survey cost and performance; (e) that there is presently available little information to assess these effects in opinion or attitude studies; and (f) that criteria other than "fixed cost, maximum precision" and "fixed precision, minimum cost" may prove more realistic in this field of investigation.

The lack of experience already noted becomes particularly apparent if we consider a sampling unit used in a very high proportion of opinion and attitude surveys, namely the household. If a sample of households is selected for a particular investigation, we may draw a single person from each household or may interview all individuals in the household. In most situations, practical considerations have supported the former course of action. We expect that interaction between individuals within the same household at the time of interview will affect the answers that would be given if each person were interviewed alone. Control of this disturbing factor by having simultaneous interviews in separate rooms is not always practical. Moreover, there is little published experience on the homogeneity of opinions and attitudes within the same household. From our general knowledge of families, we expect that members of a family tend to be similar in most attitudes, but there is no good way to measure the similarity except from survey data.

The effect of clustering will occasionally operate to increase the precision of sampling for variables on which members of a cluster tend to be *dissimilar*. One of the well-known examples is the sex of the members. To simplify the example, suppose we are sampling a population made up of four-person households—a husband, a wife, and two children—for the purpose of estimating the sex ratio in the population. It can be shown that the sampling variability of a sample estimate based on a random sample of households (with complete enumeration on the members of each household in the sample) is one-half what would be obtained if a sample of equal size were chosen by selecting

individuals at random. The husbands and wives also tend to hav
attitudes and opinions on some issues that are opposite to those
the children.

The only way to make progress on all these points is for surve
organizations to provide means for obtaining material on these aspect
of sampling for all possible types of surveys, and to make these dat
generally available to all people interested in sampling. With th
body of data available we could at least benefit from past experienc
and it might be possible to build up general models of cost and varianc
for differing types of sampling units. As it is, we are now forced t
resort to intuitive judgment and general theory instead of definite in
formation about the major factors that affect the accuracy of a pai
ticular survey.

REFERENCES

1. Lester R. Frankel, "Establishment Sampling," unpublished paper presente
   before the American Statistical Association, Dec. 28, 1947.
2. A. J. King and R. J. Jessen, "The Master Sample of Agriculture," *J. Am
   Statist. Assoc.*, 40 (1945), 38–56.
3. William G. Cochran, *Sampling Techniques,* John Wiley & Sons, New Yor
   1953.
4. Morris H. Hansen, William N. Hurwitz, and William G. Madow, *Sampl
   Survey Methods and Theory,* Vol. I, John Wiley & Sons, New York, 1953.
5. W. Edwards Deming, *Some Theory of Sampling,* John Wiley & Sons, Ne
   York, 1950.
6. Margaret J. Hagood and Eleanor H. Bernert, "Component Indexes as a Bas
   for Stratification in Sampling," *J. Am. Statist. Assoc.*, 40 (1945), 330–341.
7. Lester R. Frankel and J. Stevens Stock, "The Allocation of Samplings amon
   Several Strata," *Ann. Math. Statist.*, 10 (1939), 288–293.
8. Philip J. McCarthy, *Sampling: Elementary Principles,* Bulletin 15, New Yor
   State School of Industrial and Labor Relations, Ithaca, 1951 (reissued 1956
9. Leslie Kish, "A Two-Stage Sample of a City," *Am. Sociol. Rev.*, 17 (1952
   761–769.

CHAPTER 21

Control of Deviations
from and Changes
in Original Sample Design
during Operations

21.1 NO SAMPLING PLAN IS FINAL
UNTIL THE FIELD WORK IS COMPLETED

The final results, and therefore the performance, of a survey depend not only upon the survey design but also on its execution. Consequently little is accomplished in setting up an "ideal" design that can never be satisfied in the field. Rather, we should recognize that both known and unknown difficulties will occur in execution and, as far as possible, we should make provisions in the design so that they can be reduced and kept under control. We shall discuss in this chapter some of the considerations, with particular reference to the sampling portion of the design, that enter into this phase of survey operations.

Three different but related aspects of this design problem will be distinguished.

1. As successive portions of the field work are completed and are examined, it may become apparent that some of the survey goals are not going to be satisfied. For example, the time allowed may not have been sufficient to permit completion of the survey as planned; cost estimates may have been in error so that the field work cannot be completed within the survey budget; poor estimates of the size of clusters may mean that the sample being obtained will be too small for analysis purposes; or poor advance estimates of certain population quantities may mean that the desired precision will not be attained. When these facts become known, it may be possible to institute changes in the sampling plan that will aid in circumventing these difficulties, or at least in minimizing their effects.

2. During the course of an attitude survey, events may occur sud denly that have highly emotional effects on the members of the popula tion and change the attitudes that are being studied. Examples ar provided by strikes, international incidents, race riots, and the like When such an event occurs, responses obtained before the event from part of the sample may not be directly comparable with those ob tained for the remainder of the sample after the event. It may b necessary to adjust the survey plan quickly, not only to take ad vantage of the change in attitudes but also to insure that the survey results can be analyzed meaningfully.

3. Finally we note that some deviation from the prepared sample design must be expected during the actual operation of a survey There will be people who refuse to be interviewed; there will be others who cannot be reached because of personal characteristics and habits or because of special conditions arising from illness, weather, and the like; mistakes will occur in reading instructions or consulting maps and there may be deliberate cheating and falsification by interviewers In any particular situation the problem becomes one of discovering what deviations will occur or have occurred, of devising means for re ducing or controlling the anticipated deviations, and of making ap propriate corrections in the analysis for those that have not been eliminated.

The following discussion will treat some of the points that may be considered in connection with these three items. This will be done in an illustrative manner since each special situation will probably re quire its own modifications of sample design and theory and its own use of practical devices to control and reduce deviations from design.

21.2 PROPERLY DESIGNED SAMPLING PROCEDURES CAN FREQUENTLY BE ADJUSTED TO ACCUMULATING EXPERIENCE

Surveys are often conducted by individuals and organizations with insufficient data regarding the money and time costs involved in con ducting field work—e.g., prelisting, interviewing, and the like. If these costs are underestimated in the original survey design, then it may happen that such an undertaking will find itself with an exhausted budget and with an incomplete sample (at least in terms of the original design). Moreover, if no attention has been paid to this eventuality, the completed portion of the sample may not be "representative" of the entire population, or, indeed, of any specifiable subpopulation.

We should most certainly attempt to guard against such a situation.

Perhaps the simplest design approach to this difficulty is to consider the possibility of a two-sample procedure. That is, a first sample is drawn which is smaller than the one original estimates indicate as feasible and which can certainly be completed within the stated time and cost limitations. As field experience accumulates for this sample, reasonably accurate time and cost estimates can be obtained and a second sample, similar to but independent of the first sample, can then be drawn so that the combined first and second samples will be completed according to specifications. A different but related approach is to select originally a number of similar but independent samples (instead of one large sample), complete the field work for each of these samples in succession, and simply stop when time or money runs out. The stopping will of course be done at the end of one of the subsamples.

Two-sample procedures have applicability in many situations where there is insufficient information concerning certain features of the population or of the survey process. For example, the first sample may be used to estimate variability in the population or in strata in order to determine the total sample size required to achieve stated precision or to achieve "optimum" allocation of a sample among strata. Thus, for simple random sampling, Cochran (1, p. 59) describes a two-sample procedure originated by Stein which will produce results having at least a stated degree of precision, even though the population variance is unknown. Such procedures may also be helpful if cluster sizes are not known accurately and a sample of fixed size must be obtained.

The use of procedures similar to those already mentioned will have certain effects on survey costs in relation to the precision of survey results. These may become particularly important if "optimum" solutions are desired, and some general comments on the problem will be given in the final section of this chapter.

21.3 SURVEY DESIGNS MAY HAVE TO BE REVISED QUICKLY IN REACTION TO UNEXPECTED HAPPENINGS

If the attitudes under investigation in a particular survey are such that they might be markedly influenced by the occurrence, during the course of the survey, of some possible and probable events, it may be advisable to include some provision for such happenings in the survey design. Thus attitudes toward air travel might be influenced by a disastrous accident; community attitudes toward a company might be influenced by strikes or government suits; or attitudes toward government policies might be influenced by domestic developments and by international incidents. Under such circumstances, if that portion of the sample interviewed before the event comes from one segment of the population, then it may not be possible to analyze and interpret the survey results meaningfully.

There are various ways in which a survey can be adjusted to such circumstances as these, all of which must involve to some extent the sampling procedure and its execution. For example,

1. If the portion of the sample interviewed before the event comes from a known subpopulation, then it might be desirable to draw a second and independent sample from this subpopulation and thus insure a "before" and "after" comparison for a known group.

2. If the portion of the sample interviewed before the event comes from a known subpopulation, it may be possible to reinterview the same individuals after the event and thus convert to a panel study.

3. Finally, if the "before" sample is "representative" of the entire population, either of these two possibilities may be used with respect to the population as a whole rather than to a subpopulation.

These possibilities aim to insure that the portion of the sample completed at any point in time will be reasonably "representative," either of the entire population or of some known subpopulation. Such a condition might be obtained by successively completing field work in each of a number of strata, or by successively completing the field work in a number of independent samples of the entire population. The remarks given in the preceding section have a bearing on this problem. Such designs may well have their effects on survey costs.

21.4 MANY DEVIATIONS FROM A SAMPLING PLAN CAN BE REDUCED OR CORRECTED BY DESIGN

During the actual operation of a survey plan, it must be expected that deviations from design will occur. These deviations may be of widely varying types, but the net effect will always be that the sample actually obtained differs in some respects from the sample that would be obtained under ideal circumstances. Provision for deviations can and should be made in the survey plan.

The first step in this process is to examine the sample design to see what deviations may be expected in our own particular circumstances. Some of these will be readily apparent, others will be pointed out by past experience in similar situations, and still others may be detected only with special design features, or not at all. The more patent forms of deviation are: refusal of certain kinds of designated sample elements to submit to the measurement process, inability of interviewers to find certain kinds of designated sample elements, failure of interviewers to fill their quotas exactly, failure of field workers or interviewers to follow their instructions on definitions and other details, errors in map reading and prelisting, deliberate cheating and falsification of interviews. Though a number of these items may appear to pertain only to a certain type of sample design (e.g., people who cannot be reached for interview with a fixed address form of sampling), it may well be that one form of design points them up whereas another more or less ignores them. The quota interviewer who does his interviewing in the home does not have any better chance of picking up individuals who are rarely at home than does the interviewer who must make his interviews at fixed addresses.

Following this initial inventory, we must take steps in the survey plan to reduce and control the effects of deviations from design, recognizing that such action is going to affect not only the precision and accuracy of the survey but also its costs.

The action taken in the survey plan may take any of several forms. Among the more usual are:

Reduction in the amount of deviation by appropriate instructions, supervision, and check procedures. There is little that can be said here about reducing deviation from design by instructions, supervision, and check procedures that has not already been considered in other portions of this book. This aspect of the problem cuts across many phases of survey design, including the definition of the population, the determination of the measurement process and its actual application, and the specification and identification of units of sampling. It is only when we can foresee a specific type of deviation that plans can be made to attack it directly, and the procedures devised may vary widely from situation to situation. Many specific comments and references on this are summarized in Chapters 17, 18, and 19, and a detailed discussion of the refusal and not-at-home problem is given in Chapter 11.

Gathering of data on the deviant cases for use in adjustment and correction. Provided the proper data have been gathered during the course of the survey, it is frequently possible either to state that a certain type of deviation from design can have little or no effect on the survey results or to introduce a correction for it into the estimation process. In order to do this, it is essential that each deviation be recognized as it occurs and that a record be made of all the associated circumstances. This is, of course, simply another phase of the maintenance of systematic records throughout the actual survey operation. It costs little to record a number of factual characteristics for those individuals who refuse to be interviewed, or to obtain equivalent information on not-at-homes from the neighbor. The principal point is that the need for these data must be recognized, their importance made clear to the interviewers and other field workers, and appropriate forms prepared on which to enter the information.

At the very least, this type of record will give the extent of sample losses, and an appropriate adjustment factor can be incorporated into the mean square error of the survey results. Moreover, it may be possible to introduce some of this material into the estimation process itself. Some of the possibilities in this respect are discussed in Section 11.2C.

Incorporation of special features into the design itself. In some instances the sample design itself may be built around deviations that are certain to occur. An excellent example of a design specifically prepared for treating problems of this type has been given by Hansen and Hurwitz (2). In essence, the population being sampled is divided into two strata, those people from whom a response can be secured on the first attempt and those who will not respond to the first attempt. It is assumed that a follow-up attempt to obtain information from a non-respondent is more costly than the original attempt, for example, where the original attempt is by mail and the follow-up attempt is by personal interview. The sample design is set up in the following manner. A sample is selected from the population, and each sample element is approached for the first attempt. A fraction of those who do not respond is selected for the follow-up attempt, and the results obtained from these two samples are combined in such a manner as to give an unbiased estimate of the desired population quantity. The problem considered by Hansen and Hurwitz consists of determining the number of first attempts and the number of follow-up attempts to be made in order that the sample estimate will have a prescribed amount of sampling variation at minimum cost. In other words, it is known that there will be non-response among the first attempts, and this is taken into account in the sample design. If the response to mail questionnaires is not equivalent to the response to interviewing, there will still be a residual effect of the deviation from design.

Other design modifications for detecting and measuring differences between interviewers have been suggested and applied, for example, the system of interpenetrating samples as applied by Mahalanobis (3, 4). (See also page 369.)

A recent paper by Sukhatme (5) considers ". . . the methods of measuring observational errors based upon a mathematical model which is general enough to cover conditions commonly met in agricultural and socio-economic surveys. . . ." After describing the model and its application, Sukhatme discusses its use as a regular feature in surveys and concludes that replication is not a satisfactory alternative to adequate supervision for the control of field work..

21.5 CONSEQUENCES FOR SURVEY COSTS

Additional features which are incorporated into a survey design for any of the purposes discussed in this chapter will usually lead to an increase in costs. It is quite clear that this will be the case if extra information is collected during the course of the survey, or if supervisory check procedures are instituted. However, it will also be true that some of the special sampling procedures mentioned in the preceding sections will tend to have the same effect.

This latter point can be illustrated through the use of two-stage cluster sampling and the cost function that was given in Section 20.4C. It will be recalled that this function was

$$C = C_0\sqrt{m} + C_1 m + C_2 m\bar{n}$$

where the symbols are defined on page 422. Suppose we now consider a two-sample procedure such as was discussed in Section 21.2. Assume that the first sample is obtained by taking $m/2$ randomly selected clusters and that the second sample consists of the remaining $m/2$ clusters. We can now compare the cost of doing the field work for the entire group of m clusters with the corresponding cost when the two-sample procedure is used. In the second case this will be equal to

$$2[C_0\sqrt{m/2} + C_1(m/2) + C_2(m/2)\bar{n}] = 1.4C_0\sqrt{m} + C_1 m + C_2 m\bar{n}$$

It will be noted that the first term in this expression (referring to travel between clusters) has been increased by 40 per cent through the use of the two-sample procedure, but the other parts of the cost are not changed.

This increase in cost is not surprising when we consider that a field worker traveling between two clusters in the first sample may well have to cover approximately the same route again when traveling between two clusters in the second sample. If more than two samples

were used, the cost of travel between clusters would be still furth
increased over what it is when the entire sample is covered at o
time. Of course it should be recognized that the total cost of the su
vey does not increase in the same proportion as does the travel co
The relative increase in total cost depends also on the magnitude
travel costs in relation to the cost of drawing clusters and of carryi
out the field work in the chosen clusters.

In closing this chapter, we emphasize again that the choice of ho
far to go in the use of special design features involves balancing tl
increased cost and complexity of the survey against gains in precisi
and accuracy, and that the practical difficulties raised in the introdu
tion to Chapter 19 still apply. Important deviations that are n
controlled by modification of the sample should be corrected by a
propriate adjustment of the estimating process if that is feasible.
they cannot be controlled or corrected, they should be recognized ar
described in the presentation of results of the survey so that the use
of the results can make allowance for them.

REFERENCES

1. William G. Cochran, *Sampling Techniques,* John Wiley & Sons, New Yor
   1953.
2. Morris H. Hansen and William N. Hurwitz, "The Problem of Non-Respons
   in Sample Surveys," *J. Am. Statist. Assoc.,* 41 (1946), 517–528.
3. P. C. Mahalanobis, "On Large-Scale Sample Surveys," *Phil. Trans. Roy. So*
   B231 (1944), 329–451.
4. P. C. Mahalanobis, "Recent Experiments in Statistical Sampling in the India
   Statistical Institute," *J. Roy. Statist. Soc.,* 109 (1946), 325–370.
5. P. V. Sukhatme, "Measurement of Observational Errors in Surveys," *Revu
   de l'institut international de statistique,* 20 (1952), 121–134.

CHAPTER 22

Appraising the Performance
to Be Expected
from Survey Designs

22.1 GENERAL METHODS OF APPRAISAL

The purpose of a survey design is not only to guide and control the actual operations of survey taking but to permit the researcher to form a sound judgment in advance about whether or not the survey is worth taking. Frequently several designs are prepared and examined so that the most advantageous of them may be selected for use. Each presents its own variety of sampling problems. Often one or more prospective users of the survey's results participates in the examination of the design and the decision to select and operate it. Many techniques have been developed for use in the detailed operations of surveys but few for the appraisal of a whole design. At this late stage, then, we find that all the problems that have not been mastered in the development of the designs present themselves for a summary judgment of their probable effect on the performance of the survey. In terms of the model in Chapter 4, the sampler must analyze the situation and decide how he will play the game.

Since the sampler seldom knows all that he needs to know about the situation, he must make various assumptions about some of the more important factors that will affect the results of the survey. He must also make assumptions about the manner in which the survey will actually be operated and the extent to which deviations from the design will occur. From these assumptions and the information that he has about the situation, he must then construct as sound a forecast as he can of the way in which the survey will fulfill the purposes for which it is undertaken. When he does so, his uncertainty about the conditions and the lack of complete control of the survey process compel him to admit that his forecast can only approximate the re-

sults that he will obtain later, i.e., that he is attempting to ascertain b
an imperfect means the distribution of results that would be produce
by the survey process. In the present state of the art, it is difficul
even to attempt such forecasts of performance. The reliability of th
forecasts is not likely to be high, and the assurance with which com
parisons can be made between forecasts is still more dubious. None
theless, this is the direction in which improvement in the operatio
of surveys and the use of survey results must go.

There are still greater obstacles in the way of determining wha
the user stands to gain or lose from each of the possible outcomes o
a given survey design in a given situation. Here too, it is necessary
to learn more than is now known about how to make a valid appraisal
However, even very rough approximations can be quite useful. The
ultimate user of the survey results may have to make these estimates
of gain or loss, but frequently the sampler, when he is not the ultimate
user, will have to participate in the task.

From the two preceding steps we should like to work out an estimate of the probability distribution of gains and losses of various
magnitudes that would result from the use of a given survey design.
This and similar estimates for alternative designs could then form a
basis for a comprehensive judgment of the desirability of using any
particular design among those that are under consideration. Even
though this approach may not always lead us to a unique choice, it
would provide a general method of appraisal. It appears to be superior, even when it is quite approximate, to the methods that rely to a
greater extent on unguided judgment on the one hand, or on arbitrary
conventions on the other.

There are many ways of producing a forecast of the distribution of
results to be expected from a given survey design. Among them we
may distinguish three: (a) an appeal to the experience provided by
similar surveys in the past, (b) combination into a composite or syn-
thetic example of the results of experience with each of the various
components of the design (drawn from a larger group of past surveys
not necessarily comparable in respects other than the components in
question), and (c) the formulation of a model of the survey process
from which an appraisal can be deduced.

The use of similar surveys taken in the past seems to be a direct
and dependable method of appraisal, but it actually has several serious
weaknesses. In the first place it is usually difficult or even impossible
to find a survey that is strictly comparable with the design under consideration. In the second place, the results of past surveys are seldom
known in terms of their relation to the true values that are to be

estimated or the precise purposes they were to serve. In the third place, each example may provide only one result and not a distribuion of results. Offsetting these weaknesses is the fact that the past surveys may incorporate the effects of important factors that are overooked in the other approaches and may in this sense be more "realstic." They are also easier to present and discuss with persons who are not versed in sampling.

The second approach pieces together information from whatever sources are available about the separate parts of the survey and even includes, when necessary, the use of models for some of the parts. In this way it can attain a greater degree of comparability with the design that is being appraised, though this may be accomplished at the expense of overlooking some important relations of the parts to each other, especially in instances where they may be incompatible in actual operations. It may be the only feasible approach in many instances.

The third method of forecasting the distribution of results for a given design is to go the whole way in constructing a model from which the performance of the survey can be derived. The model should be sufficiently complex so that it can be made to fit rather closely the actual operations that will be carried out under its general guidance. The construction of such models requires more information about the details of past surveys than is required by the other approaches, and its suitability for the design in question may be difficult to determine. When it is feasible to set up such models, they offer the advantage of leading more directly to estimates of the distribution of results that is needed for the full use of the general method already outlined. The current texts and articles on sampling techniques provide much of, but by no means all, the abstract theory that is needed to construct models for the whole survey design.

The performance of a survey design may be appraised with respect to more than one assumed set of circumstances, i.e., in different situations. It may also be judged with respect to several sets of results when it is to serve different purposes, or when the purposes to be served have not yet been determined in full. Several different sets of assumptions may be made alternatively when there is uncertainty about essential facts or about the magnitude of the parameters of the model. These explorations tend to "cover" the region of uncertainty and thus facilitate the decisions that must be made in the final judgment of the design. In the absence of a sufficient amount of information, the appraisal becomes highly speculative and of lesser depend-

ability, either for the improvement of the design or for the choic
between competing designs.

Though a sound appraisal of a survey design is difficult, it is stil
well worth the effort for the prospective user of the survey results, a
well as for the sampler who is contemplating the use of the desig
in actual survey operations. For the user, it clears up questions tha
are raised by the conflicting claims of various samplers or surve
organizations. It tends to protect him from faulty judgments abou
the most advantageous manner of expending his funds, either to pur
chase information or to apply it. It gives him some insurance agains
major flaws in the surveys on which he depends and provides a furthe
check of his previous judgments about the survey. It can usuall
assist him in using the results of the survey to greater advantage onc
they are obtained.

22.2 COST ANALYSIS AND BUDGETING

A full discussion of the technical problems of cost analysis an
budgeting is well beyond the scope of this book, but some aspects o
them are sufficiently important to receive special attention at thi
point. Cost factors enter into the appraisal of a design and are formu
lated in a budget to assure that the operation of the survey conserve
the value of the design. In addition, the budget fulfills its norma
functions within the administrative organization responsible for th
survey. Schedules of the rates of completion of the various phases o
the work are also important for the operation of a survey. They tend
to assure the fulfillment of the design and its related objectives, ir
addition to their normal purposes in management of the enterprise
The customary practices in scheduling and budgeting, however, may
not suffice for these purposes.

One of the major problems in the development of sound cost analysis
for surveys is how to allocate expenditures that benefit more than one
survey. Many of them are essentially capital expenditures, invest-
ments in anticipation of future work as well as the current program.
They include the preparation of material for the selection of the sample and for the selection and training of the headquarters organization and the field personnel. They include the development of procedures of estimation and the many techniques to be used in the
processing of the results. If all these expenditures are allocated to the
first few surveys in a longer program of surveys, the costs of these
surveys will often appear prohibitive. Once a particular design is put
into operation, it will probably appear more economical than other

signs because it carries no share of the preparatory expense. At
e same time, an organization that has adopted a second design will
d it more costly to change to the first design than to continue the use
the second. Any third party that chooses a sample design on the
sis of the recommendations of one of these organizations, without
coming aware of the lack of comparability in the cost analyses of
e two designs, will be misled. The first two organizations appear
have a current advantage in continuing the use of their old designs,
d this judgment may persist even when there is a long-run advantage
adopting a new design immediately.

Similar to the problem of allocating costs of work that benefits more
an one survey is the problem of allocating costs of work that also
nefits non-survey activities. For example, the *Literary Digest* poll
as part of a direct-mail solicitation of subscriptions. In judging
hether it was an expensive or a relatively inexpensive survey, we
ust solve difficult problems of allocating the expense of the mailings
nd of the preparation of lists of the millions of persons to whom the
allots were sent. The answer to the question of the expensiveness
f mail surveys depends on answers to questions about cost allocations,
hich in the end may be arbitrary.

In the analysis of field costs, travel time is commonly regarded as
art of the sampling expense. Actually time spent in travel may con-
ibute to the quality of interviews by providing some relaxation of
he tension and fatigue of continuous interviewing and thus serve to
ome extent the functions of a rest period. This is only one of sev-
ral ways in which various kinds of costs turn out to be joint costs for
everal wholly distinct functions.

Ordinarily it would be impractical to distinguish the costs of survey
perations in such detail as that suggested by the preceding paragraph.
'or the purposes of appraising designs and improving them, however,
he effects of various changes in operations on the costs of the survey
ust be examined carefully from the point of view of their relations
o the joint costs of the operations.

A thorough cost analysis of a survey operation would show many
nterrelations between the cost of one part of the survey and the costs
f other parts. In other words, whenever the mode of operation of any
ne part is changed, it affects the costs of some other parts. These
nterrelationships affecting costs are difficult to discover and evaluate,
ut they can be of major importance. The degree of control exercised
ver the field work and other matters of the actual operating pro-
edures may have quite general effects on most of the other parts of
he survey. This is true in part because they affect the morale and

motivation of the personnel involved in training and interviewing. Moreover, the actual effect of a change on costs may turn out to be quite different from the anticipated effect because of human reaction to the changes. These interrelations and human factors make the development of models for the cost analysis of a survey design both complex and difficult. Little is known about the degree to which simple linear models approximate the more complex system of cost relationships, even when the changes are quite moderate, but it is a fair assumption that the use of simple cost functions is valid and helpful when they are not extrapolated beyond the region of experience on which they are based.

It must always be recognized that the costs of survey operation depend not only on the design and its execution but also on the current prices of various goods and services in the market. As these prices change, the relative costs of different designs change with them. Moreover, the relative prices vary between geographic regions. Transportation costs are a case in point.

The need is greater for an accurate determination of the comparative cost of different designs under different conditions than for detailed accounting for the costs of all the elements of a single survey.

22.3 THE VALUE OF SURVEY RESULTS

Much of what has been said about costs applies in a similar manner to the determination of the value or utility of the results of a survey. Both the gains and the losses that may result must be combined with the estimates of costs to determine the worth of the survey and thus form the basis for both planning and operating decisions. Many of the potential values of survey results are contingent on the availability of other information and the existence of certain conditions favorable to their use. This leads to a problem of joint value analogous to the problem of joint cost. There is also a problem of the effect of the survey results on the value of other material. The problem of appraising a design extends outward in this manner to link up with other problems of appraisal and choice. It is clear that the intelligent design and use of surveys cannot be accomplished except in conjunction with the development of rational analyses of the activities of obtaining and using information that are closely related to the use of the results.

It is likely that various projects that involve experimentation and testing, such as those aimed at objective research into consumer tastes, will increasingly be linked with surveys in the study of opinion, attitudes, and consumer wants. They may be introduced into the field work as well as being conducted as separate, though parallel, studies. When this occurs, the kinds of information that can best be obtained by experimentation will be less valuable as subjects for surveys, and surveys will be directed toward the kinds of information for which they prove to be best suited.

These shifts may have important effects on design and operations. They may have implications for the selection and training of survey personnel, both in the central offices and in the field. They may shift the emphasis on various sampling problems and modify the choices of sampling techniques. This prospect reinforces the general conclusion of the Study of Sampling, namely that sampling procedures differ greatly in their dependabilty, not only according to the type of procedure that is used and the size of the sample taken, but also according to their purposes, the mode of operation of the survey, the situation in which they are employed, and the manner in which the results are processed. There is no simple rule book that suffices to guide the user of sampling procedures and their survey results. A patient analysis of the whole survey process is needed for the appraisal of the results. Though this may be done, at least in part, by users, it can only be done thoroughly and accurately by the producers of survey results. Hence it should be the regular practice of survey organizations to make a full analysis of the accuracy and characteristics of their results and to publish the principal facts developed by the analysis for the guidance of the users.

Index of Persons and Organizations

Subject Index